ALLE · ZEIT · WACH · 1842

Fritz Oberhettinger

Tables of Mellin Transforms

Springer-Verlag
Berlin Heidelberg New York 1974

Fritz Oberhettinger
Professor of Mathematics, Oregon State University, Corvallis, Oregon, U.S.A.

AMS Subject Classification (1970): 44-02, 44A10, 44A15

ISBN 3-540-06942-9 Springer-Verlag Berlin Heidelberg New York
ISBN 0-387-06942-9 Springer-Verlag New York Heidelberg Berlin

Library of Congress Cataloging in Publication Data

Oberhettinger, Fritz. Tables of Mellin transforms.
Bibliography: p.
1. Mellin transform-Tables. I. Title.
QA432.024 515'.723 74-16456

Preface

This book contains tables of integrals of the Mellin transform type

$$\text{(a)} \qquad \Phi(z) = \int_0^\infty \phi(x) x^{z-1} dx$$

Since the substitution $x = e^{-t}$ transforms (a) into

$$\text{(b)} \qquad \Phi(z) = \int_{-\infty}^\infty \phi(e^{-t}) e^{-tz} dt$$

the Mellin transform is sometimes referred to as the two sided Laplace transform. The use of the Mellin transform in various problems in mathematical analysis is well established. Particularly widespread and effective is its application to problems arising in analytic number theory. This is partially due to the fact that if $\Phi(z)$ corresponding to a given $\phi(x)$ by (a) is known, then $\Phi(z)$ belonging to $x^\alpha \phi(x)$ or more general to $x^\alpha \phi(x^p)$ (p real) is likewise known. (See particularly the rules in sections 1.1 and 2.1 of this book.)

A list of major contributions concerning Mellin transforms is added at the end of the introduction. Latin letters (unless otherwise stated) denote real positive numbers while Greek letters denote complex parameters within the given range of validity. The author is indebted to Mrs. Jolan Eröss for her tireless effort and patience while typing this manuscript.

Oregon State University
Corvallis, Oregon
May 1974

Fritz Oberhettinger

Contents

Part I. Mellin Transforms

Part II. Inverse Mellin Transforms

Part I. Mellin Transforms

Introduction

The integral

$$(1) \qquad M[\phi(x),z] = \Phi(z) = \int_0^\infty x^{z-1} \phi(x)dx$$

is called the Mellin transform of the function $\phi(x)$ with respect to the complex parameter

$$(2) \qquad z = \sigma+i\tau$$

The substitution $x = e^{-t}$ transforms (1) into a two-sided Laplace integral

$$(3) \qquad \Phi(z) = \int_{-\infty}^\infty \phi(e^{-t})e^{-tz}dt$$

or into the sum of two one-sided Laplace integrals of parameter z and $-z$

$$(3') \qquad \Phi(z) = \int_0^\infty \phi(e^{-t})e^{-tz}\,dt + \int_0^\infty \phi(e^{t})e^{-t(-z)}dt.$$

Denote the abscissas of absolute and ordinary convergence by β and α respectively for the first integral in (3) and by β' and α' for the second integral. Then it is evident that the domains of absolute and ordinary convergence of the integral (1) consist of the respective strips.

$$\beta < \text{Re } z < -\beta'; \qquad \alpha < \text{Re } z < -\alpha'$$

For the inversion of the integral (1)

(4) $\phi(x) = M^{-1}[\Phi(z);x]$

exists the following theorem.

Let $\Phi(z)$ be a function of the complex variable $z = \sigma+i\tau$, regular in the strip $S = \{z:a < \sigma < b\}$ such that $\Phi(z) \to 0$ as $|\tau| \to \infty$ uniformly in the strip $a+\eta \leqq \sigma \leqq b-\eta$ for any arbitrary small $\eta > 0$.

Then if

$$\int_{-\infty}^{\infty} |\Phi(\sigma+i\tau)|d\tau < \infty$$

for each σ in the open interval (a,b) and if a function $\phi(x)$ is defined by

(5) $$\phi(x) = \frac{1}{2\pi i} \int_{c-i\infty}^{c+i\infty} x^{-z} \Phi(z)dz$$

for $x > 0$ and a fixed $c\varepsilon(a,b)$ then

$$\Phi(z) = \int_{0}^{\infty} \phi(x)x^{z-1} dx$$

<u>Some relation between the Mellin transform and other integral transform types</u>.

Consider the following integral transforms of a given function $\phi(x)$.

(a) $$F_s[\phi(t);x] = (2/\pi)^{\frac{1}{2}} \int_{0}^{\infty} \phi(t)\sin(xt)dt$$

Fourier sine transform

(b) $$F_c[\phi(t);x] = (2/\pi)^{\frac{1}{2}} \int_{0}^{\infty} \phi(t)\cos(xt)dt$$

Fourier cosine transform

(c) $$L[\phi(t);x] = \int_0^\infty \phi(t)e^{-xt}\,dt$$

Laplace transform

(d) $$h_\nu[\phi(t);x] = \int_0^\infty \phi(t)(xt)^{\frac{1}{2}}J_\nu(xt)\,dt$$

Hankel transform

(e) $$k_\nu[\phi(t);x] = \int_0^\infty \phi(t)(xt)^{\frac{1}{2}}K_\nu(xt)\,dt$$

K_ν transform

(f) $$y_\nu[\phi(t);x] = \int_0^\infty \phi(t)(xt)^{\frac{1}{2}}Y_\nu(xt)\,dt$$

Y_ν transform

(g) $$\mathbf{h}_\nu[\phi(t);x] = \int_0^\infty \phi(t)(xt)^{\frac{1}{2}}\mathbf{H}_\nu(xt)\,dt$$

(h) $$S_\nu[\phi(t);x] = \int_0^\infty \phi(t)(x+t)^{-\nu}\,dt$$

Generalized Stieltjes transform

(i) $$W_\nu[\phi(t);x] = \frac{1}{\Gamma(\nu)}\int_x^\infty \phi(t)(t-x)^{\nu-1}\,dt$$

Weyl's fractional integral.

Then the corresponding relations are valid.

(a') $$M\{F_s[\phi(t);x];z\} = (2/\pi)^{\frac{1}{2}}\Gamma(z)\sin(\tfrac{1}{2}\pi z)M[\phi(x);1-z]$$

(b') $$M\{F_c[\phi(t);x];z\} = (2/\pi)^{\frac{1}{2}}\Gamma(z)\cos(\tfrac{1}{2}\pi z)M[\phi(x);1-z]$$

(c') $$M\{L[\phi(t);x];z\} = \Gamma(z)M[\phi(x);1-z]$$

(d') $$M\{h_\nu[\phi(t);x];z\} = 2^{z-\frac{1}{2}}\,\frac{\Gamma(\frac{1}{4}+\frac{1}{2}\nu+\frac{1}{2}z)}{\Gamma(\frac{3}{4}+\frac{1}{2}\nu-\frac{1}{2}z)}\,M[\phi(x);1-z]$$

(e') $M\{k_\nu[\phi(t);x];z\}=2^{z-3/2}\Gamma(\frac14+\frac12\nu+\frac12 z)\Gamma(\frac14-\frac12\nu+\frac12 z)M[\phi(x);1-z]$

(f') $M\{y_\nu[\phi(t);x];z\}$

$$= 2^{z-\frac12}\pi^{-1}\sin[\tfrac12\pi(\nu-z-\tfrac32)]\Gamma(\tfrac14+\tfrac12 z+\tfrac12\nu)\Gamma(\tfrac14+\tfrac12 z-\tfrac12\nu)$$

$$\cdot\ M[\phi(x);1-z]$$

(g') $M\{\mathbf{h}_\nu[\phi(t);x];z\}$

$$= 2^{z-\frac12}\tan[\tfrac12\pi(\tfrac12+z+\nu)]\ \frac{\Gamma(\frac14+\frac12\nu+\frac12 z)}{\Gamma(\frac34+\frac12\nu-\frac12 z)}\ M[\phi(x);1-z]$$

(h') $M\{S_\nu[\phi(t);x];z\} = B(z,\nu-z)M[\phi(x);1-\nu+z]$

(i') $M\{W_\nu[\phi(t);x];z\} = \dfrac{\Gamma(z)}{\Gamma(\nu+z)}\ M[\phi(x);\nu+z]$

If, for instance $\phi(x)$ is such that both, its Hankel transform (d) and its Mellin transform (1) is known, then the relation listed under (d') gives an additional result. For tables of integral transforms of the types (a) - (i) see list of references at the end of this introduction.

Laplace and finite Mellin transforms.

Tables of Laplace and inverse Laplace transforms (see list of references at the end of this introduction) can be used to obtain additions to the transform tables presented here. Let

(6) $$\Phi(z) = \int_0^\infty f(t)e^{-tz}dt$$

Then the substitution $t = -\log x$ transforms (6) into a finite Mellin transform

$$(7) \qquad \Phi(z) = \int_0^1 x^{z-1}\phi(x)\,dx \quad \text{with} \quad \phi(x) = f(\log \tfrac{1}{x}).$$

For instance, the Laplace transform pair

$$f(t) = J_\nu(a \sinh t) \qquad \Phi(z) = I_{\frac{1}{2}\nu+\frac{1}{2}z}(a)K_{\frac{1}{2}\nu-\frac{1}{2}z}(a)$$

leads to the Mellin transform pair

$$\phi(x) = \begin{cases} J_\nu[\tfrac{1}{2}a(x^{-1}-x)], & x < 1 \\ 0 & x > 1 \end{cases} \qquad \Phi(z) = I_{\frac{1}{2}\nu+\frac{1}{2}z}(a)K_{\frac{1}{2}\nu+\frac{1}{2}z}(a)$$

which is listed in Part I, under 10.77. Vice versa, the pair of the inverse Laplace transform type

$$\Phi(z) = I_\nu(bz^{\frac{1}{2}})K_\nu(az^{\frac{1}{2}}), \qquad f(t) = \tfrac{1}{2}t^{-1}\exp[-\tfrac{1}{4}(a^2+b^2)/t]I_\nu(\tfrac{1}{2}ab/t)$$

yields the inverse Mellin transform type pair

$$\Phi(z) = I_\nu(bz^{\frac{1}{2}})K_\nu(az^{\frac{1}{2}}), \qquad \phi(x) = \begin{cases} \tfrac{1}{2}y^{-1}\exp[-\tfrac{1}{4}(a^2+b^2)/y]I_\nu(\tfrac{1}{2}ab/y) & x < 1 \\ 0 & x > 1 \end{cases}$$

with $y = \log(\frac{1}{x})$. This result is listed in Part II under 7.85.

Some Applications of the Mellin Transform Analysis.

Only a few examples will be singled out.

(A) Application to certain integral equations (Sneddon, p. 277, Titchmarsh, p. 303). Solutions of the following integral equations can be given in the form of the inverse Mellin Transform.

(a)
$$g(y) = \int_0^\infty f(x)K(xy)\,dx, \quad y > 0, \text{ with}$$
$$M[f(t);z] = \frac{M[g(t);z]}{M[K(t);z]}$$

(b)
$$g(y) + \int_0^\infty f(x)K(xy)\,dx = f(y), \quad y > 0, \quad \text{with}$$
$$M[f(t);z] = \frac{M[g(t);z]+M[K(t);z]M[g(t);1-z]}{1-M[K(t);z]M(K(t);1-z]}$$

(c)
$$g(y) = \int_0^\infty K(y/x)\,dx, \quad y > 0, \quad \text{with}$$
$$M[f(t);z] = \frac{M[g(t);z-1]}{M[K(t);z-1]}$$

where in (a) and (b) the kernel function K depends on the product xy and in (c) on the quotient x/y.

(B) A summation formula. An infinite series of the form

$\sum_{n=0}^{\infty} f(n+a)$ can be transformed into an integral expression of the inverse Mellin type (Sneddon, p. 283).

$$\sum_{n=0}^{\infty} f(n+a) = (2\pi i)^{-1} \int_{\sigma-i\infty}^{\sigma+i\infty} \zeta(z,a)\Phi(z)\,dz$$

[$\zeta(z,a)$ is the Hurwitz zeta function] with $\max(1,\delta)<\sigma<\gamma$ where δ and γ are the abscissas of absolute convergence of the Mellin transform of the function $f(x)$ involved in the above infinite sum. Hence

$$\Phi(z) = \int_0^{\infty} f(x)x^{z-1}dx$$

(C) An asymptotic expansion theorem (Doetsch, Vol.2, p. 115). This theorem is widely used in problems of analytic number theory.

<u>Theorem</u> 1 Let $\phi(x) = \int_{c-i\infty}^{c+i\infty} x^{-z}\Phi(z)\,dz$ such that

(a) $\Phi(z)$ is analytic in a left half plane $\operatorname{Re} z \leqq c$ except for singular points of one-calued character (poles or essential singularities) $\lambda_0, \lambda_1, \lambda_2, \cdots$; $c > \operatorname{Re}\lambda_0 > \operatorname{Re}\lambda_1 > \cdots \to -\infty$

The principle part of the Laurent expansion of $\Phi(z)$ at $z = \lambda_\nu$ be

$$b_1^{(\nu)}(z-\lambda_\nu)^{-1} + b_2^{(\nu)}(z-\lambda_\nu)^{-2} + \cdots + b_{r_\nu}^{(\nu)}(z-\lambda_\nu)^{-r_\nu} + \cdots$$

(b) In every strip of finite width $c_0 \leqq \sigma \leqq c$, $\phi(\sigma+i\tau) \to 0$ as $|\tau| \to \infty$ uniformly in σ.

(c) Between two singularities λ_ν and $\lambda_{\nu+1}$ there exists a β_ν (real) with $\mathrm{Re}(\lambda_{\nu+1}) < \beta < \mathrm{Re}(\lambda_\nu)$, $(\nu=0,1,2,\cdots)$ such that the integral

$$\int_{-\infty}^{\infty} x^{-i\tau}\Phi(\beta_\nu+i\tau)\,d\tau$$

converges uniformly for $0 < x \leqq X_\nu$. [This is the case, for example, if $\Phi(z) = \Phi(\sigma+i\tau) = O(|\tau|^\alpha)$ for fixed σ with $\mathrm{Re}\,\alpha < 0$.]

Then

$$(2\pi i)^{-1}\int_{c-i\infty}^{c+i\infty} x^{-z}\Phi(z)\,dz = \phi(x)$$

converges for $0 < x \leqq X_0$ and

$$\phi(x) = \sum_{\nu=0}^{n}\left[b_1^{(\nu)} + \frac{b_2^{(\nu)}}{1!}(-\log x) + \cdots + \frac{b_{r_\nu}^{(\nu)}}{(r_\nu-1)!}(-\log x)^{r_\nu-1} + \cdots\right]x^{-\lambda_\nu}$$

$$+ (2\pi i)^{-1}\int_{\beta_n-i\infty}^{\beta_n+i\infty} x^{-z}\Phi(z)\,dz$$

as $x \to 0$ (through positive values), where the last term (the integral) is $o(x^{-\beta_n})$.

Similarly

<u>Theorem</u> 2

(a) The function $\Phi(z)$ be analytic in a right half plane $\mathrm{Re}\,z \geqq a$ except for singular points of one valued character $\lambda_0, \lambda_1, \cdots$ with $a<\mathrm{Re}\,\lambda_0<\mathrm{Re}\,\lambda_1<\cdots \to +\infty$.

The principal part of the Laurent expansion of $\Phi(z)$ at $z = \lambda_\nu$ be

$$b_1^{(\nu)}(z-\lambda_\nu)^{-1} + b_2^{(\nu)}(z-\lambda_\nu)^{-2} + \cdots + b_{r_\nu}^{(\nu)}(z-\lambda_\nu)^{-r_\nu} + \cdots$$

(b) In every strip of finite width $a \leqq \operatorname{Re} z \leqq a_0$, $\Phi(\sigma+i\tau) \to 0$ as $|\tau| \to \infty$ uniformly in σ.

(c) Between two singularities λ_ν and $\lambda_{\nu+1}$ there exists a β_ν (real) with $\operatorname{Re} \lambda_\nu < \beta_\nu < \operatorname{Re} \lambda_{\nu+1}$, $(\nu=0,1,2,\cdots)$ such that the integral

$$\int_{-\infty}^{\infty} x^{-i\tau}\Phi(\beta_\nu+i\tau)\,d\tau$$

converges uniformly for $x \geqq X_\nu > 0$.

Then $(2\pi i)^{-1} \int_{a-i\infty}^{a+i\infty} x^{-z}\Phi(z)\,dz$ converges for $x \geqq X_0$ and

$$\begin{aligned}\phi(x) = -\sum_{\nu=0}^{n} \Big[b_1^{(\nu)} - \frac{b_2^{(\nu)}}{1!}\log x + \cdots \\ + (-1)^{r_\nu-1}\frac{b_{r_\nu}^{(\nu)}}{(r_\nu-1)!}(\log x)^{r_\nu-1} + \cdots\Big]x^{-\lambda_\nu} \\ + (2\pi i)^{-1}\int_{\beta_n-i\infty}^{\beta_n+i\infty} x^{-z}\Phi(z)\,dz\end{aligned}$$

as $x \to +\infty$ where the last term is $o(x^{-\beta_n})$.

These theorems can be applied as follows: If the asymptotic behavior for $x \to 0$ or $x \to +\infty$ of a given function $\phi(x)$ is to be investigated one forms its

Mellin transform $\Phi(z)$ by (1) and represents $\phi(x)$ by the Mellin inversion formula (5) in the form required in the above theorems 1 and 2.

References

Churchill, R. V., 1958: Operational Mathematics, McGraw-Hill, New York.

Doetsch, G., 1950-1956: Handbuch der Laplace Transformation, 3 vols. Birkhauser Verlag, Basel.

Erdélyi, A. et.al., 1954: Tables of Integral Transforms, 2 vols. McGraw-Hill, New York.

Oberhettinger, F., 1957: Tabellen zur Fourier Transformation, Springer Verlag, Berlin.

Oberhettinger, F., 1972: Tables of Bessel Transforms, Springer Verlag, Berlin.

Oberhettinger, F. and Badii, L., 1973: Tables of Laplace Transforms, Springer Verlag, Berlin.

Van der Pol, B. and Bremmer, H., 1950: Operational Calculus based on the Two-Sided Laplace Integral, Cambridge University Press, London.

Sneddon, I. N., 1972: The Use of Integral Transforms, McGraw-Hill, New York.

Titchmarsh, E. C., 1948: Theory of Fourier Integrals, Oxford University Press.

Widder, D. V., 1971: An Instroduction to Transform Theory, Academic Press, New York.

1.1 General Formulas

	$\phi(x)$		$\Phi(z) = \int_0^\infty \phi(x) x^{z-1} dx$
1.1	$(2\pi i)^{-1} \int_{c-i\infty}^{c+i\infty} \Phi(z) x^{-z} dz$		$\Phi(z)$
1.2	$\phi(ax)$	$a > 0$	$a^{-z}\Phi(z)$
1.3	$x^{\alpha}\phi(x)$		$\Phi(z+\alpha)$
1.4	$\phi(x^{p})$	$p > 0$	$p^{-1}\Phi(z/p)$
1.5	$\phi(x^{-p})$	$p > 0$	$p^{-1}\Phi(-z/p)$
1.6	$x^{\nu}\phi(ax^{p})$	$a,p > 0$	$p^{-1}a^{-(z+\nu)/p}\Phi[(z+\nu)/p]$
1.7	$x^{\nu}\phi(ax^{-p})$	$a,p > 0$	$p^{-1}a^{(z+\nu)/p}\Phi[-(z+\nu)/p]$
1.8	$\phi(x)(\log x)^{n}$		$\Phi^{(n)}(z)$
1.9	$\phi^{(n)}(x)$ provided that $\lim_{x\to 0} x^{z-k-1}\phi^{(k)}(x) = 0$, $k = 0,1,\cdots n-1$		$(-1)^{n} \frac{\Gamma(z)}{\Gamma(z-n)} \Phi(z-n)$ $= \frac{\Gamma(n+1-z)}{\Gamma(1-z)} \Phi(z-n)$

	$\phi(x)$	$\Phi(z) = \int_0^\infty \phi(x) x^{z-1} dx$
1.10	$(x \frac{d}{dx})^n \phi(x)$	$(-z)^n \Phi(z)$
1.11	$(\frac{d}{dx} x)^n \phi(x)$	$(1-z)^n \Phi(z)$
1.12	$(x^{1-\alpha} \frac{d}{dx})^n \phi(x)$ $\alpha \neq 0$	$\frac{(-\alpha)^n \Gamma(z/\alpha)}{\Gamma(-n+z/\alpha)} \Phi(z-n\alpha)$ $= \alpha^n \frac{\Gamma(n+1-z/\alpha)}{\Gamma(1-z/\alpha)} \Phi(z-n\alpha)$
1.13	$\phi_1(x) \phi_2(x)$	$(2\pi i)^{-1} \int_{c-i\infty}^{c+i\infty} \Phi_1(s) \Phi_2(z-s) ds$
1.14	$x^\alpha \int_0^\infty t^\beta \phi_1(xt) \phi_2(t) dt$	$\Phi_1(z+\alpha) \Phi_2(1-z-\alpha+\beta)$
1.15	$x^\alpha \int_0^\infty t^\beta \phi_1(x/t) \phi_2(t) dt$	$\Phi_1(z+\alpha) \Phi_2(z+\alpha+\beta+1)$
1.16	$\int_0^x \phi(t) dt$	$-z^{-1} \Phi(z+1)$
1.17	$\int_x^\infty \phi(t) dt$	$z^{-1} \Phi(z+1)$

1.2 Algebraic Functions and Powers of Arbitrary Order

	$\phi(x)$	$\Phi(z) = \int_0^\infty \phi(x)x^{z-1}dx$
2.1	x^ν $x < a$ 0 $x > a$	$(\nu+z)^{-1}a^{\nu+z}$ $\text{Re } z > -\text{Re } \nu$
2.2	0 $x < a$ x^ν $x > a$	$-(\nu+z)^{-1}a^{\nu+1}$ $\text{Re } z < -\text{Re } \nu$
2.3	x $x < a$ $b-x$ $a<x<b$ 0 $x > b$	$(z+1)^{-1}(1+a^{z+1}-b^{z+1})+bz^{-1}(b^z-a^z)$ $\text{Re } z > -1$ $= 1+a-b+b \log(b/a)$ for $z = 0$
2.4	$(a + x)^{-1}$	$\pi a^{z-1}\csc(\pi z)$ $0<\text{Re } z < 1$
2.5	$(a + x)^{-n}$ $n = 2,3,4,\cdots$	$(-1)^{n+1}\pi[(n-1)!]^{-1}(z-1)(z-2)\cdots(z-n+1)$ $\cdot a^{z-n}\csc(\pi z)$ $0<\text{Re } z < n$
2.6	$(a + x)^{-\frac{1}{2}}$	$(\pi a)^{-\frac{1}{2}}a^z\Gamma(z)\Gamma(\frac{1}{2}-z)$ $0<\text{Re } z < \frac{1}{2}$

	$\phi(x)$		$\Phi(x) = \int_0^\infty \phi(x) x^{z-1} dx$
2.7	$(a+x)^{-1}$ 0	$x < a$ $x > a$	$\frac{1}{2} a^{z-1} [\psi(\frac{1}{2}+\frac{1}{2}z) - \psi(\frac{1}{2}z)$ $\mathrm{Re}\, z > 0$
2.8	$(a+x)^{-1}$ 0	$x < b$ $x > b$	$a^{-1} b^z Y(-b/a, 1, z)$ $\mathrm{Re}\, z > 0$
2.9	0 $(c+x)^{-1}$ 0	$x < a$ $a<x<b$ $x > b$	$c^{-1} b^z Y(-b/c,1,z) - c^{-1} a^z Y(-a/c,1,z)$ $\mathrm{Re}\, z > 0$
2.10	$[(c+bx)(d+ax)]^{-1}$		$\pi(ac-bd)^{-1}(ab)^{1-z} \csc(\pi z)$ $\cdot [(bd)^{z-1} - (ac)^{z-1}], \; 0 < \mathrm{Re}\, z < 2$
2.11	$(x+a)[(x+b)(x+c)]^{-1}$		$\pi \csc(\pi z) [(\frac{b-a}{b-c}) b^{z-1} + (\frac{c-a}{c-b}) c^{z-1}]$ $0 < \mathrm{Re}\, z < 1$
2.12	$(a-x)^{-1}$ Principal value		$\pi a^{z-1} \cot(\pi z)$ $0 < \mathrm{Re}\, z < 1$
2.13	$[(a+x)(b-x)]^{-1}$ Principal value		$\pi(a+b)^{-1}[a^{z-1}\csc(\pi z) + b^{z-1}\cot(\pi z)]$ $0 < \mathrm{Re}\, z < 2$

	$\phi(x)$	$\Phi(z) = \int_0^\infty \phi(x) x^{z-1} dx$
2.14	$[(a-x)(b-x)]^{-1}$ Principal value	$\pi\cot(\pi z)(a^{z-1}-b^{z-1})/(b-a)$ $0<\mathrm{Re}\ z<2$
2.15	$[(x+a)(x^2+b^2)]^{-1}$	$\frac{1}{2}\pi(a^2+b^2)^{-1}[2a^{z-1}\csc(\pi z)$ $-b^{z-1}\sec(\frac{1}{2}\pi z)+ab^{z-2}\csc(\frac{1}{2}\pi z)]$ $0<\mathrm{Re}\ z<3$
2.16	$(x^2+a^2+b^2)$ $\cdot\{[x^2+(a+b)^2][x^2+(a-b)^2]\}^{-1}$	$\frac{1}{4}\pi\csc(\frac{1}{2}\pi z)$ $\cdot[\lvert b-a\rvert^{z-2} + (b+a)^{z-2}]$ $0<\mathrm{Re}\ z<2$
2.17	$(x^2-b^2+a^2)$ $\cdot\{[x^2+(a+b)^2][x^2+(a-b)^2]\}^{-1}$	$\frac{1}{4}\pi a^{-1}\csc(\frac{1}{2}\pi z)$ $\cdot[(a+b)^{z-1}+\mathrm{sgn}(a-b)\lvert a-b\rvert^{z-1}]$ $0<\mathrm{Re}\ z<2$
2.18	$(1-x^{\alpha})(1-x^{n\alpha})^{-1}$ $n = 2,3,\cdots$	$\pi(n\alpha)^{-1}\sin(\pi/n)$ $\cdot\csc[\pi z/(n\alpha)]\csc(\frac{\pi z+\pi\alpha}{n\alpha})$ $0<\mathrm{Re}\ z<(n-1)\alpha$
2.19	$(b+ax)^{-\nu}$	$(b/a)^z b^{-\nu} B(z,\nu-z)$ $0<\mathrm{Re}\ z<\nu$

	$\phi(x)$	$\Phi(z) = \int_0^\infty \phi(x)x^{z-1}dx$
2.20	$(a-x)^\nu$ $\quad x < a$ 0 $\quad x > a$ $\mathrm{Re}\ \nu > -1$	$a^{\nu+z}B(\nu+1,z)$ $\mathrm{Re}\ z > 0$
2.21	0 $\quad x < a$ $(x-a)^\nu$ $\quad x > a$ $\mathrm{Re}\ \nu > -1$	$a^{\nu+z}B(-\nu-z,\nu+1)$ $\mathrm{Re}\ z<-\mathrm{Re}\ \nu$
2.22	0 $\quad x < b$ $(c+ax)^{-\nu}$ $\quad x > b$	$(ab)^{-\nu}b^z(\nu-z)^{-1}$ $\cdot {}_2F_1[\nu,\nu-z;1+\nu-z;-c/(ab)]$ $= b^z(\nu-z)^{-1}(c+ab)^{-\nu}$ $\cdot {}_2F_1[1,\nu;1+\nu-z;(ab)(c+ab)^{-1}]$ $\mathrm{Re}\ z < \mathrm{Re}\ \nu$
2.23	$(c+ax)^{-\nu}$ $\quad x < b$ 0 $\quad x > b$	$c^{-\nu}z^{-1}b^{-z}{}_2F_1(1,z;1+z;-ab/c)$ $\mathrm{Re}\ z > 0$
2.24	$[(c+bx)(d+ax)]^\nu$	$2^{-2\nu-1}\Gamma(\frac{1}{2}-\nu)\|bd-ac\|^{\frac{1}{2}+\nu}(ab)^{-\frac{1}{2}\nu-\frac{1}{2}z-\frac{1}{4}}$ $\cdot(cd)^{\frac{1}{2}\nu+\frac{1}{2}z-\frac{1}{4}}B(z_1-2\nu-z)$ $\cdot P_{z+\nu-\frac{1}{2}}^{\frac{1}{2}+\nu}[\frac{1}{2}(abcd)^{-\frac{1}{2}}(ac+bd)]$ $0<\mathrm{Re}\ z<-2\mathrm{Re}\ \nu$

	$\phi(x)$	$\Phi(z) = \int_0^\infty \phi(x) x^{z-1} dx$
2.25	$(c+bx)^{\nu}(d+ax)^{-\nu-\frac{1}{2}}$ $ac < bd$	$2^{\nu+z-1} b^{\frac{1}{2}\nu-\frac{1}{2}z} (c/a)^{\frac{1}{2}\nu+\frac{1}{2}z} d^{\frac{1}{2}z-\frac{1}{2}\nu-\frac{1}{2}}$ $\cdot\Gamma(1+\nu)\Gamma(z)\sec(\pi z)$ $\cdot\{P^{-\nu-z}_{\nu-z}[(1-\frac{ac}{bd})^{\frac{1}{2}}+P^{-\nu-z}_{\nu-z}[-(1-\frac{ac}{bd})^{\frac{1}{2}}]\}$ $0 < \mathrm{Re}\ z < \frac{1}{2}$
2.26	$(c+bx)^{\nu}(d+ax)^{-\nu-\frac{1}{2}}$ $ac > bd$	$2^{z-\nu-3/2} a^{-\frac{1}{4}-\frac{1}{2}\nu-\frac{1}{2}z} b^{\frac{1}{4}+\frac{1}{2}\nu-\frac{1}{2}z} c^{\frac{1}{2}\nu+\frac{1}{2}z-\frac{1}{4}}$ $\cdot d^{\frac{1}{2}z-\frac{1}{2}\nu-\frac{1}{4}}\Gamma(\frac{1}{2}-\nu)\Gamma(z)\sec(\pi z)$ $\cdot\{P^{\nu-z+\frac{1}{2}}_{\nu+z-\frac{1}{2}}[(1-\frac{bd}{dc})^{\frac{1}{2}}]+P^{\nu-z+\frac{1}{2}}_{\nu+z-\frac{1}{2}}[-(1-\frac{bd}{ac})^{\frac{1}{2}}]\}$ $0 < \mathrm{Re}\ z < \frac{1}{2}$
2.27	$(c+bx)^{\nu}(d+ax)^{-\nu-3/2}$ $ac < bd$	$2^{\nu+z-1}(bd-ac)^{-\frac{1}{2}} a^{-\frac{1}{2}\nu-\frac{1}{2}z} b^{\frac{1}{2}+\frac{1}{2}\nu-\frac{1}{2}z}$ $\cdot c^{\frac{1}{2}z+\frac{1}{2}\nu} d^{\frac{1}{2}z-\frac{1}{2}\nu-1}\Gamma(1+\nu)\Gamma(z)\sec(\pi z)$ $\cdot\{P^{-\nu-z}_{1+\nu-z}[(1-\frac{ac}{bd})^{\frac{1}{2}}]-P^{-\nu-z}_{1+\nu-z}[-(1-\frac{ac}{bd})^{\frac{1}{2}}]\}$ $0 < \mathrm{Re}\ z < 3/2$
2.28	$(c+bx)^{\nu}(d+ax)^{-\nu-3/2}$ $ac > bd$	$2^{z-\nu-3/2}(ac-bd)^{-\frac{1}{2}} a^{-\frac{1}{4}-\frac{1}{2}\nu-\frac{1}{2}z} b^{3/4+\frac{1}{2}\nu-\frac{1}{2}z}$ $\cdot c^{\frac{1}{2}\nu+\frac{1}{2}z-\frac{1}{4}} d^{\frac{1}{2}z-\frac{1}{2}\nu-3/4}$ $\cdot\Gamma(-\nu-\frac{1}{2})\Gamma(z)\sec(\pi z)$ $\cdot[P^{3/2+\nu-z}_{\nu+z-\frac{1}{2}}[(1-\frac{bd}{ac})^{\frac{1}{2}}]-P^{3/2+\nu-z}_{\nu+z-\frac{1}{2}}[-(1-\frac{bd}{ac})^{\frac{1}{2}}]\}$ $0 < \mathrm{Re}\ z < 3/2$

	$\phi(x)$	$\Phi(z) = \int_0^\infty \phi(x) x^{z-1} dx$
2.29	$(c+bx)^{\nu}(d+ax)^{\mu}$	$d^{\mu} c^{\nu+z} b^{-z} B(z,-\mu-\nu-z)$ $\cdot {}_2F_1(-\mu,z;-\mu-\nu;1-\frac{ac}{bd})$ $= c^{\nu} a^{-z} d^{\mu+z} B(z,-\mu-\nu-z)$ $\cdot {}_2F_1(-\nu,z;-\mu-\nu;1-\frac{bd}{ac})$ $0<\mathrm{Re}\ z<-\mathrm{Re}(\nu+\mu)$
2.30	$[(a-x)(c+bx)]^{\nu}$ $x<a$ 0 $x>a$ $\mathrm{Re}\ \nu > -1$	$\Gamma(1+\nu)(c+ab)^{\nu}(ac/b)^{\frac{1}{2}\nu+\frac{1}{2}z}$ $\cdot\Gamma(z) P_{\nu}^{-\nu-z}[(c-ab)/(c+ab)]$ $\mathrm{Re}\ z > 0$
2.31	$(a-x)^{\nu}(c+bx)^{-\nu-\frac{1}{2}}$ $x<a$ 0 $x>a$ $\mathrm{Re}\ \nu > -1$	$\Gamma(1+\nu) 2^{\nu+z} \Gamma(z) c^{\frac{1}{2}z-\frac{1}{2}\nu-\frac{1}{2}}$ $\cdot(a/b)^{\frac{1}{2}\nu+\frac{1}{2}z} P_{\nu-z}^{-\nu-z}[(1+ab/c)^{\frac{1}{2}}]$ $\mathrm{Re}\ z > 0$
2.32	$(a-x)^{\nu}(c+bx)^{-\nu-3/2}$ $x<a$ 0 $x>a$ $\mathrm{Re}\ \nu > -1$	$\Gamma(1+\nu) 2^{\nu+z}(c+ab)^{-\frac{1}{2}}\Gamma(z)$ $\cdot c^{\frac{1}{2}z-\frac{1}{2}\nu-1}(a/b)^{\frac{1}{2}\nu+\frac{1}{2}z}$ $\cdot P_{\nu-z+1}^{-\nu-z}[(1+ab/c)^{\frac{1}{2}}]$ $\mathrm{Re}\ z > 0$

	$\phi(x)$	$\Phi(z) = \int_0^\infty \phi(x) x^{z-1} dx$
2.33	$(a-x)^\nu (c+bx)^\mu \quad x < a$ $0 \quad x > a$ $\mathrm{Re}\ \nu > -1,\ b > -c/a$	$a^{\nu+z} c^\mu B(\nu+1,z)$ $\cdot {}_2F_1(-\mu,z;\nu+1+z;-ab/c)$ $= c^{\nu+z}(c+ab)^\mu B(\nu+1,z)$ $\cdot {}_2F_1(-\mu,1+\nu;1+\nu+z;\frac{ab}{c+ab})$ $\mathrm{Re}\ z > 0$
2.34	$0 \quad x < a$ $[(x-a)(c+bx)]^\nu \quad x > a$ $\mathrm{Re}\ \nu > -1$	$\Gamma(1+\nu)(c+ab)^\nu (ac/b)^{\frac{1}{2}\nu+\frac{1}{2}z}$ $\cdot \Gamma(-2\nu-z) P_\nu^{\nu+z}[(ab-c)/(ab+c)]$ $\mathrm{Re}\ z < -2\ \mathrm{Re}\ \nu$
2.35	$0 \quad x < a$ $(x-a)^\nu (c+bx)^{-\nu-\frac{1}{2}} \quad x > a$ $\mathrm{Re}\ \nu > -1$	$\Gamma(1+\nu)\Gamma(\frac{1}{2}-z) 2^{\frac{1}{2}+\nu-z} a^{\frac{1}{2}z+\frac{1}{2}\nu-\frac{1}{4}}$ $\cdot b^{-\frac{1}{4}-\frac{1}{2}\nu-\frac{1}{2}z} c^{\frac{1}{2}z-\frac{1}{2}\nu-\frac{1}{4}}$ $\cdot P_{\nu+z-\frac{1}{2}}^{z-\nu-\frac{1}{2}}[(1+\frac{c}{ab})^{\frac{1}{2}}]$ $\mathrm{Re}\ z < \frac{1}{2}$
2.36	$0 \quad x < a$ $(x-a)^\nu (c+bx)^{-\nu-3/2} \quad x > a$ $\mathrm{Re}\ \nu > -1$	$\Gamma(1+\nu)\Gamma(3/2-z)(c+ab)^{-\frac{1}{2}}$ $\cdot 2^{3/2+\nu-z} a^{\frac{1}{2}z+\frac{1}{2}\nu-\frac{1}{4}} b^{-\frac{1}{2}\nu-\frac{1}{2}z-\frac{1}{4}}$ $\cdot c^{\frac{1}{2}z-\frac{1}{2}\nu-3/4} P_{\nu+z-\frac{1}{2}}^{z-\nu-3/2}[(1+\frac{c}{ab})^{\frac{1}{2}}]$ $\mathrm{Re}\ z < 3/2$

	$\phi(x)$	$\Phi(z) = \int_0^\infty \phi(x)x^{z-1}dx$
2.37	$0 \quad x < a$ $(x-a)^{\nu}(c+bx)^{\mu} \quad x > a$ $\mathrm{Re}\ \nu > -1$	$(c+ab)^{\mu}a^{z+\nu}B(1+\nu,-\mu-\nu-z)$ $\cdot {}_2F_1[-\mu,1+\nu;1-\mu-z;c/(c+ab)]$ $= b^{\mu}a^{z+\nu+\mu}B(1+\nu;-\mu-\nu-z)$ $\cdot {}_2F_1[-\mu,-\nu-\mu-z;1-\mu-z;-c/(ab)]$ $\mathrm{Re}\ z<-\mathrm{Re}(\nu+\mu)$
2.38	$[x+(a^2+x^2)^{\frac{1}{2}}]^{-\nu}$	$\nu(\nu+z)^{-1}a^{-\nu}(\frac{1}{2}a)^{z}B(z,\frac{1}{2}\nu-\frac{1}{2}z)$ $0 < \mathrm{Re}\ z < \mathrm{Re}\ \nu$
2.39	$(a^2+x^2)^{-\frac{1}{2}}[(a^2+x^2)^{\frac{1}{2}}+a]^{\nu}$	$(2a)^{\nu+z-1}B(\frac{1}{2}z,1-\nu-z)$ $0 < \mathrm{Re}\ z<1-\mathrm{Re}\ \nu$
2.40	$(a^2+x^2)^{-\frac{1}{2}}[(a^2+x^2)^{\frac{1}{2}}+x]^{\nu}$	$2^{-z}a^{\nu+z-1}B(z,\frac{1}{2}-\frac{1}{2}z-\frac{1}{2}\nu)$ $0 < \mathrm{Re}\ z<1-\mathrm{Re}\ \nu$
2.41	$(a^2+x^2)^{-\frac{1}{2}}[(a^2+x^2)^{\frac{1}{2}}-x]^{\nu}$	$2^{-z}a^{\nu+z-1}B(z,\frac{1}{2}-\frac{1}{2}z+\frac{1}{2}\nu)$
2.42	$(b-x)^{\nu-1}$ $\cdot(x^k+a^k)^{\lambda} \quad x < b$ $0 \quad x > b$ $k=1,2,3,\cdots;\ \mathrm{Re}(\nu,z)>0$	$\Gamma(\nu)^{k\lambda}b^{\nu+1+z}\Gamma(z)[\Gamma(\nu+z)]^{-1}$ $\cdot{}_{k+1}F_k(-\lambda,\frac{z}{k},\frac{z+1}{k},\cdots,\frac{z+k-1}{k};$ $\frac{z+\nu}{k},\frac{z+\nu+1}{k},\cdots,\frac{z+\nu+k-1}{k};-b^k/a^k)$

	$\phi(x)$	$\Phi(z) = \int_0^\infty \phi(x) x^{z-1} dx$
2.43	$(a^2+x^2)^{-\frac{1}{2}}[(a^2+x^2)^{\frac{1}{2}}+b]^{\nu}$	$(2a)^{\frac{1}{2}z-1}\Gamma(\frac{1}{2}z)\Gamma(1-\nu-z)$ $\cdot(b^2-a^2)^{\frac{1}{2}\nu+\frac{1}{4}z}P_{\frac{1}{2}\nu-1}^{\nu+\frac{1}{2}z}(b/a)$ $= (2a)^{\nu+z-1}[\Gamma(1-\nu-\frac{1}{2}z)]^{-1}$ $\cdot {}_2F_1(-\frac{1}{2}\nu-\frac{1}{2}z, 1-\frac{3}{2}\nu-\frac{1}{2}z; 1-\nu-\frac{1}{2}z; \frac{1}{2}-\frac{1}{2}b/a)$ $0<\mathrm{Re}\ z<1-\mathrm{Re}\ \nu$
2.44	$(a^2+x^2)^{-\frac{1}{2}}[(a^2+x^2)^{\frac{1}{2}}+bx]^{\nu}$	$2^{-\frac{1}{2}-\frac{1}{2}\nu-\frac{1}{2}z}a^{\nu+z-1}\Gamma(z)\Gamma(\frac{1}{2}-\frac{1}{2}\nu-\frac{1}{2}z)$ $\cdot(b^2-1)^{\frac{1}{4}+\frac{1}{4}\nu-\frac{1}{4}z}P_{\frac{1}{2}\nu-1}^{\frac{1}{2}+\frac{1}{2}\nu-\frac{1}{2}z}(b)$ $= 2^{-z}a^{\nu+z-1}B(z,\frac{1}{2}-\frac{1}{2}\nu-\frac{1}{2}z)$ $\cdot {}_2F_1(\frac{1}{2}z-\frac{1}{2}, \frac{1}{2}-\nu+\frac{1}{2}z; \frac{1}{2}-\frac{1}{2}\nu+\frac{1}{2}z; \frac{1}{2}-\frac{1}{2}b)$ $0<\mathrm{Re}\ z<1-\mathrm{Re}\ \nu$
2.45	$0 \quad x<a$ $(x^2-a^2)^{-\frac{1}{2}}\{[x+(x^2-a^2)^{\frac{1}{2}}]^{\nu}$ $+[x-(x^2-a^2)^{\frac{1}{2}}]^{\nu}\} \quad x>a$	$a^{\nu+z-1}2^{-z}B(\frac{1}{2}-\frac{1}{2}z+\frac{1}{2}\nu, \frac{1}{2}-\frac{1}{2}z-\frac{1}{2}\nu)$ $\mathrm{Re}\ z<1\pm\mathrm{Re}\ \nu$
2.46	$(a^2-x^2)^{-\frac{1}{2}}[a+(a^2-x^2)^{\frac{1}{2}}]^{\nu}$ $+[a-(a^2-x^2)^{\frac{1}{2}}]^{\nu} \quad x<a$ $0 \quad x>a$	$a^{\nu+z-1}2^{\nu+z-1}B(\nu+\frac{1}{2}z,\frac{1}{2}z)$ $\mathrm{Re}\ z>\mathrm{Max}(0,-2\ \mathrm{Re}\ \nu)$

	$\phi(x)$	$\Phi(x) = \int_0^\infty \phi(x) x^{z-1} dx$
2.47	$[x+a+(x^2+2ax)^{\frac{1}{2}}]^{-\nu}$	$2\nu a^{-\nu}(\frac{1}{2}a)^z [\Gamma(1+\nu+z)]^{-1} \Gamma(2z)\Gamma(\nu-z)$ $0 < \mathrm{Re}\, z < \nu$
2.48	$(x+a)[(x+a)^2+b^2]^{-1}$	$\pi \csc(\pi z)(a^2+b^2)^{\frac{1}{2}z-\frac{1}{2}}$ $\cdot \cos[(1-z)\arctan(b/a)]$ $0 < \mathrm{Re}\, z < 1$
2.49	$(x+a)(x^2+2ax)^{-\nu-3/2}$	$\frac{1}{2}(z-1)(2a)^{z-2\nu-2}[\Gamma(3/2+\nu)]^{-1}$ $\cdot \Gamma(z-\nu-3/2)\Gamma(2+2\nu-z)$ $3/2+\mathrm{Re}\,\nu < \mathrm{Re}\, z < 2+2\mathrm{Re}\,\nu$
2.50	$[(x+a)^2-b^2]^{-\nu-\frac{1}{2}}$ $a > b$	$\pi^{\frac{1}{2}}(2b)^{-\nu}[\Gamma(\frac{1}{2}+\nu)]^{-1}(a^2-b^2)^{\frac{1}{2}z-\frac{1}{2}\nu-\frac{1}{2}}$ $\cdot \Gamma(z)\Gamma(1+2\nu-z) P_{\nu-z}^{-\nu}[a(a^2-b^2)^{-\frac{1}{2}}]$ $0 < \mathrm{Re}\, z < 1+2\,\mathrm{Re}\,\nu$
2.51	$(x+a)[(x+a)^2-b^2]^{-\nu-3/2}$ $a > b$	$(\frac{1}{2}b)^{-\nu}\Gamma(1+\nu)B(z,2\nu+2-z)$ $\cdot (a^2-b^2)^{\frac{1}{2}z-\frac{1}{2}\nu-1} P_{1+\nu-z}^{-\nu}[a(a^2-b^2)^{-\frac{1}{2}}]$ $0 < \mathrm{Re}\, z < 2+2\,\mathrm{Re}\,\nu$

	$\phi(x)$	$\Phi(z) = \int_0^\infty \phi(x) x^{z-1} dx$
2.52	$\{x+a+[(x+a)^2-b^2]^{\frac{1}{2}}\}^{-\nu}$ $a > b$	$\nu b^{-\nu}\Gamma(z)\Gamma(\nu-z)$ $\cdot(a^2-b^2)^{\frac{1}{2}z} P_z^{-\nu}[a(a^2-b^2)^{-\frac{1}{2}}]$ $0 < \mathrm{Re}\, z < \nu$
2.53	$(x^2+2ax)^{-\frac{1}{2}}$ $\cdot[x+a+(x^2+2ax)^{\frac{1}{2}}]^{-\nu}$	$\pi^{-\frac{1}{2}} a^{-\nu}(2a)^{z-1}\Gamma(\frac{1}{2}-z)\Gamma(z)$ $\cdot\Gamma(1+\nu-z)[\Gamma(z+\nu)]^{-1}$ $\frac{1}{2} < \mathrm{Re}\, z < 1 + \mathrm{Re}\, \nu$
2.54	$(x^2+2ax\cos\theta+a^2)^{-1}$ $-\pi < \theta < \pi$	$\pi\csc\theta a^{z-2}\csc(\pi z)\sin[(1-z)\theta]$ $0 < \mathrm{Re}\, z < 2$
2.55	$(x^2+2ax\cosh y+a^2)^{-1}$	$\pi\,\mathrm{csch}\, y\; a^{z-2}\csc(\pi z)\sinh[(1-z)y]$ $0 < \mathrm{Re}\, z < 2$
2.56	$(x^2+2ax\cos\theta+a^2)^{-\frac{1}{2}}$ $-\pi < \theta < \pi$	$\pi a^{z-1}\csc(\pi z) P_{z-1}(\cos\theta)$ $0 < \mathrm{Re}\, z < 1$
2.57	$(x^2+2ax\cosh y+a^2)^{-\frac{1}{2}}$	$\pi a^{z-1}\csc(\pi z) P_{z-1}(\cosh y)$ $0 < \mathrm{Re}\, z < 1$

	$\phi(x)$	$\Phi(z) = \int_0^\infty \phi(x) x^{z-1} dx$
2.58	$(x^2+2ax\cos\theta + a^2)^{-\nu}$ $-\pi < \theta < \pi$	$2^{\nu-\frac{1}{2}}\Gamma(\frac{1}{2}+\nu)(\sin\theta)^{\frac{1}{2}-\nu} a^{z-2\nu}$ $\cdot B(z,2\nu-z) P^{\frac{1}{2}-\nu}_{z-\nu-\frac{1}{2}}(\cos\theta)$ $0 < \mathrm{Re}\, z < 2\mathrm{Re}\,\nu$
2.59	$(x^2+2ax\cosh y + a^2)^{-\nu}$	$2^{\nu-\frac{1}{2}}\Gamma(\frac{1}{2}+\nu)(\sinh y)^{\frac{1}{2}-\nu} a^{z-2\nu}$ $\cdot B(z,2\nu-z) P^{\frac{1}{2}-\nu}_{z-\nu-\frac{1}{2}}(\cosh y)$ $0 < \mathrm{Re}\, z < 2\,\mathrm{Re}\,\nu$
2.60	$x\sin\theta$ $\cdot(x^2+2ax\cos\theta+a^2)^{-1}$ $-\pi < \theta < \pi$	$\pi a^{z-1}\csc(\pi z)\sin(z\theta)$ $-1 < \mathrm{Re}\, z < 1$
2.61	$x\sinh y$ $\cdot(x^2+2ax\cosh y+a^2)^{-1}$	$\pi a^{z-1}\csc(\pi z)\sinh(yz)$ $-1 < \mathrm{Re}\, z < 1$
2.62	$(a+x\cos\theta)$ $\cdot(x^2+2ax\cos\theta+a^2)^{-1}$ $-\pi < \theta < \pi$	$\pi a^{z-1}\csc(\pi z)\cos(z\theta)$ $0 < \mathrm{Re}\, z < 1$

	$\phi(x)$	$\Phi(z) = \int_0^\infty \phi(x) x^{z-1} dx$
2.63	$(a+x \cosh y)$ $\cdot(x^2+2ax \cosh y + a^2)^{-1}$	$\pi a^{z-1} \csc(\pi z) \cosh(zy)$ $0 < \mathrm{Re}\, z < 1$
2.64	$(x^2+2a\zeta x+a^2)^{-\nu}$ ζ not on the real axis between -1 and $-\infty$.	$2^{\nu-\frac{1}{2}}\Gamma(\frac{1}{2}+\nu) a^{z-2\nu} B(z,2\nu-z)$ $\cdot(\zeta^2-1)^{\frac{1}{4}-\frac{1}{2}\nu} P^{\frac{1}{2}-\nu}_{z-\nu-\frac{1}{2}}(\zeta)$ $= a^{z-2\nu} B(z,2\nu-z)$ $\cdot {}_2F_1(z,2\nu-z;\frac{1}{2}+\nu;\frac{1}{2}-\frac{1}{2}\zeta)$ $0 < \mathrm{Re}\, z < 2\, \mathrm{Re}\, \nu$

1.3 Exponential Functions

3.1	e^{-ax}	$a^{-z}\Gamma(z)$ $\mathrm{Re}\, z > 0$
3.2	e^{-bx} $x < a$ 0 $x > a$	$b^{-z}\gamma(z,ab)$ $\mathrm{Re}\, z > 0$
3.3	0 $x < a$ e^{-bx} $x > a$	$b^{-z}\Gamma(z,ab)$
3.4	$(b+x)^{-1}e^{-ax}$	$e^{ab}b^{z-1}\Gamma(z)\Gamma(1-z,ab)$ $\mathrm{Re}\, z > 0$

	$\phi(x)$	$\Phi(z) = \int_0^\infty \phi(x) x^{z-1} dx$
3.5	$(b+x)^{\nu} e^{-ax}$	$e^{\frac{1}{2}ab} b^{\frac{1}{2}z+\frac{1}{2}\nu-\frac{1}{2}} a^{-\frac{1}{2}-\frac{1}{2}\nu-\frac{1}{2}z}$ $\cdot \Gamma(z) W_{\frac{1}{2}+\frac{1}{2}\nu-\frac{1}{2}z, \frac{1}{2}\nu+\frac{1}{2}z}(ab)$ $\operatorname{Re} z > 0$
3.6	$(b^2+x^2)^{-1} e^{-ax}$	$a^{\frac{1}{2}} b^{z-3/2} \Gamma(z) S_{\frac{1}{2}-z, \frac{1}{2}}(ab)$ $\operatorname{Re} z > 0$
3.7	$(b-x)^{\nu} e^{ax}$ $x < b$; 0 $x > b$; $\operatorname{Re} \nu > -1$	$b^{\nu+z} B(z, 1+\nu)$ $\cdot {}_1F_1(z; z+\nu+1; ab)$ $\operatorname{Re} z > 0$
3.8	$(b^2-x^2)^{-\frac{1}{2}} \exp[-a(b^2-x^2)^{\frac{1}{2}}]$ $x < b$; 0 $x > b$	$\frac{1}{2}\pi^{\frac{1}{2}} (2b/a)^{\frac{1}{2}z-\frac{1}{2}} \Gamma(\frac{1}{2}z)$ $\cdot [I_{\frac{1}{2}z-\frac{1}{2}}(ab) - \mathbf{L}_{\frac{1}{2}z-\frac{1}{2}}(ab)]$ $\operatorname{Re} z > 0$
3.9	0 $x < b$; $\exp[-a(x^2-b^2)^{\frac{1}{2}}]$ $x > b$	$(ab)^{\frac{1}{2}} (b/a)^{\frac{1}{2}z} S_{\frac{1}{2}z-1-\nu, \frac{1}{2}z+\nu}(ab)$
3.10	0 $x < b$; $(x^2-b^2)^{-\frac{1}{2}} \exp[-a(x^2-b^2)^{\frac{1}{2}}]$ $x > b$	$\frac{1}{2}\pi^{\frac{1}{2}} (2b/a)^{\frac{1}{2}z-\frac{1}{2}} \Gamma(\frac{1}{2}z)$ $\cdot [\mathbf{H}_{\frac{1}{2}z-\frac{1}{2}}(ab) - Y_{\frac{1}{2}z-\frac{1}{2}}(ab)]$

	$\phi(x)$	$\Phi(z) = \int_0^\infty \phi(x) x^{z-1} dx$
3.11	$(a^2+x^2)^{-\frac{1}{2}}$ $\cdot \exp[-b(a^2+x^2)^{\frac{1}{2}}]$	$\pi^{-\frac{1}{2}}(2a/b)^{\frac{1}{2}z-\frac{1}{2}}\Gamma(\frac{1}{2}z)K_{\frac{1}{2}-\frac{1}{2}z}(ab)$ Re $z > 0$
3.12	$\exp[-b(a^2+x^2)^{\frac{1}{2}}]$	$\pi^{-\frac{1}{2}}a(2a/b)^{\frac{1}{2}z-\frac{1}{2}}\Gamma(\frac{1}{2}z)K_{\frac{1}{2}+\frac{1}{2}z}(ab)$ Re $z > 0$
3.13	e^{-ax^2-bx}	$(2a)^{-\frac{1}{2}z}\Gamma(z)\exp(\frac{1}{8}b^2/a)$ $\cdot D_{-z}[b(2a)^{-\frac{1}{2}}]$ Re $z > 0$
3.14	$\exp(-ax^p)$ $p > 0$	$p^{-1}a^{-z/p}\Gamma(z/p)$ Re $z > 0$
3.15	$\exp(-ax^{-p})$ $p > 0$	$p^{-1}a^{z/p}\Gamma(-z/p)$ Re $z < 0$
3.16	$\exp(-ax^p-bx^{-p})$ $p > 0$	$2p^{-1}(b/a)^{\frac{1}{2}z/p}K_{z/p}[2(ab)^{\frac{1}{2}}]$
3.17	$1-\exp(-ax^p)$ $p > 0$	$-p^{-1}a^{-z/p}\Gamma(z/p)$ $-p <$ Re $z < 0$
3.18	$1-\exp(-ax^{-p})$ $p > 0$	$-p^{-1}a^{z/p}\Gamma(-z/p)$ $0 <$ Re $z < p$

	$\phi(x)$	$\Phi(z) = \int_0^\infty \phi(x) x^{z-1} dx$
3.19	$(e^{ax}-1)^{-1}$	$a^{-z}\Gamma(z)\zeta(z)$ Re $z > 1$
3.20	$(e^{ax}-1)^{-1}e^{-bx}$	$a^{-z}\Gamma(z)\zeta(z,1+b/a)$ Re $z > 1$
3.21	$e^{-px}(e^{ax}+1)^{-1}$ Re $z>0$, $p>-a$	$(2a)^{-z}\Gamma(z)[\zeta(z,\tfrac{1}{2}+\tfrac{1}{2}p/a)-\zeta(z,1+\tfrac{1}{2}p/a)]$
3.22	$e^{-ax}[\tfrac{1}{2}-x^{-1}$ $+(e^{x}-1)^{-1}]$	$\Gamma(z)[\zeta(z,a)-\tfrac{1}{2}a^{-z}+(1-z)^{-1}a^{1-z}]$ Re $z > -1$
3.23	$(e^{x}-1)^{-2}$	$\Gamma(z)[\zeta(z-1)-\zeta(z)]$ Re $z > 2$
3.24	$e^{-ax}(1-e^{-x})^{-2}$	$\Gamma(z)[\zeta(z-1,a-1)-(a-1)\zeta(z,a-1)]$ Re $z > 2$
3.25	$(e^{x}-\zeta)^{-1}e^{-ax}$ $\lvert\arg\zeta\rvert < \pi$	$\Gamma(z)\Upsilon(\zeta,z,a+1)$ Re $z > 0$
3.26	$(e^{ax}+1)^{-1}$	$a^{-z}\Gamma(z)(1-2^{1-z})\zeta(z)$ Re $z > 0$

	$\phi(x)$	$\Phi(z) = \int_0^\infty \phi(x) x^{z-1} dx$
3.27	$0 \quad x < 1$ $(x^2-1)^{-\frac{1}{2}}$ $\cdot \exp[-a(x-x^{-1})] \quad x > 1$	$\frac{1}{2}\pi^{3/2} a^{\frac{1}{2}} [J_{\frac{1}{2}-\frac{1}{2}z}(a) Y_{-\frac{1}{2}z}(a)$ $-J_{-\frac{1}{2}z}(a) Y_{\frac{1}{2}-\frac{1}{2}z}(a)]$
3.28	$(1-t^2)^{-\frac{1}{2}} \quad x < 1$ $\cdot \exp[-a(x^{-1}-x)]$ $0 \quad x > 1$	$\frac{1}{2}\pi^{3/2} a^{\frac{1}{2}} [J_{\frac{1}{2}z}(a) Y_{\frac{1}{2}z-\frac{1}{2}}(a)$ $-J_{\frac{1}{2}z-\frac{1}{2}}(a) Y_{\frac{1}{2}z}(a)]$
3.29	$e^{-ax}[(e^x-1)^{-1}-x^{-1}]$ $\mathrm{Re}\, z > 0$	$\Gamma(z)[\zeta(z,a)+(1-z)^{-1}a^{1-z}-a^{-z}]$
3.30	$(1+x^2)^{-\nu-1}$ $\cdot \exp[-a(\frac{1-x^2}{1+x^2})]$	$\frac{1}{2}(2a)^{-\frac{1}{2}\nu-\frac{1}{2}}[\Gamma(1+\nu)]^{-1}\Gamma(\frac{1}{2}z)$ $\cdot \Gamma(1+\nu-\frac{1}{2}z) M_{\frac{1}{2}-\frac{1}{2}z+\frac{1}{2}\nu, \frac{1}{2}\nu}(2a)$ $0 < \mathrm{Re}\, z < 2+2\, \mathrm{Re}\, \nu$
3.31	$(1+x^2)^{-\nu-1}$ $\cdot \exp(-\frac{2a}{1+x^2})$	$\frac{1}{2}(2a)^{-\frac{1}{2}\nu-\frac{1}{2}}e^{-a}[\Gamma(1+\nu)]^{-1}\Gamma(\frac{1}{2}z)$ $\cdot \Gamma(1+\nu-\frac{1}{2}z) M_{\frac{1}{2}-\frac{1}{2}z+\frac{1}{2}\nu, \frac{1}{2}\nu}(2a)$ $0 < \mathrm{Re}\, z < 2+2\, \mathrm{Re}\, \nu$

	$\phi(x)$	$\Phi(z) = \int_0^\infty (x) x^{z-1} dx$
3.32	$(1-x^2)^{-\nu-1}$ $\cdot \exp[-a(\frac{1+x^2}{1-x^2})]$, $x < 1$ 0 $x > 1$	$\frac{1}{2}(2a)^{-\frac{1}{2}\nu-\frac{1}{2}}\Gamma(\frac{1}{2}z) W_{\frac{1}{2}\nu+\frac{1}{2}-\frac{1}{2}z, \frac{1}{2}\nu}(2a)$ $\mathrm{Re}\, z > 0$
3.33	$(1-x^2)^{-\nu-1}$ $\cdot \exp(-\frac{2ax^2}{1-x^2})$ $x < 1$ 0 $x > 1$	$\frac{1}{2}(2a)^{-\frac{1}{2}\nu-\frac{1}{2}} e^a \Gamma(\frac{1}{2}z) W_{\frac{1}{2}\nu+\frac{1}{2}-\frac{1}{2}z, \frac{1}{2}\nu}(2a)$ $\mathrm{Re}\, z > 0$
3.34	$(1-x^2)^{-\frac{1}{2}}$ $\cdot \exp[-a(\frac{1+x}{1-x})]$, $x < 1$ 0 $x > 1$	$\Gamma(z) D_{-z}^2[(2a)^{\frac{1}{2}}]$ $\mathrm{Re}\, z > 0$
3.35	$(1-x^2)^{-\frac{1}{2}}$ $\cdot \exp(-\frac{2ax}{1-x})$, $x < 1$ 0 $x > 1$	$e^a \Gamma(z) D_{-z}^2[(2a)^{\frac{1}{2}}]$ $\mathrm{Re}\, z > 0$
3.36	$(1-x^2)^{-\frac{1}{2}}$ $\cdot \exp[-(\frac{2ax+b+bx^2}{1-x^2})]$ $x < 1$ 0 $x > 1$	$\Gamma(z) D_{-z}\{2^{\frac{1}{2}}[b+(b^2-a^2)^{\frac{1}{2}}]^{\frac{1}{2}}\}$ $\cdot D_{-z}\{2^{\frac{1}{2}}[b-(b^2-a^2)^{\frac{1}{2}}]^{\frac{1}{2}}\}$ $\mathrm{Re}\, z > 0$

	$\phi(x)$	$\Phi(z) = \int_0^\infty \phi(x) x^{z-1} dx$
3.37	$(1-x^2)^{-\frac{1}{2}}$ $\cdot\exp[-2(\frac{ax+bx^2}{1-x^2})]$ $x < 1$ 0 $x > 1$	$e^b\Gamma(z)D_{-z}\{2^{\frac{1}{2}}[b+(b^2-a^2)^{\frac{1}{2}}]^{\frac{1}{2}}\}$ $\cdot D_{-z}\{2^{\frac{1}{2}}[b-(b^2-a^2)^{\frac{1}{2}}]\}$ $\mathrm{Re}\ z > 0$
3.38	$(b-x)^{\nu-1}$ $\cdot\exp(ax^k)$ $x < b$ 0 $x > b$ $k = 2,3,4,\cdots$	$\Gamma(\nu)\Gamma(z)[\Gamma(\nu+z)]^{-1}b^{\nu-1+z}$ $\cdot{}_kF_k(\frac{z}{k},\frac{z+1}{k},\cdots,\frac{z+k-1}{k};\frac{z+\nu}{k},\frac{z+\nu+1}{k},\cdots\frac{z+\nu+k-1}{k};\ ab^k)$ $\mathrm{Re}(\nu,z) > 0$
3.39	$(b-x)^{\nu-1}$ $\cdot\exp[a(b-x)^k]$ $x < b$ 0 $x > b$ $k = 2,3,4,\cdots$	$\Gamma(\nu)\Gamma(z)[\Gamma(\nu+z)]^{-1}b^{\nu-1+z}$ $\cdot{}_kF_k(\frac{\nu}{k},\frac{\nu+1}{k},\cdots,\frac{\nu+k-1}{k},\frac{z+\nu}{k},\frac{z+\nu+1}{k},\cdots\frac{z+\nu+k-1}{k};\ ab^k)$ $\mathrm{Re}(\nu,z) > 0$
3.40	$(b-x)^{\nu-1}$ $x < b$ $\cdot\exp(ax^{\frac{1}{2}})$ 0 $x > b$ $\mathrm{Re}(\nu,z) > 0$	$\Gamma(z)\Gamma(\nu)[\Gamma(\nu+z)]^{-1}b^{\nu-1+z}$ $\cdot{}_1F_2(z;\frac{1}{2},\nu+z;\frac{1}{4}a^2b)+a\Gamma(\frac{1}{2}+z)\Gamma(\nu)$ $\cdot[\Gamma(\frac{1}{2}+\nu+z)]^{-1}b^{\nu-\frac{1}{2}+z}$ $\cdot{}_1F_2(\frac{1}{2}+z;\frac{3}{2},\frac{1}{2}+\nu+z;\frac{1}{4}a^2b)$

	$\phi(x)$	$\Phi(z) = \int_0^\infty \phi(x) x^{z-1} dx$
3.41	$0 \qquad x < 1$ $(x^2-1)^{-\frac{1}{2}}$ $\cdot \exp[-a(\frac{x+1}{x-1})]$ $x > 1$	$D^2_{z-1}[(2a)^{\frac{1}{2}}]\Gamma(1-z)$ $\mathrm{Re}\, z < 1$
3.42	$0 \qquad x < 1$ $(x^2-1)^{-\frac{1}{2}}$ $\cdot \exp(-\frac{2a}{x-1}) \qquad x > 1$	$e^a\Gamma(1-z)D^2_{z-1}[(2a)^{\frac{1}{2}}]$ $\mathrm{Re}\, z < 1$
3.43	$0 \qquad x < 1$ $(x^2-1)^{-\frac{1}{2}}$ $\cdot \exp[-(\frac{2ax+b+bx^2}{x^2-1})]$ $x > 1$	$\Gamma(1-z)D_{z-1}\{2^{\frac{1}{2}}[b+(b^2-a^2)^{\frac{1}{2}}]^{\frac{1}{2}}\}$ $\cdot D_{z-1}\{2^{\frac{1}{2}}[b-(b^2-a^2)^{\frac{1}{2}}]^{\frac{1}{2}}\}$ $\mathrm{Re}\, z < 1$
3.44	$0 \qquad x < 1$ $(x^2-1)^{-\frac{1}{2}}$ $\cdot \exp[-2(\frac{ax+b}{x^2-1})]$ $x > 1$	$e^b\Gamma(1-z)D_{z-1}\{2^{\frac{1}{2}}[b+(b^2-a^2)^{\frac{1}{2}}]^{\frac{1}{2}}\}$ $\cdot D_{z-1}\{2^{\frac{1}{2}}[b-(b^2-a^2)^{\frac{1}{2}}]\}$ $\mathrm{Re}\, z < 1$

	$\phi(x)$	$\Phi(z) = \int_0^\infty \phi(x) x^{z-1} dx$
3.45	$(1+x^2)^{-\nu-1}$ $\cdot \exp[-a(\frac{1-x^2}{1+x^2})]$	$\frac{1}{2}(2a)^{-\frac{1}{2}\nu-\frac{1}{2}}[\Gamma(1+\nu)]^{-1}\Gamma(1+\nu-\frac{1}{2}z)$ $\cdot\Gamma(\frac{1}{2}z) M_{\frac{1}{2}-\frac{1}{2}z+\frac{1}{2}\nu,\frac{1}{2}}(2a)$ $0 < \mathrm{Re}\ z < 2+2\ \mathrm{Re}\ \nu$
3.46	$(1+x^2)^{-\nu-1}$ $\cdot \exp(-\frac{2a}{1+x^2})$	$\frac{1}{2}(2a)^{-\frac{1}{2}\nu-\frac{1}{2}}e^{-a}[\Gamma(1+\nu)]^{-1}\Gamma(1+\nu-\frac{1}{2}z)$ $\cdot\Gamma(\frac{1}{2}z) M_{\frac{1}{2}-\frac{1}{2}z+\frac{1}{2}\nu,\frac{1}{2}\nu}(2a)$ $0 < \mathrm{Re}\ z < 2+2\ \mathrm{Re}\ \nu$
3.47	$0 \quad x < 1$ $(x^2-1)^{-\nu-1}$ $\cdot \exp[-a(\frac{x^2+1}{x^2-1})]$ $x > 1$	$\frac{1}{2}(2a)^{-\frac{1}{2}\nu-\frac{1}{2}}\Gamma(1+\nu-\frac{1}{2}z) W_{\frac{1}{2}z-\frac{1}{2}\nu-\frac{1}{2},\frac{1}{2}\nu}(2a)$ $\mathrm{Re}\ z < 2+2\ \mathrm{Re}\ \nu$
3.48	$0 \quad x < 1$ $(x^2-1)^{-\nu-1}$ $\cdot \exp(-\frac{2a}{x^2-1})$ $x > 1$	$\frac{1}{2}(2a)^{-\frac{1}{2}\nu-\frac{1}{2}}e^{a}\Gamma(1+\nu-\frac{1}{2}z)$ $\cdot W_{\frac{1}{2}z-\frac{1}{2}\nu-\frac{1}{2},\frac{1}{2}\nu}(2a)$ $\mathrm{Re}\ z < 2+2\ \mathrm{Re}\ \nu$

1.4 Logarithmic Functions

	$\phi(x)$	$\Phi(z) = \int_0^\infty \Phi(x) x^{z-1} dx$
4.1	$\log x \quad x < a$ $0 \quad x > a$	$z^{-1} a^z (\log a - z^{-1}) \quad \text{Re } z > 0$
4.2	$0 \quad x < a$ $\log(x/a) \quad x > a$	$a^z z^{-2} \quad \text{Re } z < 0$
4.3	$(x+a)^{-1} \log x$	$\pi a^{z-1} \csc(\pi z)[\log a - \pi \cot(\pi z)]$ $0 < \text{Re } z < 1$
4.4	$x^{-1} \log(1+x)$	$\pi \csc(\pi z)(1-z)^{-1} \quad 0 < \text{Re } z < 1$
4.5	$\log(1+a/x)$	$\pi z^{-1} a^z \csc(\pi z) \quad 0 < \text{Re } z < 1$
4.6	$\log(1+ax)$	$\pi z^{-1} a^{-z} \csc(\pi z) \quad -1 < \text{Re } z < 0$
4.7	$(b+ax)^{-1} \log(b+ax)$ $0 < \text{Re } z < 1$	$-b^{1-\nu} a^{-z} \pi \csc(\pi z)$ $\cdot [\gamma + \psi(1-z) - \log b]$
4.8	$\log[(a+cx)/(b+cx)]$	$\pi z^{-1} c^{-z} \csc(\pi z)(a^z - b^z)$ $0 < \text{Re } z < 1$

	$\phi(x)$	$\Phi(x) = \int_0^\infty \Phi(x) x^{z-1} dx$
4.9	$\log\|1-ax\|$	$\pi z^{-1} a^{-z} \cot(\pi z)$ $-1<\mathrm{Re}\ z<0$
4.10	$\log\left\|\frac{1+x}{1-x}\right\|$	$\pi z^{-1} \tan(\frac{1}{2}\pi z)$ $-1<\mathrm{Re}\ z<1$
4.11	$\log\ \left\|\frac{a+x}{b-x}\right\|$	$\pi z^{-1} \csc(\pi z)[a^2-b^2\cos(\pi z)]$ $0<\mathrm{Re}\ z<1$
4.12	$(x+1)^{-1}\log x$ $x<1$ 0 $x>1$	$\frac{1}{4}[\psi'(\frac{1}{2}+\frac{1}{2}z)-\psi'(\frac{1}{2}z)]$ $\mathrm{Re}\ z>0$
4.13	$(x-1)^{-1}\log x$ $x<1$ 0 $x>1$	$\psi'(z)$ $\mathrm{Re}\ z>0$
4.14	$(a-x)^{-1}\log x$ Principal value	$\pi a^{z-1}[\log a \cot(\pi z)-\pi\csc^2(\pi z)]$ $0<\mathrm{Re}\ z<1$
4.15	$(a+x)^{-1}(b+x)^{-1}\log x$ $0<\mathrm{Re}\ z<1$	$\pi(b-a)^{-1}\csc(\pi z)[a^{z-1}\log a$ $-b^{z-1}\log b-\pi\cot(\pi z)(a^{z-1}-b^{z-1})]$
4.16	$\log(1+x)$ $x<1$ 0 $x>1$ $\mathrm{Re}\ z>-1$	$z^{-1}[\log 2-\frac{1}{2}\psi(1+\frac{1}{2}z)$ $+\frac{1}{2}\psi(\frac{1}{2}+\frac{1}{2}z)]$

	$\phi(x)$	$\Phi(z) = \int_0^\infty \phi(x) x^{z-1} dx$
4.17	$0 \qquad x < 1$ $\log(1+x) \qquad x > 1$	$\pi z^{-1} \cot(\pi z)$ $-1 < \mathrm{Re}\, z < 0$
4.18	$\log(1-x) \qquad x < 1$ $0 \qquad x > 1$	$-z^{-1}[\gamma+\psi(z+1)] \qquad \mathrm{Re}\, z > -1$
4.19	$0 \qquad x < 1$ $\log(x-1) \qquad x > 1$	$z^{-1}[z^{-1}+\psi(1-z)+\gamma]$ $\mathrm{Re}\, z < 0$
4.20	$(1+x)^{-1}(\log x)^2$	$\pi^3 \csc^3(\pi z)[2-\sin^2(\pi z)] \quad 0 < \mathrm{Re}\, z < 1$
4.21	$(1+x)^{-1}\log(1+x^2)$ $-2 < \mathrm{Re}\, z < 1$	$\tfrac{1}{2}\pi\csc(\pi z)\{\log 4+(1-z)\sin(\tfrac{1}{2}\pi z)$ $\cdot[\psi(\tfrac{3}{4}-\tfrac{1}{4}z)-\psi(\tfrac{1}{4}-\tfrac{1}{4}z)]$ $-(2-z)\cos(\tfrac{1}{2}\pi z)[\psi(1-\tfrac{1}{4}z)-\psi(\tfrac{1}{2}-\tfrac{1}{4}z)]\}$
4.22	$\log(1+2x\cos\theta+x^2)$ $-\pi < \theta < \pi$	$2\pi z^{-1}\cos(z\theta)\csc(\pi z)$ $-1 < \mathrm{Re}\, z < 0$
4.23	$\log[x+(a^2+x^2)^{\frac{1}{2}}]$	$-(\tfrac{1}{2}a)^z z^{-1} B(z,\tfrac{1}{2}\nu-\tfrac{1}{2}z) \qquad -1 < \mathrm{Re}\, z < 0$

	$\phi(x)$	$\Phi(z) = \int_0^\infty \phi(x)x^{z-1}dx$
4.24	$(a^2+x^2)^{-\frac{1}{2}}$ $\cdot\log[(a^2+x^2)^{\frac{1}{2}}-x]$	$2^{-z}a^{z-1}B(z,\frac{1}{2}-\frac{1}{2}z)[\log a-\frac{1}{2}\pi\tan(\frac{1}{2}\pi z)]$ $0 < \operatorname{Re} z < 1$
4.25	$(a^2+x^2)^{-\frac{1}{2}}$ $\cdot\log[(a^2+x^2)^{\frac{1}{2}}+x]$	$2^{-z}a^{z-1}B(z,\frac{1}{2}-\frac{1}{2}z)[\log a+\frac{1}{2}\pi\tan(\frac{1}{2}\pi z)]$ $0 < \operatorname{Re} z < 1$
4.26	$\log[1+(1-x^2)^{\frac{1}{2}}]$ $x < 1$ 0 $x > 1$	$z^{-1}[\frac{1}{2}\pi^{\frac{1}{2}}\frac{\Gamma(\frac{1}{2}z)}{\Gamma(\frac{1}{2}+\frac{1}{2}z)} - z^{-1}]$ $\operatorname{Re} z > 0$
4.27	0 $x < 1$ $\log[x+(x^2-1)^{\frac{1}{2}}]$ $x > 1$	$-\frac{1}{2}\pi^{\frac{1}{2}}z^{-1}\Gamma(-\frac{1}{2}z)[\Gamma(\frac{1}{2}-\frac{1}{2}z)]^{-1}$ $\operatorname{Re} z < 0$
4.28	0 $x < a$ $\log[\frac{x+(x^2-a^2)^{\frac{1}{2}}}{x-(x^2-a^2)^{\frac{1}{2}}}]$ $x > a$	$-\pi^{\frac{1}{2}}z^{-1}a^z\Gamma(-\frac{1}{2}z)[\Gamma(\frac{1}{2}-\frac{1}{2}z)]^{-1}$ $\operatorname{Re} z < 0$
4.29	$(y^2-1)^{-\frac{1}{2}}$ $\cdot\log[y+(y^2-1)^{\frac{1}{2}}]$ $y = (2ab)^{-1}(a^2+b^2+x^2)$	$2a^{3/2}b^{3/2}\Gamma(\frac{1}{2}z)[\Gamma(3/2-\frac{1}{2}z)]^2$ $\cdot[\Gamma(1-\frac{1}{2}z)]^{-1}(\lvert b^2-a^2\rvert)^{-\frac{1}{2}z}q_{-\frac{1}{2}z-1}(\frac{b^2+a^2}{\lvert b^2-a^2\rvert})$ $0 < \operatorname{Re} z < 1$

	$\phi(x)$	$\Phi(z) = \int_0^\infty \phi(x)x^{z-1}dx$
4.30	$\log[x+(1+x^2)^{\frac{1}{2}}]$	$-\frac{1}{2}z^{-1}B(\frac{1}{2}+\frac{1}{2}z,-\frac{1}{2}z)$ $\quad -1<\text{Re } z<0$
4.31	$(a^2+x^2+2ax\cos\theta)^{-1} \cdot (\log x)^n$ $\quad -\pi<\theta<\pi$	$-\pi\cos\theta \frac{d^n}{dz^n}[a^{z-2}\csc(\pi z)\sin(z-1)\theta]$ $\quad 0<\text{Re } z<2$
4.32	$e^{-x}(\log x)^n$	$\frac{d^n}{dz^n}\Gamma(z)$ $\quad \text{Re } z>0$
4.33	$\log(1+e^{-ax})$	$a^{-z}(1-2^{-z})\Gamma(z)\zeta(1+z)$ $\quad \text{Re } z>0$
4.34	$\log(1-e^{-ax})$	$-a^{-z}\Gamma(z)\zeta(1+z)$ $\quad \text{Re } z>0$
4.35	$\log[\tanh(ax)]$	$-(2a)^{-z}\Gamma(z)\zeta(1+z)(2-2^{-z})$ $\quad \text{Re } z>0$
4.36	$\log(1-2ae^{-x}\cos\theta+a^2e^{-2x})$ $\quad 0<a<1,\ -\pi<\theta<\pi$	$-2\Gamma(z)\sum_1^\infty (n+1)^{-1}a^n\cos(n\theta)$ $\quad \text{Re } z>0$
4.37	$(\log 1/x)^\nu$ $\quad x<1$; 0 $\quad x>1$	$z^{-\nu-1}\Gamma(\nu+1)$ $\quad \text{Re } \nu>-1$ $\quad \text{Re } z>0$

	$\phi(x)$	$\Phi(z) = \int_0^\infty \phi(x)x^{z-1}dx$
4.38	$(a-x)^\nu \log(a-x)$ $x < a$ 0 $x > a$ $\mathrm{Re}\,\nu > -1$	$a^{\nu+z}B(\nu+1,z)$ $\cdot[\log a+\psi(1+\nu)-\psi(1+\nu+z)]$ $\mathrm{Re}\,z > 0$
4.39	0 $x < a$ $(x-a)^\nu \log(x-a)$ $x > a$ $\mathrm{Re}\,\nu > -1$	$a^{\nu+z}B(\nu+1,-\nu-z)$ $\cdot[\log a+\psi(1+\nu)-\psi(-\nu-z)]$ $\mathrm{Re}\,z < -\mathrm{Re}\,\nu$
4.40	$(b+ax)^{-\nu}$ $\cdot\log(b+ax)$	$b^{-\nu}(b/a)^z B(z,\nu-z)$ $\cdot[\log b+\psi(\nu)-\psi(\nu-z)]$ $0 < \mathrm{Re}\,z < \mathrm{Re}\,\nu$
4.41	$(1-x)^\nu \log x$ $x < 1$ 0 $x > 1$ $\mathrm{Re}\,\nu > -2$	$B(\nu+1,z)[\psi(z)-\psi(\nu+1+z)]$ $\mathrm{Re}\,z > 0$
4.42	$(1-x)^\nu (\log x)^2$ $x < 1$ 0 $x > 1$ $\mathrm{Re}\,\nu > -1$	$B(\nu+1,z)\{[\psi(z)-\psi(\nu+1+z)]^2$ $+\psi'(z)-\psi'(\nu+1+z)\}$ $\mathrm{Re}\,z > 0$
4.43	$e^{-at}(\log t)^2$	$\Gamma(z)a^{-z}\{[\psi(z)-\log a]^2+\psi'(z)\}$ $\mathrm{Re}\,z > 0$

	$\phi(x)$	$\Phi(z) = \int_0^\infty \phi(x) x^{z-1} dx$
4.44	$e^{-at}\log t$	$\Gamma(z) a^{-z}[\psi(z)-\log a]$ Re $z > 0$
4.45	$[x+(a^2+x^2)^{\frac{1}{2}}]^{-\nu}$ $\cdot\log[x+(a^2+x^2)^{\frac{1}{2}}]$ For $a=1$, $-1<\text{Re } z<\text{Re }\nu$	$-a^{-\nu}(\nu+z)^{-1}(\frac{1}{2}a)^z B(z,\frac{1}{2}\nu-\frac{1}{2}z)$ $\cdot[1-\nu\log a-\frac{1}{2}\nu\psi(1+\frac{1}{2}\nu+\frac{1}{2}z)$ $+\frac{1}{2}\nu\psi(\frac{1}{2}\nu-\frac{1}{2}z)]$, $0<\text{Re } z<\text{Re }\nu$
4.46	$(a^2+x^2)^{-\frac{1}{2}}[(a^2+x^2)^{\frac{1}{2}}-x]^{\nu}$ $\cdot\log[(a^2+x^2)^{\frac{1}{2}}-x]$ $0<\text{Re } z<1+\text{Re }\nu$	$a^{\nu-1}(\frac{1}{2}a)^z B(z,\frac{1}{2}-\frac{1}{2}z+\frac{1}{2}\nu)$ $\cdot[\log a+\frac{1}{2}\psi(\frac{1}{2}-\frac{1}{2}z+\frac{1}{2}\nu)$ $-\frac{1}{2}\psi(\frac{1}{2}+\frac{1}{2}z+\frac{1}{2}\nu)]$
4.47	$(a^2+x^2)^{-\frac{1}{2}}[(a^2+x^2)^{\frac{1}{2}}+x]^{\nu}$ $\cdot\log[(a^2+x^2)^{\frac{1}{2}}+x]$ $0<\text{Re } z<1-\text{Re }\nu$	$a^{\nu-1}(\frac{1}{2}a)^z B(z,\frac{1}{2}-\frac{1}{2}z-\frac{1}{2}\nu)$ $\cdot[\log a-\frac{1}{2}\psi(\frac{1}{2}-\frac{1}{2}z-\frac{1}{2}\nu)$ $+\frac{1}{2}\psi(\frac{1}{2}+\frac{1}{2}z-\frac{1}{2}\nu)]$
4.48	$(a^2-x^2)^{-\frac{1}{2}}$ $\cdot\{[a+(a^2-x^2)^{\frac{1}{2}}]^{\nu}\log(a+(a^2-x^2)^{\frac{1}{2}}]$ $+[a-(a^2-x^2)^{\frac{1}{2}}]^{\nu}\log[a-(a^2-x^2)^{\frac{1}{2}}]$ $x < a$ 0 $x > a$	$(2a)^{\nu+z-1}B(\nu+\frac{1}{2}z,\frac{1}{2}z)$ $\cdot[\log(2a)+\psi(\nu+\frac{1}{2}z)$ $-\psi(\nu+z)]$ Re $z>\text{Max}(0,-2\text{ Re }\nu)$

	$\phi(x)$	$\Phi(z) = \int_0^\infty \phi(x) x^{z-1} dx$
4.49	$0 \qquad x < a$ $[x+(x^2-a^2)^{\frac{1}{2}}]^\nu$ $\cdot\log[x+(x^2-a^2)^{\frac{1}{2}}]$ $-[x-(x^2-a^2)^{\frac{1}{2}}]^\nu$ $\cdot\log[x-(x^2-a^2)^{\frac{1}{2}}] \quad x > a$	$\frac{1}{2}a^\nu[\Gamma(1-z)]^{-1}(\frac{1}{2}a)^z$ $\Gamma(\frac{1}{2}\nu-\frac{1}{2}z)\Gamma(-\frac{1}{2}\nu-\frac{1}{2}z)$ $\cdot[1+\nu\log a+\frac{1}{2}\nu\psi(\frac{1}{2}\nu-\frac{1}{2}z)$ $-\frac{1}{2}\nu\psi(-\frac{1}{2}\nu-\frac{1}{2}z)]$ $\operatorname{Re} z < \pm \operatorname{Re} \nu$
4.50	$0 \qquad x < a$ $(x^2-a^2)^{-\frac{1}{2}}\{[x+(x^2-a^2)^{\frac{1}{2}}]^\nu$ $\cdot\log[x+(x^2-a^2)^{\frac{1}{2}}]$ $+[x-(x^2-a^2)^{\frac{1}{2}}]^\nu$ $\cdot\log[x-(x^2-a^2)^{\frac{1}{2}}]\} \quad x > a$	$a^{\nu-1}(\frac{1}{2}a)^z B(\frac{1}{2}-\frac{1}{2}z+\frac{1}{2}\nu, \frac{1}{2}-\frac{1}{2}z-\frac{1}{2}\nu)$ $\cdot[\log a+\frac{1}{2}\psi(\frac{1}{2}-\frac{1}{2}z+\frac{1}{2}\nu)$ $-\frac{1}{2}\psi(\frac{1}{2}-\frac{1}{2}z-\frac{1}{2}\nu)]$ $\operatorname{Re} z < 1 \pm \operatorname{Re} \nu$
4.51	$\log\left[\frac{c^2+(x+a)^2}{c^2+(x+b)^2}\right]$ $0<\operatorname{Re} z<1$	$-2z^{-1}\{(b^2+c^2)^{\frac{1}{2}z}\cos[z\arctan(c/b)]$ $-(a^2+c^2)^{\frac{1}{2}z}\cos[z\arctan(c/a)]\}$

1.5 Trigonometric Functions

	$\phi(x)$	$\Phi(z) = \int_0^\infty \phi(x) x^{z-1} dx$
5.1	$\sin(ax)$	$a^{-z}\Gamma(z)\sin(\frac{1}{2}\pi z)$ $-1 < \text{Re } z < 1$
5.2	$\cos(ax)$	$a^{-z}\Gamma(z)\cos(\frac{1}{2}\pi z)$ $0 < \text{Re } z < 1$
5.3	$\sin(bx) \quad x < a$ $0 \quad x > a$	$\frac{1}{2}iz^{-1}a^{z}[{}_1F_1(z;1+z;-iab)$ $- {}_1F_1(z;1+z;iab)] \quad \text{Re } z > -1$
5.4	$\cos(bx) \quad x < a$ $0 \quad x > a$	$\frac{1}{2}z^{-1}a^{z}[{}_1F_1(z;1+z;iab)$ $+ {}_1F_1(2;1+z;-iab)]$ $\text{Re } z > 0$
5.5	$(a+x)^{-1}\sin(bx)$ $-1 < \text{Re } z < 2$	$(\pi ab)^{\frac{1}{2}}(2a)^{z-1}[\frac{\Gamma(\frac{1}{2}+\frac{1}{2}z)}{\Gamma(1-\frac{1}{2}z)}$ $\cdot S_{\frac{1}{2}-z,\frac{1}{2}}(ab) - \frac{2\Gamma(1+\frac{1}{2}z)}{\Gamma(\frac{1}{2}-\frac{1}{2}z)} S_{-\frac{1}{2}-z,\frac{1}{2}}(ab)]$
5.6	$(a+x)^{-1}\cos(bx)$ $0 < \text{Re } z < 2$	$(\pi ab)^{\frac{1}{2}}(2a)^{z-1}[\frac{\Gamma(\frac{1}{2}z)}{\Gamma(\frac{1}{2}-\frac{1}{2}z)} S_{\frac{1}{2}-z,\frac{1}{2}}(ab)$ $- 2\frac{\Gamma(\frac{1}{2}+\frac{1}{2}z)}{\Gamma(-\frac{1}{2}z)} S_{-\frac{1}{2}-z,\frac{1}{2}}(ab)]$

	$\phi(x)$	$\Phi(z) = \int_0^\infty \phi(x) x^{z-1} dx$
5.7	$(a^2+x^2)^{-1}$ $\cdot\sin(bx)$ $-1 < \mathrm{Re}\, z < 3$	$2^{z-3}\pi^{\frac{1}{2}}\Gamma(\frac{1}{2}z-\frac{1}{2})[\Gamma(2-\frac{1}{2}z)]^{-1}b^{-z}$ $\cdot {}_1F_2(1;2-\frac{1}{2}z,\frac{3}{2}-\frac{1}{2}z;\frac{1}{4}a^2b^2)$ $+\frac{1}{2}\pi a^{z-2}\sec(\frac{1}{2}\pi z)\sinh(ab)$
5.8	$(a^2+x^2)^{-1}$ $\cdot\cos(bx)$ $0 < \mathrm{Re}\, z < 3$	$2^{z-3}\pi^{\frac{1}{2}}\Gamma(\frac{1}{2}z-1)[\Gamma(\frac{3}{2}-\frac{1}{2}z)]^{-1}b^{-z}$ $\cdot {}_1F_2(1;\frac{3}{2}-\frac{1}{2}z,2-\frac{1}{2}z;\frac{1}{4}a^2b^2)$ $+\frac{1}{2}\pi a^{z-2}\csc(\frac{1}{2}\pi z)\cosh(ab)$
5.9	$(a^2+x^2)^{-\nu}$ $\cdot\sin(bx)$ $-1<\mathrm{Re}\, z<1+2\,\mathrm{Re}\,\nu$	$\frac{1}{2}a^{z+1-2\nu}b\, B(\frac{1}{2}+\frac{1}{2}z,\nu-\frac{1}{2}-\frac{1}{2}z)$ $\cdot {}_1F_2(\frac{1}{2}+\frac{1}{2}z;\frac{3}{2}+\frac{1}{2}z-\nu,\frac{3}{2};\frac{1}{4}a^2b^2)$ $+2^{z-2\nu-1}\pi^{\frac{1}{2}}\Gamma(\frac{3}{2}-\frac{1}{2}z-\nu)[\Gamma(1+\nu-\frac{1}{2}z)]^{-1}$ $\cdot b^{2\nu-2}\,{}_1F_2(\nu;1+\nu-\frac{1}{2}z,\frac{1}{2}+\nu-\frac{1}{2}z;\frac{1}{4}a^2b^2)$
5.10	$(a^2+x^2)^{-\nu}$ $\cdot\cos(bx)$ $0<\mathrm{Re}\, z<1+2\,\mathrm{Re}\,\nu$	$\frac{1}{2}a^{z-2\nu}B(\frac{1}{2}z,\nu-\frac{1}{2}z)$ $\cdot {}_1F_2(\frac{1}{2}z;1+\frac{1}{2}z-\nu,\frac{1}{2};\frac{1}{4}a^2b^2)$ $+\pi^{\frac{1}{2}}2^{z-2\nu-1}b^{2\nu-z}[\Gamma(\frac{1}{2}+\nu-\frac{1}{2}z)]^{-1}\Gamma(\frac{1}{2}z-\nu)$ $\cdot {}_1F_2(\nu;\frac{1}{2}+\nu-\frac{1}{2}z,1+\nu-\frac{1}{2}z;\frac{1}{4}a^2b^2)$

	$\phi(x)$	$\Phi(z) = \int_0^\infty \phi(x)x^{z-1}dx$
5.11	$(a-x)^{\nu-1}$ $x < a$ $\cdot\sin(bx)$ 0 $x > a$ $\mathrm{Re}\,\nu > 0$	$\frac{1}{2}iB(z,\nu)a^{z+\nu-1}[{}_1F_1(z,\nu+z;-i\,ab)$ $-{}_1F_1(z,\nu+z;iab]$ $\mathrm{Re}\,z > -1$
5.12	$(a-x)^{\nu-1}$ $x < a$ $\cdot\cos(bx)$ 0 $x > a$ $\mathrm{Re}\,\nu > 0$	$\frac{1}{2}B(z,\nu)a^{z+\nu-1}[{}_1F_1(z,\nu+z;iab)$ $+{}_1F_1(z,\nu+z;-iab)]$ $\mathrm{Re}\,z > 0$
5.13	$(a^2-x^2)^{\nu}$ $\cdot\sin(bx)$ $x < a$ 0 $x > a$	$\frac{1}{2}a^{z+2\nu+1}bB(\nu+1,\frac{1}{2}+\frac{1}{2}z)$ $\cdot\,{}_1F_2(\frac{1}{2}+\frac{1}{2}z;\,^3/_2,\,^3/_2+\nu+\frac{1}{2}z;-\frac{1}{4}a^2b^2)$ $\mathrm{Re}\,z > -1,$ $\mathrm{Re}\,\nu > -1$
5.14	$(a^2-x^2)^{\nu}$ $\cdot\cos(bx)$ $x < a$ 0 $x > a$	$\frac{1}{2}a^{z+2\nu}B(\frac{1}{2}z,\nu+1)\,{}_1F_2(\frac{1}{2}z;\frac{1}{2},1+\nu+\frac{1}{2}z;$ $-\frac{1}{4}a^2b^2)$ $\mathrm{Re}\,z > 0,$ $\mathrm{Re}\,\nu > 1$
5.15	$(a^2-x^2)^{-1}$ $\cdot\sin(bx)$ (Principal value)	$-\frac{1}{2}\pi a^{z-2}\cos(ab)+(\pi b/a)^{\frac{1}{2}}(2a)^{z-1}$ $\cdot\Gamma(\frac{1}{2}+\frac{1}{2}z)[\Gamma(1-\frac{1}{2}z)]^{-1}S_{\frac{1}{2}-z,\frac{1}{2}}(ab)$ $-1 < \mathrm{Re}\,z < 3$

	$\phi(x)$	$\Phi(z) = \int_0^\infty \phi(x) x^{z-1} dx$
5.16	$(a^2-x^2)^{-1}$ $\cdot\cos(bx)$ Principal value	$\frac{1}{2}\pi\sin(ab)a^{z-2}+(2\pi b)^{\frac{1}{2}}(2a)^{z-\frac{3}{2}}$ $\cdot\Gamma(\frac{1}{2}z)[\Gamma(\frac{1}{2}-\frac{1}{2}z)]^{-1}S_{\frac{1}{2}+z,\frac{1}{2}}(ab)$ $0 < \mathrm{Re}\, z < 3$
5.17	$(a+x)^{-1}$ $\cdot\cos[b(a+x)^{\frac{1}{2}}]$ $0<\mathrm{Re}\, z < \frac{3}{2}$	$\pi\csc(\pi z)a^{z}\{1-\frac{1}{2}\pi a^{\frac{1}{2}}b$ $\cdot[J_{\frac{1}{2}-z}(ba^{\frac{1}{2}})\mathbf{H}_{-\frac{1}{2}-z}(ba^{\frac{1}{2}})-\mathbf{H}_{\frac{1}{2}-z}(ba^{\frac{1}{2}})J_{-\frac{1}{2}-z}(ba^{\frac{1}{2}})]\}$
5.18	$(a-x)^{\nu}$ $x < a$ $\cdot\cos[b(a-x)^{\frac{1}{2}}]$ 0 $x > a$	$a^{\nu+1}B(z,\nu+1)$ $\cdot {}_1F_2(\nu+1;1+\nu+z,\frac{1}{2};-\frac{1}{4}ab^2)$ $\mathrm{Re}\, z > 0,$ $\mathrm{Re}\,\nu > -1$
5.19	$e^{-bx}\sin(ax)$	$\Gamma(z)(b^2+a^2)^{-\frac{1}{2}z}\sin[z\arctan(a/b)]$ $\mathrm{Re}\, z > -1$
5.20	$e^{-bx}\cos(ax)$	$\Gamma(z)(b^2+a^2)^{-\frac{1}{2}z}\cos[z\arctan(a/b)]$
5.21	$e^{-bx}\sin(ax)$ $x < c$ 0 $x > c$	$\frac{1}{2}i(b+ia)^{-z}\gamma[z,c(b+ia)]$ $-\frac{1}{2}i(b-ia)^{-z}\gamma[z,c(b-ia)]$ $\mathrm{Re}\, z > -1$

	$\phi(x)$	$\Phi(z) = \int_0^\infty \phi(x) x^{z-1} dx$
5.22	$e^{-bx}\cos(ax)$ $x < c$ 0 $x > c$	$\frac{1}{2}(b+ia)^{-z}\gamma[z,c(b+ia)]$ $+\frac{1}{2}(b-ia)^{-z}\gamma[z,c(b-ia)]$ $\mathrm{Re}\ z > 0$
5.23	0 $x < c$ $e^{-bx}\sin(ax)$ $x > c$	$\frac{1}{2}i(b+ia)^{-z}\Gamma[z,c(b+ia)]$ $-\frac{1}{2}i(b-ia)^{-z}\Gamma[z,c(b-ia)]$
5.24	0 $x < c$ $e^{-bx}\cos(ax)$ $x > c$	$\frac{1}{2}(b+ia)^{-z}\Gamma[z,c(b+ia)]$ $+\frac{1}{2}(b-ia)^{-z}\Gamma[z,c(b-ia)]$
5.25	$(e^{ax}-1)^{-1}$ $\cdot\sin(bx)$	$\frac{1}{2}ia^{-z}\Gamma(z)[\zeta(z,1+ib/a)-\zeta(z,1-ib/a)]$ $\mathrm{Re}\ z > 0$
5.26	$(e^{ax}-1)^{-1}$ $\cdot\cos(bx)$	$\frac{1}{2}a^{-z}\Gamma(z)[\zeta(z,1+ib/a)+\zeta(z,1-ib/a)]$ $\mathrm{Re}\ z > 1$
5.27	$(e^{ax}+1)^{-1}$ $\cdot\sin(bx)$ $\mathrm{Re}\ z > -1$	$\Gamma(z)\{b^{-z}\sin(\frac{1}{2}\pi z)+\frac{1}{2}i(2a)^{-z}$ $\cdot[\zeta(z,\frac{1}{2}+\frac{1}{2}ib/a)-\zeta(z,\frac{1}{2}-\frac{1}{2}ib/a)$ $-\zeta(z,\frac{1}{2}ib/a)+\zeta(z,-\frac{1}{2}ib/a)]\}$

	$\phi(x)$	$\Phi(z) = \int_0^\infty \phi(x) x^{z-1} dx$
5.28	$(e^{ax}+1)^{-1}$ $\cdot\cos(bx)$ Re $z > 0$	$\Gamma(z)\{b^{-z}\cos(\frac{1}{2}\pi z)+\frac{1}{2}(2a)^{-z}$ $\cdot[\zeta(z,\frac{1}{2}+\frac{1}{2}ib/a)+\zeta(z,\frac{1}{2}-\frac{1}{2}ib/a)$ $-\zeta(z,\frac{1}{2}ib/a)-\zeta(z,-\frac{1}{2}ib/a)]$
5.29	$e^{-ax^2}\sin(bx)$ Re $z > -1$	$\frac{1}{2}ba^{-\frac{1}{2}-\frac{1}{2}z}\Gamma(\frac{1}{2}+\frac{1}{2}z)\exp(-\frac{1}{4}b^2/a)$ $\cdot {}_1F_1(-\frac{1}{2}z;\frac{3}{2};\frac{1}{4}b^2/a)$
5.30	$e^{-ax^2}\cos(bx)$ Re $z > 0$	$\frac{1}{2}a^{-\frac{1}{2}z}\Gamma(\frac{1}{2}z)\exp(-\frac{1}{4}b^2/a)$ $\cdot {}_1F_1(-\frac{1}{2}z+\frac{1}{2};\frac{1}{2};\frac{1}{4}b^2/a)$
5.31	e^{-ax^2-bx} $\cdot\sin(cx)$ Re $z > -1$	$-\frac{1}{2}i\Gamma(z)(2a)^{-\frac{1}{2}z}\exp[\frac{1}{8}a^{-1}(b^2-c^2)]$ $\cdot\{\exp(-\frac{1}{4}ibc/a)D_{-z}[(2a)^{-\frac{1}{2}}(b-ic)]$ $-\exp(\frac{1}{4}ibc/a)D_{-z}[(2a)^{-\frac{1}{2}}(b+ic)]\}$
5.32	e^{-ax^2-bx} $\cdot\cos(cx)$ Re $z > 0$	$\frac{1}{2}(2a)^{-\frac{1}{2}z}\Gamma(z)\exp[\frac{1}{8}a^{-1}(b^2-c^2)]$ $\cdot\{\exp(-\frac{1}{4}ibc/a)D_{-z}[(2a)^{-\frac{1}{2}}(b-ic)]$ $+\exp(\frac{1}{4}ibc/a)D_{-z}[(2a)^{-\frac{1}{2}}(b+ic)]\}$
5.33	$e^{-ax-b^2/x}$ $\cdot\sin(cx)$	$ib^z\{(a+ic)^{-\frac{1}{2}z}K_z[2b(a+ic)^{\frac{1}{2}}]$ $-(a-ic)^{-\frac{1}{2}z}K_z[2b(a-ic)^{\frac{1}{2}}]\}$

	$\phi(x)$	$\Phi(z) = \int_0^\infty \phi(x) x^{z-1} dx$
5.34	$e^{-ax-b^2/x}$ $\cdot\cos(cx)$	$b^z\{(a+ic)^{-\frac{1}{2}z}K_z[2b(a+ic)^{\frac{1}{2}}]$ $+(a-ic)^{-\frac{1}{2}z}K_z[2b(a+c)^{\frac{1}{2}}]\}$
5.35	$e^{-ax^{\frac{1}{2}}}\sin(bx)$ $\mathrm{Re}\ z > -1$	$i(2b)^{-z}\Gamma(2z)\{\exp[-i(\frac{1}{2}\pi z+{}^1/_8a^2/b)]$ $\cdot D_{-2z}[\frac{1}{2}ab^{-\frac{1}{2}}(1-i] - \exp[i(\frac{1}{2}\pi z+{}^1/_8a^2/b)]$ $\cdot D_{-2z}[\frac{1}{2}ab^{-\frac{1}{2}}(1+i)]\}$
5.36	$e^{-ax^{\frac{1}{2}}}\cos(bx)$ $\mathrm{Re}\ z > 0$	$(2b)^{-z}\Gamma(2z)\{\exp[-i(\frac{1}{2}\pi z+{}^1/_8a^2/b)]$ $\cdot D_{-2z}[\frac{1}{2}ab^{-\frac{1}{2}}(1-i)]+\exp[i(\frac{1}{2}\pi z+{}^1/_8a^2/b)]$ $\cdot D_{-2z}[\frac{1}{2}ab^{-\frac{1}{2}}(1+i)]\}$
5.37	$\sin[b(a^2+x^2)^{\frac{1}{2}}]$ $0 < \mathrm{Re}\ z < 1$	$-\frac{1}{2}\pi^{\frac{1}{2}}a(2a/b)^{\frac{1}{2}z-\frac{1}{2}}\Gamma(\frac{1}{2}z)$ $\cdot[Y_{\frac{1}{2}+\frac{1}{2}z}(ab)\sin(\frac{1}{2}\pi z)$ $-J_{\frac{1}{2}+\frac{1}{2}z}(ab)\cos(\frac{1}{2}\pi z)]$
5.38	$\cos[b(a^2+x^2)^{\frac{1}{2}}]$ $0 < \mathrm{Re}\ z < 1$	$-\frac{1}{2}\pi^{\frac{1}{2}}a(2a/b)^{\frac{1}{2}z-\frac{1}{2}}\Gamma(\frac{1}{2}z)$ $\cdot[J_{\frac{1}{2}+\frac{1}{2}z}(ab)\sin(\frac{1}{2}\pi z)+Y_{\frac{1}{2}+\frac{1}{2}z}(ab)\cos(\frac{1}{2}\pi z)]$

	$\phi(x)$	$\Phi(z) = \int_0^\infty \phi(x) x^{z-1} dx$
5.39	$(a^2+x^2)^{-\frac{1}{2}}$ $\cdot\sin[b(a^2+x^2)^{\frac{1}{2}}]$	$\frac{1}{2}(\frac{1}{2}\pi b/a)^{\frac{1}{2}}(2a/b)^{\frac{1}{2}z}\Gamma(\frac{1}{2}z)J_{\frac{1}{2}-\frac{1}{2}z}(ab)$ $0 < \mathrm{Re}\, z < 2$
5.40	$(a^2+x^2)^{-\frac{1}{2}}$ $\cdot\cos[b(a^2+x^2)^{\frac{1}{2}}]$	$-\frac{1}{2}(\frac{1}{2}\pi b/a)^{\frac{1}{2}}(2a/b)^{\frac{1}{2}z}\Gamma(\frac{1}{2}z)Y_{\frac{1}{2}-\frac{1}{2}z}(ab)$ $0 < \mathrm{Re}\, z < 2$
5.41	$\sin[b(a^2-x^2)^{\frac{1}{2}}]$ $x < a$; 0 $x > a$	$\frac{1}{2}(\frac{1}{2}\pi ab)^{\frac{1}{2}}(2a/b)^{\frac{1}{2}z}J_{\frac{1}{2}+\frac{1}{2}z}(ab)$ $\mathrm{Re}\, z > 0$
5.42	$\cos[b(a^2-x^2)^{\frac{1}{2}}]$ $x < a$; 0 $x > a$	$-a(a/b)^{\frac{1}{2}z-\frac{1}{2}}s_{\frac{1}{2}z-3/2,\frac{1}{2}z+\frac{1}{2}}(ab)$ $\mathrm{Re}\, z > 0$
5.43	$(a^2-x^2)^{-\frac{1}{2}}$ $\cdot\sin[b(a^2-x^2)^{\frac{1}{2}}]$ $x < a$; 0 $x > a$	$\frac{1}{2}\pi^{\frac{1}{2}}(2a/b)^{\frac{1}{2}z-3/2}\Gamma(\frac{1}{2}z)\mathbf{H}_{\frac{1}{2}z-\frac{1}{2}}(ab)$ $\mathrm{Re}\, z > 0$
5.44	$(a^2-x^2)^{-\frac{1}{2}}$ $\cdot\cos[(a^2-x^2)^{\frac{1}{2}}]$ $x < a$; 0 $x > a$	$\frac{1}{2}(\frac{1}{2}\pi b/a)^{\frac{1}{2}}(2a/b)^{\frac{1}{2}z}J_{\frac{1}{2}z-\frac{1}{2}}(ab)$ $\mathrm{Re}\, z > 0$

	$\phi(x)$	$\Phi(z) = \int_0^\infty \phi(x) x^{z-1} dx$
5.45	$0 \qquad x < a$ $\sin[b(x^2-a^2)^{\frac{1}{2}}] \quad x > a$	$(\frac{1}{2}\pi ab)^{\frac{1}{2}}(2a/b)^{\frac{1}{2}z}[\Gamma(1-\frac{1}{2}z)]^{-1}$ $\cdot K_{\frac{1}{2}z+\frac{1}{2}}(ab) \qquad \operatorname{Re} z < 1$
5.46	$0 \qquad x < a$ $\cos[b(x^2-a^2)^{\frac{1}{2}} \quad x > a$ $\operatorname{Re} z < 1$	$(\frac{1}{2}ab)^{\frac{1}{2}}(a/b)^{\frac{1}{2}z}[2^{\frac{1}{2}z-1}\Gamma(\frac{1}{2}z)I_{-\frac{1}{2}-\frac{1}{2}z}(ab)$ $+2^{\frac{1}{2}}e^{-i\frac{\pi}{4}(z-1)} s_{\frac{1}{2}z-3/2,\frac{1}{2}z+\frac{1}{2}}(iab)]$
5.47	$0 \qquad x < a$ $(x^2-a^2)^{-\frac{1}{2}}$ $\cdot\sin[b(x^2-a^2)^{\frac{1}{2}}]$ $x > a$	$\frac{1}{2}\pi^{\frac{1}{2}}(\frac{1}{2}a/b)^{\frac{1}{2}z-\frac{1}{2}}\Gamma(\frac{1}{2}z)[I_{\frac{1}{2}-\frac{1}{2}z}(ab)-\mathbf{L}_{\frac{1}{2}z-\frac{1}{2}}(ab)]$ $\operatorname{Re} z < 2$
5.48	$0 \qquad x < a$ $(x^2-a^2)^{-\frac{1}{2}}$ $\cdot\cos[b(x^2-a^2)^{\frac{1}{2}}]$ $x > a$	$(\frac{1}{2}\pi b/a)^{\frac{1}{2}}(2a/b)^{\frac{1}{2}z}[\Gamma(1-\frac{1}{2}z)]^{-1}K_{\frac{1}{2}z-\frac{1}{2}}(ab)$ $\operatorname{Re} z < 2$
5.49	$e^{-u}\cos v$ $\frac{u}{v}=(2a)^{\frac{1}{2}}[(b^2+x^2)^{\frac{1}{2}}\pm b]^{\frac{1}{2}}$	$(2a)^{\frac{1}{2}}2^{z-\frac{1}{2}}\Gamma(\frac{1}{2}z)[\Gamma(\frac{1}{2}-\frac{1}{2}z)]^{-1}$ $\cdot(a/b)^{-\frac{1}{2}z-\frac{1}{4}}K_{\frac{1}{2}+z}[2(ab)^{\frac{1}{2}}]$ $\operatorname{Re} z > 0$

	$\phi(x)$	$\Phi(z) = \int_0^\infty \phi(x) x^{z-1} dx$
5.50	$e^{-u}\sin v$ $\begin{matrix} u \\ v \end{matrix} = (2a)^{\frac{1}{2}}[(b^2+x^2)^{\frac{1}{2}} \pm b]$	$(2a)^{\frac{1}{2}} 2^{z-\frac{1}{2}} \Gamma(\frac{1}{2}+\frac{1}{2}z)[\Gamma(1-\frac{1}{2}z)]^{-1}$ $\cdot (a/b)^{-\frac{1}{2}z-\frac{1}{4}} K_{\frac{1}{2}+z}[2(ab)^{\frac{1}{2}}]$ $\mathrm{Re}\, z > -1$
5.51	$\sin(a/x)\sin(bx)$ $-2 < \mathrm{Re}\, z < 2$	$\frac{1}{4}\pi(a/b)^{\frac{1}{2}z}\csc(\frac{1}{2}\pi z)$ $\cdot \{J_z[2(ab)^{\frac{1}{2}}-J_{-z}[2(ab)^{\frac{1}{2}}]$ $+2\pi^{-1}\sin(\pi z)K_z[2(ab)^{\frac{1}{2}}]\}$
5.52	$\sin(a/x)\cos(bx)$ $-1 < \mathrm{Re}\, z < 2$	$\frac{1}{4}\pi(a/b)^{\frac{1}{2}z}\sec(\frac{1}{2}\pi z)$ $\cdot \{J_z[2(ab)^{\frac{1}{2}}]+J_{-z}[2(ab)^{\frac{1}{2}}]$ $-2\pi^{-1}\sin(\pi z)K_z[2(ab)^{\frac{1}{2}}]\}$
5.53	$\cos(a/x)\sin(bx)$ $-1 < \mathrm{Re}\, z < 2$	$\frac{1}{4}\pi(a/b)^{\frac{1}{2}z}\sec(\frac{1}{2}\pi z)$ $\cdot \{J_z[2(ab)^{\frac{1}{2}}]+J_{-z}[2(ab)^{\frac{1}{2}}]$ $+2\pi^{-1}\sin(\pi z)K_z[2(ab)^{\frac{1}{2}}]\}$
5.54	$\cos(a/x)\cos(bx)$ $-1 < \mathrm{Re}\, z < 1$	$\frac{1}{4}\pi(a/b)^{\frac{1}{2}z}\csc(\frac{1}{2}\pi z)$ $\cdot \{J_{-z}[2(ab)^{\frac{1}{2}}]-J_z[2(ab)^{\frac{1}{2}}]$ $+2\pi^{-1}\sin(\pi z)K_z[2(ab)^{\frac{1}{2}}]\}$

	$\phi(x)$	$\Phi(z) = \int_0^\infty \phi(x) x^{z-1} dx$
5.55	$\sin[a(x-b^2/x)]$	$2b^z \sin(\frac{1}{2}\pi z) K_z(2ab)$ $-1<\mathrm{Re}\ z<1$
5.56	$\sin[a(x+b^2/x)]$ $-1<\mathrm{Re}\ z<1$	$\frac{1}{2}\pi b^z \sec(\frac{1}{2}\pi z)[J_z(2ab)+J_{-z}(2ab)]$ $-1<\mathrm{Re}\ z<1$
5.57	$\cos[a(x-b^2/x)]$	$2b^z \cos(\frac{1}{2}\pi z) K_z(2ab)$ $-1<\mathrm{Re}\ z<1$
5.58	$\cos[a(x+b^2/x)]$	$\frac{1}{2}\pi b^z \csc(\frac{1}{2}\pi z)[J_{-z}(2ab)-J_z(2ab)]$ $-1<\mathrm{Re}\ z<1$
5.59	$(1+x^2)^{-\frac{1}{2}}$ $\cdot\sin[a(x+x^{-1})]$ $-1<\mathrm{Re}\ z<2$	$-(\frac{1}{2}\pi)^{3/2} a^{\frac{1}{2}} [J_{-\frac{1}{4}-\frac{1}{2}z}(a) Y_{3/4+\frac{1}{2}z}(a)$ $+J_{3/4+\frac{1}{2}z}(a) Y_{-\frac{1}{4}-\frac{1}{2}z}(a)]$ $=(\frac{1}{2}\pi)^{3/2} a^{\frac{1}{2}} [J_{-3/4-\frac{1}{2}z}(a) J_{\frac{1}{4}+\frac{1}{2}z}(a)$ $-Y_{-3/4-\frac{1}{2}z}(a) Y_{\frac{1}{4}+\frac{1}{2}z}(a)]$

	$\phi(x)$	$\Phi(z) = \int_0^\infty \phi(x)x^{z-1}dx$
5.60	$(1+x^2)^{-\frac{1}{2}}$ $\cdot\cos[a(x+x^{-1})]$ $-1 < \mathrm{Re}\, z < 2$	$-(\frac{1}{2}\pi)^{3/2}a^{\frac{1}{2}}[J_{-\frac{3}{4}-\frac{1}{2}z}(a)Y_{\frac{1}{4}+\frac{1}{2}z}(a)$ $+J_{\frac{1}{4}+\frac{1}{2}z}(a)Y_{-\frac{3}{4}-\frac{1}{2}z}(a)]$ $= -(\frac{1}{2}\pi)^{3/2}a^{\frac{1}{2}}[J_{-\frac{1}{4}-\frac{1}{2}z}(a)J_{\frac{3}{4}+\frac{1}{2}z}(a)$ $-Y_{-\frac{1}{4}-\frac{1}{2}z}(a)Y_{\frac{3}{4}+\frac{1}{2}z}(a)]$
5.61	$(1-x^2)^{-\frac{1}{2}}$ $\cdot\sin[a(x-x^{-1})]$ $x < 1$; 0 $x > 1$	$-(\frac{1}{2}\pi a)^{\frac{1}{2}}I_{\frac{1}{2}z}(a)K_{\frac{1}{2}z-\frac{1}{2}}(a)$ $\mathrm{Re}\, z > -1$
5.62	$(1-x^2)^{-\frac{1}{2}}$ $\cdot\cos[a(x-x^{-1})]$ $x < 1$; 0 $x > 1$	$(\frac{1}{2}\pi a)^{\frac{1}{2}}I_{\frac{1}{2}z-\frac{1}{2}}(a)K_{\frac{1}{2}z}(a)$ $\mathrm{Re}\, z > -1$
5.63	0 $x < 1$; $(x^2-1)^{-\frac{1}{2}}$ $\cdot\sin[a(x-x^{-1})]$ $x > 1$	$(\frac{1}{2}\pi a)^{\frac{1}{2}}I_{\frac{1}{2}-\frac{1}{2}z}(a)K_{\frac{1}{2}z}(a)$ $\mathrm{Re}\, z < 2$
5.64	0 $x < 1$; $(x^2-1)^{-\frac{1}{2}}$ $\cdot\cos[a(x-x^{-1})]$ $x > 1$	$(\frac{1}{2}\pi a)^{\frac{1}{2}}I_{-\frac{1}{2}z}(a)K_{\frac{1}{2}z-\frac{1}{2}}(a)$ $\mathrm{Re}\, z < 2$

	$\phi(x)$	$\Phi(z) = \int_0^\infty \phi(x) x^{z-1} dx$
5.65	$(1+x^2)^{-\frac{1}{2}} \sin(\frac{2ax}{1+x^2})$ $\cdot \exp(-\frac{2b}{1+x^2})$ $-1 < \mathrm{Re}\, z < 2$	$2(\pi a)^{-\frac{1}{2}} e^{-b} \Gamma(1-\frac{1}{2}z)\Gamma(\frac{1}{2}+\frac{1}{2}z)$ $\cdot M_{\frac{1}{4}-\frac{1}{2}z,\frac{1}{4}}[b+(b^2-a^2)^{\frac{1}{2}}]$ $\cdot M_{\frac{1}{4}-\frac{1}{2}z,\frac{1}{4}}[b-(b^2-a^2)^{\frac{1}{2}}]$
5.66	$(1+x^2)^{-\frac{1}{2}} \cos(\frac{2ax}{1+x^2})$ $\cdot \exp(-\frac{2b}{1+x^2})$ $0 < \mathrm{Re}\, z < 1$	$\frac{1}{2}(\pi a)^{-\frac{1}{2}} e^{-b} \Gamma(\frac{1}{2}-\frac{1}{2}z)\Gamma(\frac{1}{2}z)$ $\cdot M_{\frac{1}{4}-\frac{1}{2}z,-\frac{1}{4}}[b+(b^2-a^2)^{\frac{1}{2}}]$ $\cdot M_{\frac{1}{4}-\frac{1}{2}z,-\frac{1}{4}}[b-(b^2-a^2)^{\frac{1}{2}}]$
5.67	$0, \quad x < 1$ $(x^2-1)^{-\frac{1}{2}} \sin(\frac{2ax}{x^2-1}$ $\cdot \exp(-\frac{2b}{x^2-1}), \quad x > 1$	$(\frac{1}{2}a)^{-\frac{1}{2}} e^b 2^{-\frac{1}{2}z} \Gamma(1-\frac{1}{2}z)[(a^2+b^2)^{\frac{1}{2}}+b]^{\frac{1}{4}}$ $\cdot D_{z-1}\{2^{\frac{1}{2}}[(a^2+b^2)^{\frac{1}{2}}+b]^{\frac{1}{2}}\}$ $\cdot M_{\frac{1}{4}-\frac{1}{2}z,\frac{1}{4}}[(a^2+b^2)^{\frac{1}{2}}-b]$ $\mathrm{Re}\, z < 2$
5.68	$0, \quad x < 1$ $(x^2-1)^{-\frac{1}{2}} \cos(\frac{2ax}{x^2-1})$ $\cdot \exp(-\frac{2b}{x^2-1}), \quad x > 1$	$(2a)^{-\frac{1}{2}} e^b 2^{-\frac{1}{2}z} \Gamma(\frac{1}{2}-\frac{1}{2}z)[(a^2+b^2)^{\frac{1}{2}}+b]^{\frac{1}{4}}$ $\cdot D_{z-1}\{2^{\frac{1}{2}}[(a^2+b^2)^{\frac{1}{2}}+b]^{\frac{1}{2}}\}$ $\cdot M_{\frac{1}{4}-\frac{1}{2}z,-\frac{1}{4}}[(a^2+b^2)^{\frac{1}{2}}-b]$ $\mathrm{Re}\, z < 1$

	$\phi(x)$	$\Phi(z) = \int_0^\infty \phi(x) x^{z-1} dx$
5.69	$(1+x^2)^{-\frac{1}{2}} \cos\left(\frac{2ax}{1+x^2}\right)$ $\cdot \exp\left[-b\left(\frac{1-x^2}{1+x^2}\right)\right]$ $0 < \mathrm{Re}\, z < 1$	$\frac{1}{2}(\pi a)^{-\frac{1}{2}} \Gamma(\frac{1}{2}z)\Gamma(\frac{1}{2}-\frac{1}{2}z)$ $\cdot M_{\frac{1}{4}-\frac{1}{2}z,-\frac{1}{4}}[b+(b^2-a^2)^{\frac{1}{2}}]$ $\cdot M_{\frac{1}{4}-\frac{1}{2}z,-\frac{1}{4}}[b-(b^2-a^2)^{\frac{1}{2}}]$
5.70	$(1+x^2)^{-\frac{1}{2}} \sin(2ax)$ $\cdot \exp\left[-b\left(\frac{1-x^2}{1+x^2}\right)\right]$ $-1 < \mathrm{Re}\, z < 2$	$2(\pi a)^{-\frac{1}{2}} \Gamma(1-\frac{1}{2}z)\Gamma(\frac{1}{2}+\frac{1}{2}z)$ $\cdot M_{\frac{1}{4}-\frac{1}{2}z,\frac{1}{4}}[b+(b^2-a^2)^{\frac{1}{2}}]$ $\cdot M_{\frac{1}{4}-\frac{1}{2}z,\frac{1}{4}}[b-(b^2-a^2)^{\frac{1}{2}}]$
5.71	$0 \quad x < 1$ $(x^2-1)^{-\frac{1}{2}} \cos\left(\frac{2ax}{x^2-1}\right)$ $\cdot \exp\left[-b\left(\frac{x^2+1}{x^2-1}\right)\right], \quad x > 1$	$(2a)^{-\frac{1}{2}} 2^{-\frac{1}{2}z} \Gamma(\frac{1}{2}-\frac{1}{2}z)[(a^2+b^2)^{\frac{1}{2}}+b]^{\frac{1}{4}}$ $\cdot D_{z-1}\{2^{\frac{1}{2}}[(a^2+b^2)^{\frac{1}{2}}+b]^{\frac{1}{2}}\}$ $\cdot M_{\frac{1}{4}-\frac{1}{2}z,-\frac{1}{4}}[(a^2+b^2)^{\frac{1}{2}}-b]$ $\mathrm{Re}\, z < 1$
5.72	$0 \quad x < 1$ $(x^2-1)^{-\frac{1}{2}} \sin\left(\frac{2ax}{x^2-1}\right)$ $\cdot \exp\left[-b\left(\frac{x^2+1}{x^2-1}\right)\right]$ $x > 1$	$(\frac{1}{2}a)^{-\frac{1}{2}} 2^{-\frac{1}{2}z} \Gamma(1-\frac{1}{2}z)[(a^2+b^2)^{\frac{1}{2}}+b]^{\frac{1}{4}}$ $\cdot D_{z-1}\{2^{\frac{1}{2}}[(a^2+b^2)^{\frac{1}{2}}+b]^{\frac{1}{2}}\}$ $\cdot M_{\frac{1}{4}-\frac{1}{2}z,\frac{1}{4}}[(a^2+b^2)^{\frac{1}{2}}-b]$

	$\phi(x)$	$\Phi(z) = \int_0^\infty \phi(x)\, x^{z-1} dx$
5.73	$(1+x^2)^{-\frac{1}{2}}\cos(2ax)$ $\cdot\exp[b(\frac{1-x^2}{1+x^2})]$ $0 < \mathrm{Re}\, z < 1$	$\frac{1}{2}(\pi a)^{-\frac{1}{2}}\Gamma(\frac{1}{2}z)\Gamma(\frac{1}{2}-\frac{1}{2}z)$ $\cdot M_{\frac{1}{2}z-\frac{1}{4},-\frac{1}{4}}[b+(b^2-a^2)^{\frac{1}{2}}]$ $\cdot M_{\frac{1}{2}z-\frac{1}{4},-\frac{1}{4}}[b-(b^2-a^2)^{\frac{1}{2}}]$
5.74	$(1+x^2)^{-\frac{1}{2}}\sin(2ax)$ $\cdot\exp[b(\frac{1-x^2}{1+x^2})]$ $-1 < \mathrm{Re}\, z < 2$	$2(\pi a)^{-\frac{1}{2}}\Gamma(1-\frac{1}{2}z)\Gamma(\frac{1}{2}+\frac{1}{2}z)$ $\cdot M_{\frac{1}{2}z-\frac{1}{4},\frac{1}{4}}[b+(b^2-a^2)^{\frac{1}{2}}]$ $\cdot M_{\frac{1}{2}z-\frac{1}{4},\frac{1}{4}}[b-(b^2-a^2)^{\frac{1}{2}}]$
5.75	$(1-x^2)^{-\frac{1}{2}}\cos(\frac{2ax}{1-x^2})$ $\cdot\exp[-b(\frac{1+x^2}{1-x^2})]$, $x < 1$ 0 , $x > 1$	$\frac{1}{2}a^{-\frac{1}{2}}2^{\frac{1}{2}z}\Gamma(\frac{1}{2}z)[(a^2+b^2)^{\frac{1}{2}}+b]^{\frac{1}{4}}$ $\cdot D_{-z}\{2^{\frac{1}{2}}[(a^2+b^2)^{\frac{1}{2}}+b]^{\frac{1}{2}}\}$ $\cdot M_{\frac{1}{2}z-\frac{1}{4},-\frac{1}{4}}[(a^2+b^2)^{\frac{1}{2}}-b]$ $\mathrm{Re}\, z > 0$
5.76	$(1-x^2)^{-\frac{1}{2}}\sin(\frac{2ax}{1-x^2})$ $\cdot\exp[-b(\frac{1+x^2}{1-x^2})]$, $x < 1$ 0 $x > 1$	$a^{-\frac{1}{2}}2^{\frac{1}{2}z}\Gamma(\frac{1}{2}+\frac{1}{2}z)[(a^2+b^2)^{\frac{1}{2}}+b]^{\frac{1}{4}}$ $\cdot D_{-z}\{2^{\frac{1}{2}}[(a^2+b^2)^{\frac{1}{2}}+b]^{\frac{1}{2}}\}$ $\cdot M_{\frac{1}{2}z-\frac{1}{4},\frac{1}{4}}[(a^2+b^2)^{\frac{1}{2}}-b]$ $\mathrm{Re}\, z > -1$

	$\phi(x)$	$\Phi(z) = \int_0^\infty \phi(x) x^{z-1} dx$
5.77	$\log x \sin(ax)$ $-1 < \mathrm{Re}\, z < 1$	$a^{-z}\Gamma(z)\sin(\frac{1}{2}\pi z)[\psi(z)-\log a$ $+\frac{1}{2}\pi\cot(\frac{1}{2}\pi z)]$
5.78	$\cos(ax)\log x$ $0 < \mathrm{Re}\, z < 1$	$a^{-z}\Gamma(z)\cos(\frac{1}{2}\pi z)[\psi(z)-\log a$ $-\frac{1}{2}\pi\tan(\frac{1}{2}\pi z)]$
5.79	$e^{-bx}\sin(ax)$ $\cdot\log x$ $\mathrm{Re}\, z > -1$	$\Gamma(z)(a^2+b^2)^{-\frac{1}{2}z}\sin(z\arctan(a/b)]$ $\cdot\{\psi(z)-\frac{1}{2}\log(a^2+b^2)+\arctan(a/b)$ $\cdot\cot[z\arctan(a/b)]\}$
5.80	$e^{-bx}\cos(ax)$ $\cdot\log x$ $\mathrm{Re}\, z > 0$	$\Gamma(z)(a^2+b^2)^{-\frac{1}{2}z}\cos[z\arctan(a/b)]$ $\cdot\{\psi(z)-\frac{1}{2}\log(a^2+b^2)-\arctan(a/b)$ $\cdot\tan[z\arctan(a/b)]\}$
5.81	$e^{-x}\sin(x+ax^2)$ $\mathrm{Re}\, z > -1$	$(2a)^{-\frac{1}{2}z}\Gamma(z)e^{-2/a}\sin(\frac{1}{4}\pi z)$ $\cdot D_{-z}(a^{-\frac{1}{2}})$
5.82	$\sin(a\log x) \quad x < 1$ $0 \quad x > 1$	$-a(a^2+z^2)^{-1}$ $\mathrm{Re}\, z > 0$
5.83	$e^{-x}\sin(a\log x)$	$\lvert\Gamma(z+ia)\rvert\sin[\arg\Gamma(z+ia)]$ $\mathrm{Re}\, z > 0$

	$\phi(x)$	$\Phi(z) = \int_0^\infty \phi(x) x^{z-1} dx$
5.84	$e^{-x}\cos(x+ax^2)$ $\operatorname{Re} z > 0$	$(2a)^{-\frac{1}{2}z}\Gamma(z)e^{-2/a}\cos(\frac{1}{4}\pi z)$ $\cdot D_{-z}(a^{-\frac{1}{2}})$
5.85	$\cos(a \log x) \quad x < 1$ $0 \quad x > 1$	$z(a^2+z^2)^{-1}$ $\operatorname{Re} z > 0$
5.86	$e^{-x}\cos(a \log x)$	$\lvert\Gamma(z+ia)\rvert\cos[\arg\Gamma(z+ia)]$ $\operatorname{Re} z > 0$
5.87	$\arcsin(x/a) \quad x < a$ $0 \quad x > a$	$\frac{1}{2}\pi z^{-1}a^z\{1-\pi^{-\frac{1}{2}}\Gamma(\frac{1}{2}+\frac{1}{2}z)[\Gamma(1+\frac{1}{2}z)]^{-1}\}$ $\operatorname{Re} z > -1$
5.88	$\arccos(x/a) \quad x < a$ $0 \quad x > a$	$\frac{1}{2}\pi^{\frac{1}{2}}z^{-1}a^z\Gamma(\frac{1}{2}+\frac{1}{2}z)[\Gamma(1+\frac{1}{2}z)]^{-1}$ $\operatorname{Re} z > -1$
5.89	$\arctan(ax)$	$-\frac{1}{2}\pi z^{-1}a^{-z}\sec(\frac{1}{2}\pi z) \qquad -1 < \operatorname{Re} z < 0$
5.90	$\operatorname{arccot}(ax)$	$\frac{1}{2}\pi z^{-1}a^{-z}\sec(\frac{1}{2}\pi z) \qquad 0 < \operatorname{Re} z < 1$
5.91	$\arctan(x/a) \quad x < a$ $0 \quad x > a$	$\frac{1}{4}z^{-1}a^z[\pi+\psi(\frac{1}{4}+\frac{1}{4}z)-\psi(\frac{3}{4}+\frac{1}{4}z)]$ $\operatorname{Re} z > -1$

	$\phi(x)$	$\Phi(z) = \int_0^\infty \phi(x) x^{z-1} dx$
5.92	$\operatorname{arccot}(x/a)$ $x < a$ 0 $x > a$	$\frac{1}{4} z^{-1} a^z [\pi - \psi(\frac{1}{4}+\frac{1}{4}z) + \psi(\frac{3}{4}+\frac{3}{4}z)]$ $\operatorname{Re} z > 0$
5.93	$\arctan(ae^{-x})$ $\operatorname{Re} z > 0$	$2^{-z-1}\Gamma(z) a Y(-a^2, z+1, \frac{1}{2})$
5.94	$(a^2+x^2)^{-\frac{1}{2}\nu}$ $\cdot \sin[\nu \arctan(x/a)]$	$a^{z-\nu}[\Gamma(\nu)]^{-1}\Gamma(z)\Gamma(\nu-z)\sin(\frac{\pi}{2}z)$ $-1 < \operatorname{Re} z < \operatorname{Re} \nu$
5.95	$(a^2+x^2)^{-\frac{1}{2}\nu}$ $\cdot \cos[\nu \arctan(x/a)]$	$a^{z-\nu}[\Gamma(\nu)]^{-1}\Gamma(z)\Gamma(\nu-z)\cos(\frac{\pi}{2}z)$ $0 < \operatorname{Re} z < \operatorname{Re} \nu$
5.96	$(a^2-x^2)^{-\frac{1}{2}}$ $\cdot \cos[\nu \arccos(x/a)]$ $x < a$ 0 $x > a$	$\pi 2^{-z} a^{z-1}\Gamma(z)$ $\cdot [\Gamma(\frac{1}{2}+\frac{1}{2}z-\frac{1}{2}\nu)\Gamma(\frac{1}{2}+\frac{1}{2}z+\frac{1}{2}\nu)]^{-1}$ $\operatorname{Re} z > 0$
5.97	0 $x < a$ $(x^2-a^2)^{-\frac{1}{2}}$ $\cdot \cos[\nu \arccos(a/x)]$ $x > a$	$\pi 2^{z-1} a^{z-1}\Gamma(1-z)$ $\cdot [\Gamma(1-\frac{1}{2}z-\frac{1}{2}\nu)\Gamma(1-\frac{1}{2}z+\frac{1}{2}\nu)]^{-1}$ $\operatorname{Re} z < 1$

	$\phi(x)$	$\Phi(z) = \int_0^\infty \phi(x) x^{z-1} dx$
5.98	$\arcsin[a(a^2+x^2)^{\frac{1}{2}}]$	$z^{-1}a^z \sin(\frac{1}{2}\pi z)$ $0 < \mathrm{Re}\, z < 1$
5.99	$\arccos[x(a^2+x^2)^{-\frac{1}{2}}]$	$z^{-1}a^z \sin(\frac{1}{2}\pi z)$ $0 < \mathrm{Re}\, z < 1$
5.100	$(a^2-x^2)^{-\frac{1}{2}}$ $\cos[\nu \arccos(x^2/a^2-1)]$ $x < a$ $0 \quad x > a$	$\pi 2^{-z} a^{z-1} \Gamma(z)$ $\cdot [\Gamma(\frac{1}{2}+\nu+\frac{1}{2}z)\Gamma(\frac{1}{2}-\nu+\frac{1}{2}z)]^{-1}$ $\mathrm{Re}\, z > 0$

1.6 Hyperbolic Functions

	$\phi(x)$	$\Phi(z) = \int_0^\infty \phi(x) x^{z-1} dx$
6.1	$\operatorname{sech}(ax)$	$a^{-z} 2^{1-z} \Gamma(z) Y(-1,z,\tfrac{1}{2})$ $= a^{-z} 2^{1-2z} \Gamma(z)$ $\cdot [\zeta(z,\tfrac{1}{4}) - \zeta(z,\tfrac{3}{4})]$ $\quad \operatorname{Re} z > 0$
6.2	$\operatorname{csch}(ax)$	$2a^{-z}(1-2^{-z})\Gamma(z)\zeta(z)$ $\quad \operatorname{Re} z > 1$
6.3	$\tanh(ax)$	$2^{1-z} a^{-z} (2^{1-z}-1) \Gamma(z) \zeta(z)$ $-1 < \operatorname{Re} z < 0$
6.4	$1-\tanh(ax)$	$2(2a)^{-z} \Gamma(z) (1-2^{1-2}) \zeta(z)$ $\operatorname{Re} z > 0$
6.5	$\coth(bx)-1$	$2(2b)^{-z} \Gamma(z) \zeta(z)$ $\quad \operatorname{Re} z > 1$
6.6	$x^{-1} - \operatorname{csch} x$	$2(2^{-z}-1) \Gamma(z) \zeta(z)$ $\quad -1 < \operatorname{Re} z < 1$
6.7	$\operatorname{sech}^2(ax)$	$2^{2-z} a^{-z} (1-2^{2-z}) \Gamma(z) \zeta(z-1)$ $\operatorname{Re} z > 0$
6.8	$\operatorname{csch}^2(ax)$	$2^{2-z} a^{-z} \Gamma(z) \zeta(z-1)$ $\quad \operatorname{Re} z > 2$

	$\phi(x)$	$\Phi(z) = \int_0^\infty \phi(x)x^{z-1}dx$
6.9	$e^{-ax}\text{sech}(bx)$ $-a > b$	$2^{1-2z}b^{-z}\Gamma(z)$ $\cdot[\zeta(z,\frac{1}{4}+\frac{1}{4}a/b)-\zeta(z,\frac{3}{4}+\frac{1}{4}a/b)]$ Re $z > 0$
6.10	$e^{-ax}\text{csch}(bx)$ $-a > b$	$2(2b)^{-z}\Gamma(z)\zeta(z,\frac{1}{2}+\frac{1}{2}a/b)$ Re $z > 1$
6.11	$e^{-ax}\tanh(ax)$	$2^{1-2z}b^{-z}\Gamma(z)$ $\cdot[\zeta(z,\frac{1}{4}a/b)-\zeta(z,\frac{1}{2}+\frac{1}{4}a/b)]$ $-a^{-z}\Gamma(z)$ Re $z > -1$
6.12	$e^{-ax}\coth(bx)$	$b^{-z}\Gamma(z)[2^{1-z}\zeta(z,\frac{1}{2}a/b)-(b/a)^{z}]$ Re $z > 1$
6.13	$[\coth(bx)-1]e^{-ax}$	$2(2b)^{-z}\Gamma(z)\zeta(z,1+\frac{1}{2}a/b)$ Re $z > 1$
6.14	$\sinh(ax)\text{csch}(bx)$ $a < b$	$(2b)^{-z}\Gamma(z)$ $\cdot[\zeta(z,\frac{1}{2}-\frac{1}{2}a/b)-\zeta(z,\frac{1}{2}+\frac{1}{2}a/b)]$ Re $z > 0$

	$\phi(x)$	$\Phi(z) = \int_0^\infty \phi(x) x^{z-1} dx$
6.15	$\cosh(ax)\operatorname{csch}(bx)$ $a < b$	$(2b)^{-z}\Gamma(z)$ $\cdot[\zeta(z,\tfrac{1}{2}-\tfrac{1}{2}a/b)+\zeta(z,\tfrac{1}{2}+\tfrac{1}{2}a/b)]$ $\operatorname{Re} z > 1$
6.16	$\sinh(ax)\operatorname{sech}(bx)$ $a < b$	$2^{-2z}b^{-z}\Gamma(z)[\zeta(z,\tfrac{1}{4}-\tfrac{1}{4}a/b)$ $-\zeta(z,\tfrac{1}{4}+\tfrac{1}{4}a/b)+\zeta(z,\tfrac{3}{4}+\tfrac{1}{4}a/b)-\zeta(z,\tfrac{3}{4}-\tfrac{1}{4}a/b)]$ $\operatorname{Re} z > -1$
6.17	$\cosh(ax)\operatorname{sech}(bx)$ $a < b$	$2^{-2z}b^{-z}\Gamma(z)[\zeta(\tfrac{1}{4}+\tfrac{1}{4}a/b)$ $+\zeta(\tfrac{1}{4}-\tfrac{1}{4}a/b)-\zeta(\tfrac{3}{4}+\tfrac{1}{4}a/b)-\zeta(\tfrac{3}{4}-\tfrac{1}{4}a/b)]$ $\operatorname{Re} z > 0$
6.18	$[\cosh(ax)+\cos\theta]^{-1}$ $-\pi < \theta < \pi$	$\tfrac{1}{2}(2\pi/a)^z\csc\theta\csc(\tfrac{1}{2}\pi z)$ $\cdot[\zeta(1-z,\tfrac{1}{2}-\frac{\theta}{2\pi})-\zeta(1-z,\tfrac{1}{2}+\frac{\theta}{2\pi})]$ $\operatorname{Re} z > 0$
6.19	$\cosh(\tfrac{1}{2}ax)[\cosh(ax)+\cos\theta]^{-1}$ $-\pi < \theta < \pi$	$\pi^z 2^{2z-3}a^{-z}\sec(\tfrac{1}{2}\theta)\csc(\tfrac{1}{2}\pi z)$ $\cdot[\zeta(1-z,\tfrac{1}{4}+\tfrac{1}{4}\theta/\pi)+\zeta(1-z,\tfrac{1}{4}-\tfrac{1}{4}\theta/\pi)$ $-\zeta(1-z,\tfrac{3}{4}+\tfrac{1}{4}\theta/\pi)-\zeta(1-z,\tfrac{3}{4}-\tfrac{1}{4}\theta/\pi)]$ $\operatorname{Re} z > 0$

	$\phi(x)$	$\Phi(z) = \int_0^\infty \phi(x) x^{z-1} dx$
6.20	$\sinh(\tfrac{1}{2}ax)[\cosh(ax)+\cos\theta]^{-1}$ $-\pi < \theta < \pi$	$\pi^z 2^{2z-3} a^{-z} \csc(\tfrac{1}{2}\theta)\sec(\tfrac{1}{2}\pi z)$ $\cdot[\zeta(1-z,\tfrac{3}{4}+\tfrac{1}{4}\theta/\pi)-\zeta(1-z,\tfrac{3}{4}-\tfrac{1}{4}\theta/\pi)$ $+\zeta(1-z,\tfrac{1}{4}-\tfrac{1}{4}\theta/\pi)-\zeta(1-z,\tfrac{1}{4}+\tfrac{1}{4}\theta/\pi)]$ $\mathrm{Re}\, z > 0$
6.21	$e^{-ax}[\cosh(ax)+\cos\theta]^{-1}$ $-\pi < \theta < \pi$	$(2\pi/a)^z \csc\theta \csc(\pi z)$ $\cdot[\cos(\theta+\tfrac{1}{2}\pi z)](1-z,\tfrac{1}{2}+\tfrac{1}{2}\theta/\pi)$ $-\cos(\theta-\tfrac{1}{2}\pi z)\zeta(1-z,\tfrac{1}{2}-\tfrac{1}{2}\theta/\pi)]$ $\mathrm{Re}\, z > 0$
6.22	$\log\left[\frac{\cosh(\frac{\pi x}{2b})+\sin(\frac{\pi a}{2b})}{\cosh(\frac{\pi x}{2b})-\sinh(\frac{\pi a}{2b})}\right]$ $a > b$	$-\pi z^{-1}\csc(\tfrac{1}{2}\pi z)\, 2^{2z} b^z$ $\cdot[\zeta(-z,\tfrac{1}{4}-\tfrac{1}{4}a/b)-\zeta(-z,\tfrac{1}{4}+\tfrac{1}{4}a/b)$ $+\zeta(-z,\tfrac{3}{4}+\tfrac{1}{4}a/b)-\zeta(-z,\tfrac{3}{4}-\tfrac{1}{4}a/b)]$ $\mathrm{Re}\, z > 0$
6.23	$\log[\tanh(ax)]$	$\pi z^{-1}\csc(\tfrac{1}{2}\pi z)(\tfrac{1}{2}\pi/a)^z$ $\cdot(2^{1+z}-1)\zeta(-z)$ $\mathrm{Re}\, z > 0$
6.24	$\arctan[\sinh(ax)]$	$-\pi z^{-1}\sec(\tfrac{1}{2}\pi z)(2\pi/a)^z$ $\cdot[\zeta(-z,\tfrac{3}{4})-\zeta(-z,\tfrac{1}{4})]$ $-1 < \mathrm{Re}\, z < 0$

	$\phi(x)$	$\Phi(z) = \int_0^\infty \phi(x) x^{z-1} dx$
6.25	$(1+x^2)^{-\frac{1}{2}} \sinh(\frac{2ax}{1+x^2})$ $\cdot \exp(-\frac{2b}{1+x^2})$ $-1 < \text{Re } z < 2$	$2(\pi a)^{-\frac{1}{2}} e^{-b} \Gamma(1-\frac{1}{2}z) \Gamma(\frac{1}{2}+\frac{1}{2}z)$ $\cdot M_{\frac{1}{4}-\frac{1}{2}z,\frac{1}{4}}[b+(b^2+a^2)^{\frac{1}{2}}]$ $\cdot M_{\frac{1}{2}z-\frac{1}{4},\frac{1}{4}}[(b^2+a^2)^{\frac{1}{2}}-b]$
6.26	$(1+x^2)^{-\frac{1}{2}} \cosh(\frac{2ax}{1+x^2})$ $\cdot \exp(-\frac{2b}{1+x^2})$ $0 < \text{Re } z < 1$	$\frac{1}{2}(\pi a)^{-\frac{1}{2}} e^{-b} \Gamma(\frac{1}{2}-\frac{1}{2}z) \Gamma(\frac{1}{2}z)$ $\cdot M_{\frac{1}{4}-\frac{1}{2}z,-\frac{1}{4}}[(b^2+a^2)^{\frac{1}{2}}+b]$ $\cdot M_{\frac{1}{2}z-\frac{1}{4},-\frac{1}{4}}[(b^2+a^2)^{\frac{1}{2}}-b]$
6.27	$0, \quad x < 1$ $(x^2-1)^{-\frac{1}{2}} \sinh(\frac{2ax}{x^2-1})$ $\cdot \exp(-\frac{2b}{x^2-1}), \quad x > 1$ $b > a$	$(\frac{1}{2}a)^{-\frac{1}{2}} e^{b} 2^{-\frac{1}{2}z} \Gamma(1-\frac{1}{2}z) [(b^2-a^2)^{\frac{1}{2}}+b]^{\frac{1}{4}}$ $\cdot D_{z-1}\{2^{\frac{1}{2}}[b+(b^2-a^2)^{\frac{1}{2}}]^{\frac{1}{2}}\}$ $\cdot M_{\frac{1}{2}z-\frac{1}{4},\frac{1}{4}}[b-(b^2-a^2)^{\frac{1}{2}}]$ $\text{Re } z < 2$
6.28	$0, \quad x < 1$ $(x^2-1)^{-\frac{1}{2}} \cosh(\frac{2ax}{x^2-1})$ $\cdot \exp(-\frac{2b}{x^2-1}), \quad x > 1$ $b > a$	$(2a)^{-\frac{1}{2}} e^{b} 2^{-\frac{1}{2}z} \Gamma(\frac{1}{2}-\frac{1}{2}z) [(b^2-a^2)^{\frac{1}{2}}+b]^{\frac{1}{4}}$ $\cdot D_{z-1}\{2^{\frac{1}{2}}[b+(b^2-a^2)^{\frac{1}{2}}]^{\frac{1}{2}}\}$ $\cdot M_{\frac{1}{2}z-\frac{1}{4},-\frac{1}{4}}[b-(b^2-a^2)^{\frac{1}{2}}]$ $\text{Re } z < 1$

	$\phi(x)$	$\Phi(z) = \int_0^\infty \phi(x)x^{z-1}dx$
6.29	$(1+x^2)^{-\frac{1}{2}}\cosh(\frac{2ax}{1+x^2})$ $\cdot\exp[-b(\frac{1-x^2}{1+x^2})]$ $0 < \mathrm{Re}\, z < 1$	$\frac{1}{2}(\pi a)^{-\frac{1}{2}}\Gamma(\frac{1}{2}-\frac{1}{2}z)\Gamma(\frac{1}{2}z)$ $\cdot M_{\frac{1}{2}z-\frac{1}{4},-\frac{1}{4}}[(a^2+b^2)^{\frac{1}{2}}-b]$ $\cdot M_{\frac{1}{4}-\frac{1}{2}z,-\frac{1}{4}}[(a^2+b^2)^{\frac{1}{2}}+b]$
6.30	$(1+x^2)^{-\frac{1}{2}}\sinh(\frac{2ax}{1+x^2})$ $\cdot\exp[-b(\frac{1-x^2}{1+x^2})]$ $-1 < \mathrm{Re}\, z < 2$	$2(\pi a)^{-\frac{1}{2}}\Gamma(1-\frac{1}{2}z)\Gamma(\frac{1}{2}+\frac{1}{2}z)$ $\cdot M_{\frac{1}{2}z-\frac{1}{4},\frac{1}{4}}[(a^2+b^2)^{\frac{1}{2}}-b]$ $\cdot M_{\frac{1}{4}-\frac{1}{2}z,\frac{1}{4}}[(a^2+b^2)^{\frac{1}{2}}+b]$
6.31	$0 \qquad x < 1$ $(x^2-1)^{-\frac{1}{2}}\cosh(\frac{2ax}{x^2-1})$ $\cdot\exp[-b(\frac{x^2+1}{x^2-1})]$ $b > a$	$(2a)^{-\frac{1}{2}}2^{-\frac{1}{2}z}\Gamma(\frac{1}{2}-\frac{1}{2}z)[b+(b^2-a^2)^{\frac{1}{2}}]^{\frac{1}{4}}$ $\cdot D_{z-1}\{2^{\frac{1}{2}}[b+(b^2-a^2)^{\frac{1}{2}}]\}$ $\cdot M_{\frac{1}{2}z-\frac{1}{4},-\frac{1}{4}}[b-(b^2-a^2)^{\frac{1}{2}}]\}$ $\mathrm{Re}\, z < 1$
6.32	$0 \qquad x < 1$ $(x^2-1)^{-\frac{1}{2}}\sinh(\frac{2ax}{x^2-1})$ $\cdot\exp[-b(\frac{x^2+1}{x^2-1})]$ $b > a$	$(\frac{1}{2}a)^{-\frac{1}{2}}2^{-\frac{1}{2}z}\Gamma(1-\frac{1}{2}z)[b+(b^2-a^2)^{\frac{1}{2}}]^{\frac{1}{4}}$ $\cdot D_{z-1}\{2^{\frac{1}{2}}[b+(b^2-a^2)^{\frac{1}{2}}]\}$ $\cdot M_{\frac{1}{2}z-\frac{1}{4},\frac{1}{4}}[b-(b^2-a^2)^{\frac{1}{2}}]$ $\mathrm{Re}\, z < 2$

	$\phi(x)$	$\Phi(z) = \int_0^\infty \phi(x)\,x^{z-1}dx$
6.33	$(1+x^2)^{-\frac{1}{2}}\cosh\left(\frac{2ax}{1+x^2}\right)$ $\cdot\exp\left[b\left(\frac{1-x^2}{1+x^2}\right)\right]$ $0 < \mathrm{Re}\, z < 1$	$\frac{1}{2}(\pi a)^{-\frac{1}{2}}\Gamma(\frac{1}{2}z)\Gamma(\frac{1}{2}-\frac{1}{2}z)$ $\cdot M_{\frac{1}{2}-\frac{1}{2}z,-\frac{1}{4}}[(a^2+b^2)^{\frac{1}{2}}-b]$ $\cdot M_{\frac{1}{2}z-\frac{1}{4},-\frac{1}{4}}[(a^2+b^2)^{\frac{1}{2}}+b]$
6.34	$(1+x^2)^{-\frac{1}{2}}\sinh\left(\frac{2ax}{1+x^2}\right)$ $\cdot\exp\left[b\left(\frac{1-x^2}{1+x^2}\right)\right]$ $-1 < \mathrm{Re}\, z < 2$	$2(\pi a)^{-\frac{1}{2}}\Gamma(\frac{1}{2}+\frac{1}{2}z)\Gamma(1-\frac{1}{2}z)$ $\cdot M_{\frac{1}{2}-\frac{1}{2}z,\frac{1}{4}}[(a^2+b^2)^{\frac{1}{2}}-b]$ $\cdot M_{\frac{1}{2}z-\frac{1}{2},\frac{1}{4}}[(a^2+b^2)^{\frac{1}{2}}+b]$
6.35	$(1-x^2)^{-\frac{1}{2}}\cosh\left(\frac{2ax}{1-x^2}\right)$ $\cdot\exp\left[-b\left(\frac{1+x^2}{1-x^2}\right)\right]$, $x < 1$ 0 $x > 1$ $b > a$, $\mathrm{Re}\, z > 0$	$\frac{1}{2}a^{-\frac{1}{2}}2^{\frac{1}{2}z}\Gamma(\frac{1}{2}z)[b+(b^2-a^2)^{\frac{1}{2}}]^{\frac{1}{4}}$ $\cdot D_{-z}\{2^{\frac{1}{2}}[b+(b^2-a^2)^{\frac{1}{2}}]^{\frac{1}{2}}\}$ $\cdot M_{\frac{1}{4}-\frac{1}{2}z,-\frac{1}{4}}[b-(b^2-a^2)^{\frac{1}{2}}]$
6.36	$(1-x^2)^{-\frac{1}{2}}\sinh\left(\frac{2ax}{1-x^2}\right)$ $\cdot\exp\left[-b\left(\frac{1+x^2}{1-x^2}\right)\right]$ $x < 1$ 0 $x > 1$ $b > a$	$a^{-\frac{1}{2}}2^{\frac{1}{2}z}\Gamma(\frac{1}{2}+\frac{1}{2}z)[b+(b^2-a^2)^{\frac{1}{2}}]^{\frac{1}{4}}$ $\cdot D_{-z}\{2^{\frac{1}{2}}[b+(b^2-a^2)^{\frac{1}{2}}]^{\frac{1}{2}}\}$ $\cdot M_{\frac{1}{4}-\frac{1}{2}z,\frac{1}{4}}[b-(b^2-a^2)^{\frac{1}{2}}]$ $\mathrm{Re}\, z > -1$

1.7 The Gamma Function and Related Functions

	$\phi(x)$	$\Phi(z) = \int_0^\infty \phi(x)x^{z-1}dx$
7.1	$\gamma+\psi(x+1)$	$-\pi\csc(\pi z)\zeta(1-z) \quad -1<\mathrm{Re}\, z<0$
7.2	$\psi'(1+z)$	$\pi(1-z)\csc(\pi z)\zeta(2-z) \quad 0<\mathrm{Re}\, z<1$
7.3	$\psi(1+x)-\log x$	$-\pi\csc(\pi z)\zeta(1-z) \quad 0<\mathrm{Re}\, z<1$
7.4	$\psi'(1+x)-(1+x)^{-1}$	$-\pi(1-z)\csc(\pi z)[(z-1)^{-1}-\zeta(2-z)]$ $0<\mathrm{Re}\, z<2$
7.5	$\log(1+x)-\psi(1+x)$	$\pi\csc(\pi z)[\zeta(1-z)+z^{-1}] \quad 0<\mathrm{Re}\, z<1$
7.6	$\psi(\frac{1}{2}+\frac{1}{2}x)-\log(\frac{1}{2}x)$	$2\csc(\pi z)(2^{z-1}-1)\zeta(z) \quad 0<\mathrm{Re}\, z<1$
7.7	$\psi(a+x)-\psi(b+x)$	$\pi\csc(\pi z)[\zeta(1-z,b)-\zeta(1-z,a)]$ $0<\mathrm{Re}\, z<1$
7.8	$\log[x^{\frac{1}{2}} \frac{\Gamma(x)}{\Gamma(\frac{1}{2}+x)}]$	$\frac{1}{2}\pi^{-z}2^{-2z}\sec(\frac{1}{2}\pi z)\Gamma(z)$ $\cdot(2^{1+z}-1)\zeta(1+z) \quad 0<\mathrm{Re}\, z<1$
7.9	$\zeta(\nu,\alpha+ax)$ Re $\alpha > 0$, Re $\nu > 1$	$[\Gamma(\nu)]^{-1}a^{-z}\Gamma(z)\Gamma(\nu-z)\zeta(\nu-z,\alpha)$ $0 < \mathrm{Re}\, z < -1 + \mathrm{Re}\, \nu$

1.8 Legendre Functions

	$\phi(x)$	$\Phi(z) = \int_0^\infty \phi(x) x^{z-1} dx$
8.1	$P_\nu(x) \quad x < 1$ $0 \quad x > 1$	$\pi^{\frac{1}{2}} 2^{-z} \Gamma(z) [\Gamma(\frac{1}{2}-\frac{1}{2}\nu+\frac{1}{2}z) \Gamma(1+\frac{1}{2}\nu+\frac{1}{2}z)]^{-1}$ $\mathrm{Re}\, z > 0$
8.2	$P_\nu(a+x) \quad x<1-a$ $0 \quad x>1-a$ $-1 < a < 1$	$(1-a^2)^{\frac{1}{2}z} \Gamma(z) P_\nu^{-z}(a)$ $\mathrm{Re}\, z > 0$
8.3	$q_\nu(a+x)$ $a > 1$ $0 < \mathrm{Re}\, z < 1+\mathrm{Re}\, \nu$	$\Gamma(z)(a^2-1)^{\frac{1}{2}z} e^{i\pi z} q_\nu^{-z}(a)$ $= (\frac{1}{2}\pi)^{\frac{1}{2}} \Gamma(z) \Gamma(\nu+1-z)(a^2-1)^{\frac{1}{2}z-\frac{1}{4}}$ $\cdot p_{z-\frac{1}{2}}^{-\nu-\frac{1}{2}}[a(a^2-1)^{-\frac{1}{2}}]$
8.4	$p_\nu(a-x) \quad x < a-1$ $0 \quad x > a-1$ $a > 1$	$(a^2-1)^{\frac{1}{2}z} \Gamma(z) p_\nu^{-z}(a)$ $=(\frac{1}{2}\pi)^{-\frac{1}{2}} \Gamma(z)(a^2-1)^{\frac{1}{2}z-\frac{1}{4}}$ $\cdot e^{i\pi(\nu+\frac{1}{2})} q_{z-\frac{1}{2}}^{-\nu-\frac{1}{2}}[a(a^2-1)^{-\frac{1}{2}}]$ $\mathrm{Re}\, z > 0$
8.5	$q_\nu(\alpha x)$ $\lvert\arg(\alpha-1)\rvert < \pi$	$e^{i\pi z} \Gamma(z) \alpha^{-z} (\alpha^2-1)^{\frac{1}{2}z} q_\nu^{-z}(\alpha)$ $= (\frac{1}{2}\pi)^{\frac{1}{2}} \alpha^{-z} \Gamma(\nu+1-z) \Gamma(z)$ $\cdot (\alpha^2-1)^{\frac{1}{2}z-\frac{1}{4}} p_{z-\frac{1}{2}}^{-\nu-\frac{1}{2}}[\alpha(\alpha^2-1)^{-\frac{1}{2}}]$ $0 < \mathrm{Re}\, z < 1+\mathrm{Re}\, \nu$

	$\phi(x)$	$\Phi(z) = \int_0^\infty \phi(x)x^{z-1}dx$
8.6	$p_\nu(x+a)$ $a > 1$ $0 < \mathrm{Re}\ z < \begin{cases} -\mathrm{Re}\ \nu \\ 1+\mathrm{Re}\ \nu \end{cases}$	$-\pi^{-1}\sin(\pi\nu)\Gamma(z)\Gamma(-\nu-z)$ $\cdot\Gamma(1+\nu-z)(a^2-1)^{\frac{1}{2}z}p_\nu^z(a)$ $= -2^{\frac{1}{2}}\pi^{-3/2}\sin(\pi\nu)\Gamma(z)\Gamma(1+\nu-z)$ $\cdot(a^2-1)^{\frac{1}{2}z-\frac{1}{4}}e^{i\pi(\nu+\frac{1}{2})}q_{-z-\frac{1}{2}}^{-\nu-\frac{1}{2}}[a(a^2-1)^{-\frac{1}{2}}]$
8.7	$0 \qquad x < 1+a$ $(x^2+a^2-1-2ax)^{-\frac{1}{2}\mu}$ $\cdot p_\nu^\mu(x-a) \qquad x > 1+a$ $0<\mathrm{Re}\ z < \begin{cases} \mathrm{Re}(\mu-\nu) \\ 1+\mathrm{Re}(\mu+\nu) \end{cases}$	$\Gamma(\mu-\nu-z)\Gamma(1+\mu+\nu-z)[\Gamma(1-z)]^{-1}$ $\cdot(a^2-1)^{\frac{1}{2}z-\frac{1}{2}\mu}p_\nu^{z-\mu}(a)$ $= (2/\pi)^{\frac{1}{2}}\Gamma(1+\mu+\nu-z)[\Gamma(1-z)]^{-1}$ $\cdot(a^2-1)^{\frac{1}{2}z-\frac{1}{2}\mu-\frac{1}{4}}e^{i\pi(\nu+\frac{1}{2})}q_{\mu-\frac{1}{2}-z}^{-\nu-\frac{1}{2}}[a(a^2-1)^{-\frac{1}{2}}]$
8.8	$(a^2-x^2)^{-\frac{1}{2}\mu}$ $\cdot P_\nu^\mu(x/a) \qquad x < a$ $0 \qquad x > a$	$\pi^{\frac{1}{2}}2^{\mu-z}a^{z-\mu}\Gamma(z)$ $\cdot[\Gamma(\frac{1}{2}-\frac{1}{2}\nu-\frac{1}{2}\mu+\frac{1}{2}z)\Gamma(1+\frac{1}{2}\nu-\frac{1}{2}\mu+\frac{1}{2}z)]^{-1}$ $\mathrm{Re}\ z > 0$
8.9	$P_\nu(2x^2/a^2-1) \qquad x < a$ $0 \qquad x > a$	$\frac{1}{2}a^z\Gamma^2(\frac{1}{2}z)$ $\cdot[\Gamma(1+\nu+\frac{1}{2}z)\Gamma(\frac{1}{2}z-\nu)]^{-1}$ $\mathrm{Re}\ z > 0$

	$\phi(x)$	$\Phi(z) = \int_0^\infty \phi(x) x^{z-1} dx$
8.10	$(a^2-x^2)^{-\frac{1}{2}\mu}$ $\cdot P_\nu^\mu(2x^2/a^2-1) \quad x < a$ $0 \quad x > a$	$\frac{1}{2}a^{z-\mu}\Gamma(\frac{1}{2}z+\frac{1}{2}\mu)\Gamma(\frac{1}{2}z-\frac{1}{2}\mu)$ $\cdot[\Gamma(1+\nu-\frac{1}{2}\mu+\frac{1}{2}z)\Gamma(\frac{1}{2}z-\nu-\frac{1}{2}\mu)]^{-1}$ $\text{Re } z > -\text{Re } \mu$
8.11	$(a^2+x^2)^\nu$ $\cdot P_\nu^\mu(\frac{a^2-x^2}{a^2+x^2})$ $\text{Re } \mu < \text{Re } z < \text{Re}(2\nu \pm \mu)$	$\frac{1}{2}a^{z+2\nu}\Gamma(\frac{1}{2}z-\frac{1}{2}\mu)\Gamma(-\nu-\frac{1}{2}z-\frac{1}{2}\mu)$ $\cdot\Gamma(\frac{1}{2}\mu-\nu-\frac{1}{2}z)$ $\cdot[\Gamma(-\nu)\Gamma(-\nu-\mu)\Gamma(1-\frac{1}{2}\mu-\frac{1}{2}z)]^{-1}$
8.12	$(a^2+x^2)^{\frac{1}{2}\nu}$ $\cdot P_\nu(\frac{2a^2+x^2}{2a(a^2+x^2)^{\frac{1}{2}}})$	$\pi^{-\frac{1}{2}}2^{z-2}a^{\nu+z}\Gamma(\frac{1}{2}z)\Gamma(\frac{1}{2}-\frac{1}{2}z)$ $\cdot\Gamma(\frac{1}{2}-\nu-\frac{1}{2}z) \quad [\Gamma(-\nu)\Gamma(1-\frac{1}{2}z)]^{-1}$ $0 < \text{Re } z < \begin{cases} -2 \text{ Re } \nu \\ 1+2 \text{ Re } \nu \end{cases}$
8.13	$p_\nu(1+\frac{1}{2}x^2/a^2)$ $0 < \text{Re } z < \text{Re} \begin{cases} 1+\nu \\ -\nu \end{cases}$	$-\frac{1}{2}\pi^{-1}\sin(\pi\nu)(2a)^z[\Gamma(1-\frac{1}{2}z)]^{-1}$ $\cdot\Gamma(\frac{1}{2}z)\Gamma(1+\nu-\frac{1}{2}z)\Gamma(-\nu-\frac{1}{2}z)$
8.14	$q_\nu(1+\frac{1}{2}x^2/a^2)$ $0 < \text{Re } z < 1+2 \text{ Re } \nu$	$\frac{1}{4}(2a)^z\Gamma^2(\frac{1}{2}z)\Gamma(1+\nu-\frac{1}{2}z)$ $\cdot[\Gamma(1+\nu+\frac{1}{2}z)]^{-1}$

	$\phi(x)$	$\Phi(z) = \int_0^\infty \phi(x) x^{z-1} dx$
8.15	$p_\nu[(2ab)^{-1}(x^2+a^2+b^2)]$ $0 < \mathrm{Re}\, z < \begin{cases} -2\,\mathrm{Re}\,\nu \\ 2+2\,\mathrm{Re}\,\nu \end{cases}$	$-\tfrac{1}{2}\pi^{-1}\sin(\pi\nu)\Gamma(\tfrac{1}{2}z)\Gamma(1+\nu-\tfrac{1}{2}z)$ $\cdot\Gamma(-\nu-\tfrac{1}{2}z)\lvert a^2-b^2\rvert^{\frac{1}{2}z} p_\nu^{\frac{1}{2}z}\left(\frac{a^2+b^2}{2ab}\right)$ $= -\pi^{-3/2}\sin(\pi\nu)(ab)^{\frac{1}{2}}\Gamma(\tfrac{1}{2}z)$ $\Gamma(1+\nu-\tfrac{1}{2}z)\lvert a^2-b^2\rvert^{\frac{1}{2}z-\frac{1}{2}}$ $\cdot e^{i\pi(\nu+\frac{1}{2})} q_{-\frac{1}{2}-\frac{1}{2}z}^{-\nu-\frac{1}{2}}\left(\frac{a^2+b^2}{\lvert a^2-b^2\rvert}\right)$
8.16	$q_\nu[(2ab)^{-1}(x^2+a^2+b^2)]$ $0 < \mathrm{Re}\, z < 2+2\,\mathrm{Re}\,\nu$	$\tfrac{1}{2}\pi^{\frac{1}{2}}(ab)^{\frac{1}{2}}\Gamma(\tfrac{1}{2}z)\Gamma(1+\nu-\tfrac{1}{2}z)$ $\lvert a^2-b^2\rvert^{\frac{1}{2}z-\frac{1}{2}} p_{\frac{1}{2}z-\frac{1}{2}}^{-\nu-\frac{1}{2}}\left(\frac{a^2+b^2}{\lvert a^2-b^2\rvert}\right)$ $= \tfrac{1}{2}\Gamma(\tfrac{1}{2}z)\lvert a^2-b^2\rvert^{\frac{1}{2}z}$ $\cdot e^{i\frac{1}{2}\pi z} q_\nu^{-\frac{1}{2}z}\left(\frac{a^2+b^2}{2ab}\right)$
8.17	$q_\nu[(2bx)^{-1}(a^2+b^2+x^2)]$ $-1-\mathrm{Re}\,\nu < \mathrm{Re}\, z < 1+\mathrm{Re}\,\nu$	$\pi^{\frac{1}{2}} 2^{\nu-\frac{1}{2}} b^{\frac{1}{2}}\Gamma(\tfrac{1}{2}+\tfrac{1}{2}\nu+\tfrac{1}{2}z)\Gamma(\tfrac{1}{2}+\tfrac{1}{2}\nu-\tfrac{1}{2}z)$ $\cdot(a^2+b^2)^{\frac{1}{2}z-\frac{1}{4}} P_{z-\frac{1}{2}}^{-\nu-\frac{1}{2}}[a(a^2+b^2)^{-\frac{1}{2}}]$
8.18	$0 \qquad x < a+b$ $p_\nu[(2ab)^{-1}(x^2-a^2-b^2)]$ $x > a+b$	$\tfrac{1}{2}[\Gamma(1-\tfrac{1}{2}z)]^{-1}\Gamma(1+\nu-\tfrac{1}{2}z)\Gamma(-\nu-\tfrac{1}{2}z)$ $\lvert a^2-b^2\rvert^{\frac{1}{2}z} p_\nu^{\frac{1}{2}z}\left(\frac{a^2+b^2}{2ab}\right)$ $= (ab/\pi)^{\frac{1}{2}}[\Gamma(1-\tfrac{1}{2}z)]^{-1}\Gamma(1+\nu-\tfrac{1}{2}z)$ $\lvert a^2-b^2\rvert^{\frac{1}{2}z-\frac{1}{2}} e^{i\pi(\nu+\frac{1}{2})} q_{-\frac{1}{2}-\frac{1}{2}z}^{-\nu}\left(\frac{a^2+b^2}{\lvert a^2-b^2\rvert}\right)$

	$\phi(x)$	$\Phi(z) = \int_0^\infty \phi(x)x^{z-1}dx$
8.19	$(A^2-1)^{\frac{1}{2}\mu}P_\nu^\mu(A)$ $A = (2ab)^{-1}(a^2+b^2+x^2)$ $0 < \mathrm{Re}\, z < \begin{cases} 2+2\,\mathrm{Re}(\nu-\mu) \\ -2\mathrm{Re}(\nu+\mu) \end{cases}$	$2^{-\mu-1}(ab)^{-\mu}\Gamma(\tfrac{1}{2}z)\Gamma(1+\nu-\tfrac{1}{2}z-\mu)$ $\cdot\,\Gamma(-\nu-\mu-\tfrac{1}{2}z)[\Gamma(1-\mu+\nu)\Gamma(-\mu-\nu)]^{-1}$ $\cdot\,\lvert a^2-b^2\rvert^{\mu+\frac{1}{2}z}P_\nu^{\mu+\frac{1}{2}z}(\frac{a^2+b^2}{2ab})$ $= \pi^{-\frac{1}{2}}2^{-\mu}(ab)^{\frac{1}{2}-\mu}\Gamma(\tfrac{1}{2}z)\Gamma(1+\nu-\mu-\tfrac{1}{2}z)$ $\cdot[\Gamma(1-\mu+\nu)\Gamma(-\mu-\nu)]^{-1}\lvert a^2-b^2\rvert^{\mu+\frac{1}{2}z-\frac{1}{2}}$ $\cdot\,e^{i\pi(\frac{1}{2}+\nu)}Q_{-\mu-\frac{1}{2}z-\frac{1}{2}}^{-\nu-\frac{1}{2}}(\frac{a^2+b^2}{\lvert a^2-b^2\rvert})$
8.20	$e^{-i\pi\mu}(A^2-1)^{-\frac{1}{2}\mu}Q_\nu^\mu(A)$ $A = (2ab)^{-1}(a^2+b^2+x^2)$ $0 < \mathrm{Re}\, z < 1+\mathrm{Re}(\nu+\mu)$	$\pi^{\frac{1}{2}}2^{\mu-1}(ab)^{\mu+\frac{1}{2}}\Gamma(\tfrac{1}{2}z)$ $\cdot\,\Gamma(1+\nu+\mu-\tfrac{1}{2}z)\lvert a^2-b^2\rvert^{\frac{1}{2}z-\mu-\frac{1}{2}}$ $\cdot\,P_{\frac{1}{2}z-\mu-\frac{1}{2}}^{-\nu-\frac{1}{2}}(\frac{a^2+b^2}{\lvert a^2-b^2\rvert})$ $= \tfrac{1}{2}(2ab)^{\mu}\lvert a^2-b^2\rvert^{\frac{1}{2}z-\mu}\Gamma(\tfrac{1}{2}z)$ $\cdot\,e^{i\pi(\frac{1}{2}z-\mu)}Q_\nu^{\mu-\frac{1}{2}z}(\frac{a^2+b^2}{2ab})$
8.21	$(\frac{x-a}{x+a})^{\frac{1}{2}\mu}$ $\cdot\,P_\nu^\mu(\frac{x}{a})$ $0 < \mathrm{Re}\, z < \begin{cases} -\mathrm{Re}\,\nu \\ 1+\mathrm{Re}\,\nu \end{cases}$	$-\pi^{-1}\sin(\pi\nu)\Gamma(z)a^z[\Gamma(1-\mu-z)]^{-1}$ $\cdot\,\Gamma(1+\nu-z)\,\Gamma(-\nu-z)$ $\cdot\,{}_2F_1(1+\nu-z,-\nu-z;1-\mu-z;\tfrac{1}{2})$

	$\phi(x)$	$\Phi(z) = \int_0^\infty \phi(x) x^{z-1} dx$
8.22	$(2x+a)^{-\frac{1}{2}}(2x-a)^{\frac{1}{2}\mu}$ $\cdot P_\nu^\mu(\frac{1}{2}+x/a)$ $0 < \mathrm{Re}\, z < 1-\mathrm{Re}\,\mu$	$2^{\frac{3}{2}\mu-\frac{1}{2}} a^{z-\frac{1}{2}+\frac{1}{2}\mu} \Gamma(z)[\Gamma(1-z-\mu)]^{-1}$ $\cdot\ \Gamma(1-z+\frac{1}{2}\nu-\frac{1}{2}\mu)\Gamma(\frac{1}{2}-z-\frac{1}{2}\nu-\frac{1}{2}\mu)$ $\cdot\ [\Gamma(1+\frac{1}{2}\nu-\frac{1}{2}\mu)\Gamma(\frac{1}{2}-\frac{1}{2}\nu-\frac{1}{2}\mu)]^{-1}$ $\cdot\ {}_2F_1(1-z+\frac{1}{2}\nu-\frac{1}{2}\mu, \frac{1}{2}-z-\frac{1}{2}\nu-\frac{1}{2}\mu; 1-z-\mu; \frac{1}{2})$
8.23	$(a^2+x^2)^{-\frac{1}{2}\mu}$ $\cdot P_\nu^\mu(1+2x^2/a^2)$	$-\frac{1}{2}\pi^{-1}\sin(\pi\nu) a^{z-\mu}[\Gamma(1-\frac{1}{2}\mu-\frac{1}{2}z)]^{-1}$ $\cdot\ \Gamma(\frac{1}{2}z-\frac{1}{2}\mu)\Gamma(1-\frac{1}{2}z+\frac{1}{2}\mu+\nu)\Gamma(\frac{1}{2}\mu-\nu-\frac{1}{2}z)$ $\mathrm{Re}\,\mu < \mathrm{Re}\, z < \begin{cases} \mathrm{Re}(\mu-2\nu) \\ 2+\mathrm{Re}(\mu+2\nu) \end{cases}$
8.24	$(a^2+x^2)^{\frac{1}{2}\mu}$ $\cdot P_\nu^\mu(1+2x^2/a^2)$ $\mathrm{Re}\,\mu < \mathrm{Re}\, z < \begin{cases} -\mathrm{Re}(2\nu+\mu) \\ 2+\mathrm{Re}(2\nu-\mu) \end{cases}$	$\frac{1}{2}a^{z+\mu}\Gamma(\frac{1}{2}z-\frac{1}{2}\mu)\Gamma(1+\nu-\frac{1}{2}z-\frac{1}{2}\mu)$ $\cdot\ \Gamma(-\frac{1}{2}z-\frac{1}{2}\mu-\frac{1}{2}\nu)$ $\cdot\ [\Gamma(-\mu-\nu)\Gamma(1-\mu+\nu)\Gamma(1-\frac{1}{2}\mu-\frac{1}{2}z)]^{-1}$
8.25	$(a^2+x^2)^{-\frac{1}{2}\mu}$ $\cdot e^{-i\pi\mu} Q_\nu^\mu(1+2x^2/a^2)$	$\frac{1}{4}a^{z-\mu}[\Gamma(1+\nu-\frac{1}{2}\mu+\frac{1}{2}z)]^{-1}$ $\cdot\ \Gamma(\frac{1}{2}z+\frac{1}{2}\mu)\Gamma(\frac{1}{2}z-\frac{1}{2}\mu)\Gamma(1+\nu+\frac{1}{2}\mu-\frac{1}{2}z)$ $\pm\,\mathrm{Re}\,\mu < \mathrm{Re}\, z < 2+\mathrm{Re}(\mu+2\nu)$

	$\phi(x)$	$\Phi(z) = \int_0^\infty \phi(x)\,x^{z-1}dx$
8.26	$0 \quad x < a$ $(x^2-a^2)^{-\frac{1}{2}\mu}P_\nu^\mu(x/a) \quad x>a$ $\mathrm{Re}\,\mu < 1$	$\pi^{-\frac{1}{2}}2^{\mu-1-z}a^{z-\mu}[\Gamma(1-z)]^{-1}$ $\cdot\,\Gamma(\frac{1}{2}+\frac{1}{2}\mu+\frac{1}{2}\nu-\frac{1}{2}z)\Gamma(\frac{1}{2}\mu-\frac{1}{2}\nu-\frac{1}{2}z)$ $\mathrm{Re}\,z < \mathrm{Re}(\mu-\nu,\mu+\nu+1)$
8.27	$0 \quad x < a$ $(x^2-a^2)^{-\frac{1}{2}\mu}$ $\cdot P_\nu^\mu(2x^2/a^2-1) \quad x > a$ $\mathrm{Re}\,\mu < 1$	$\frac{1}{2}a^{z-\mu}\Gamma(1+\frac{1}{2}\mu+\nu-\frac{1}{2}z)\Gamma(\frac{1}{2}\mu-\frac{1}{2}z-\nu)$ $\cdot\,[\Gamma(1+\frac{1}{2}\mu-\frac{1}{2}z)\Gamma(1-\frac{1}{2}\mu-\frac{1}{2}z)]^{-1}$ $\mathrm{Re}\,z > \begin{cases}\mathrm{Re}(\mu-2\nu)\\ \mathrm{Re}(\mu+2\nu+2)\end{cases}$
8.28	$0 \quad x < a$ $(x^2-a^2)^{-\frac{1}{2}}$ $\cdot P_\nu^\mu[(x^2/a^2-1)^{\frac{1}{2}}]$ $x > a$	$2^{\mu-1}a^{z-1}\Gamma(\frac{1}{2}z-\frac{1}{2}\mu)\Gamma(1+\frac{1}{2}\nu-\frac{1}{2}z)$ $\cdot\,\Gamma(\frac{1}{2}-\frac{1}{2}\nu-\frac{1}{2}z)\,[\Gamma(\frac{1}{2}-\frac{1}{2}\mu-\frac{1}{2}\nu)$ $\cdot\,\Gamma(1-\frac{1}{2}\mu+\frac{1}{2}\nu)\Gamma(1-\frac{1}{2}\mu-\frac{1}{2}z)]^{-1}$ $-\mathrm{Re}\,\mu < \mathrm{Re}\,z < \begin{cases}1-\mathrm{Re}\,\nu\\ 2+\mathrm{Re}\,\nu\end{cases}$
8.29	$0 \quad x < a$ $(x^2-a^2)^{-\frac{1}{2}}e^{-i\pi\mu}$ $\cdot Q_\nu^\mu[(x^2/a^2-1)^{\frac{1}{2}}]$ $x > a$	$2^{\mu-z}a^{z-1}[\Gamma(1+\frac{1}{2}\nu-\frac{1}{2}\mu)\Gamma(\frac{1}{2}+\frac{1}{2}\nu+\frac{1}{2}z)]^{-1}$ $\cdot\,\Gamma(\frac{1}{2}+\frac{1}{2}\nu+\frac{1}{2}\mu)\Gamma(1+\frac{1}{2}\nu-\frac{1}{2}z)$ $\cdot\,\Gamma(\frac{1}{2}z+\frac{1}{2}\mu)\Gamma(\frac{1}{2}z-\frac{1}{2}\mu)$ $\pm\,\mathrm{Re}\,\mu < \mathrm{Re}\,z < 2+\mathrm{Re}\,\nu$

	$\phi(x)$	$\Phi(z) = \int_0^\infty \phi(x)x^{z-1}dx$
8.30	$x^{\frac{1}{2}\mu}(2+x)^{\frac{1}{2}\mu}$ $\cdot P_\nu^\mu(1+x)$	$2^{\mu+z}\Gamma(z)\Gamma(-\mu-\nu-z)\Gamma(1+\nu-\mu-z)$ $\cdot[\Gamma(1-\mu+\nu)\Gamma(-\mu-\nu)\Gamma(1-\mu-z)]^{-1}$ $0 < \mathrm{Re}\, z < \begin{cases} -\mathrm{Re}(\mu+\nu) \\ 1-\mathrm{Re}(\mu-\nu) \end{cases}$
8.31	$x^{\frac{1}{2}\mu}(2+x)^{\frac{1}{2}\mu}$ $\cdot(\alpha+x)^{-\rho}P_\nu^\mu(1+x)$ $0<\mathrm{Re}\, z<\begin{cases}\mathrm{Re}(\rho-\mu-\nu) \\ 1+\mathrm{Re}(\nu+\rho-\mu)\end{cases}$	$2^{\mu-\rho+z}\dfrac{\Gamma(z-\rho)\Gamma(\rho-\nu-\mu-z)\Gamma(1+\rho-\mu+\nu-z)}{\Gamma(1+\nu-\mu)\Gamma(-\nu-\mu)\Gamma(1+\rho-\mu-z)}$ $\cdot{}_3F_2\left({\rho,\rho-\nu-\mu-z,1+\rho+\nu-\mu-z; \atop \rho+1-z,1+\rho-\mu-z;}\tfrac{1}{2}\alpha\right)$ $+2^\mu\Gamma(z)\Gamma(\rho-z)[\Gamma(\rho)\Gamma(1-\mu)]^{-1}\alpha^{z-\rho}$ $\cdot{}_3F_2\left({z,-\mu-\nu,1+\nu-\mu; \atop 1-\rho+z,\ 1-\mu;}\tfrac{1}{2}\alpha\right)$
8.32	$x^{-\frac{1}{2}\mu}(2+x)^{-\frac{1}{2}\mu}$ $\cdot(\alpha+x)^{-\rho}P_\nu^\mu(1+x)$ $\mathrm{Re}\,\mu<\mathrm{Re}\, z<\begin{cases}\mathrm{Re}(\rho+\mu-\nu) \\ 1+\mathrm{Re}(\rho+\mu+\nu)\end{cases}$	$[\Gamma(\rho)\Gamma(1-\mu)]^{-1}\Gamma(z-\mu)\Gamma(\rho+\mu-z)\alpha^{z-\mu-\rho}$ $\cdot{}_3F_2\left({z-\mu,-\nu,1+\nu; \atop 1-\mu-\rho+z,1-\mu;}\tfrac{1}{2}\alpha\right)$ $-\pi^{-1}\sin(\pi\nu)2^{z-\rho-\mu}[\Gamma(1+\rho-z)]^{-1}$ $\cdot\Gamma(z-\mu-\rho)\Gamma(\rho+\mu-\nu-z)\Gamma(\rho+\mu+\nu+1-z)$ $\cdot{}_3F_2\left({\rho,\rho+\mu-\nu-z,\rho+\mu+\nu+1-z; \atop 1+\rho-z,1+\rho+\mu-z;}\tfrac{1}{2}\alpha\right)$

	$\phi(x)$	$\Phi(z) = \int_0^\infty \phi(x) x^{z-1} dx$
8.33	$x^{\frac{1}{2}\mu}(2+x)^{\frac{1}{2}\mu}$ $\cdot e^{-ax} p_\nu^\mu(1+x)$ Re $z > 0$	$[\Gamma(-\mu-\nu)\Gamma(1-\mu+\nu)]^{-1} a^{-\mu-z} e^{-a}$ $\cdot G_{23}^{31}\left(2a \,\middle\vert\, \begin{matrix} 1+\mu,\ 1 \\ \mu+z,-\nu,1+\nu \end{matrix}\right)$
8.34	$x^{-\frac{1}{2}\mu}(2+x)^{-\frac{1}{2}\mu}$ $\cdot e^{-ax} p_\nu^\mu(1+x)$ Re $z >$ Re μ	$-\pi^{-1}\sin(\pi\nu) a^{\mu-z} e^{-a}$ $\cdot G_{23}^{31}\left(2a \,\middle\vert\, \begin{matrix} 1,\ 1-\mu \\ z-\mu,1+\nu,-\nu \end{matrix}\right)$
8.35	$x^{\frac{1}{2}\mu}(2+x)^{\frac{1}{2}\mu}$ $\cdot e^{-ax} q_\nu^\mu(1+x)$	$\frac{1}{2}[\Gamma(1+\nu-\mu)]^{-1}\Gamma(1+\nu+\mu) e^{i\pi\mu} a^{-\mu-z} e^{-a}$ $\cdot G_{23}^{22}\left(2a \,\middle\vert\, \begin{matrix} 1+\mu,\ 1 \\ z+\mu,1+\nu,-\nu \end{matrix}\right)$
8.36	$x^{-\frac{1}{2}\mu}(2+x)^{-\frac{1}{2}\mu}$ $\cdot e^{-ax} q_\nu^\mu(1+x)$ Re $z > \begin{cases} 0 \\ \text{Re } \mu \end{cases}$	$\frac{1}{2} e^{i\pi\mu} a^{\mu-z} e^{-a}$ $\cdot G_{23}^{22}\left(2a \,\middle\vert\, \begin{matrix} 1-\mu,\ 1 \\ z-\mu,1+\nu,-\nu \end{matrix}\right)$
8.37	$(b-x)^{\mu-1}(1+\frac{1}{2}ax)^{-\frac{1}{2}}$ $\cdot p_\nu^\lambda(1+ax)$ $x < b$; 0 $x > b$; Re $\mu > 0$, $ab < 1$	$\frac{1}{2} a^{-\frac{1}{2}\lambda}[\Gamma(1-\lambda)\Gamma(\mu-\frac{1}{2}\lambda+z)]^{-1}\Gamma(\mu)$ $\cdot\Gamma(z-\frac{1}{2}\lambda) b^{\mu-\frac{1}{2}\lambda-1+z} {}_3F_2(-\nu,1+\nu,$ $z-\frac{1}{2}\lambda;1-\lambda,\mu-\frac{1}{2}\lambda+z;-\frac{1}{2}ab)$ Re $z > \frac{1}{2}$ Re λ

	$\phi(x)$	$\Phi(z) = \int_0^\infty \phi(x) x^{z-1} dx$
8.38	$(1-x^2)^{-\frac{1}{2}\mu} \cdot P_\nu^\mu(x)$ $\quad x < 1$ 0 $\quad x > 1$	$\pi^{\frac{1}{2}} 2^{\mu-z} \Gamma(z)$ $\cdot [\Gamma(\frac{1}{2}-\frac{1}{2}\nu-\frac{1}{2}\mu+\frac{1}{2}z)\Gamma(1+\frac{1}{2}\nu-\frac{1}{2}\mu+\frac{1}{2}z)]^{-1}$ $\mathrm{Re}\, z > 0, \quad \mathrm{Re}\, \mu < 1$
8.39	$(1-x^2)^{K} \cdot P_\nu^\mu(x)$ $\quad x < 1$ 0 $\quad x > 1$ $\mathrm{Re}(K-\frac{1}{2}\mu) > -1, \ \mathrm{Re}\, z > 0$	$2^{\mu-1} \Gamma(1+K-\frac{1}{2}\mu) \Gamma(\frac{1}{2}z)$ $\cdot [\Gamma(1-\mu)\Gamma(1+K-\frac{1}{2}\mu+\frac{1}{2}z)]^{-1}$ $\cdot {}_3F_2\left({\frac{1}{2}+\frac{1}{2}\nu-\frac{1}{2}\mu, -\frac{1}{2}\nu-\frac{1}{2}\mu, 1+K-\frac{1}{2}\mu; \atop 1-\mu, 1+K-\frac{1}{2}\mu+\frac{1}{2}z;} 1\right)$
8.40	$(1-x^2)^{-\frac{1}{2}\mu} \sin(ax) \cdot P_\nu^\mu(x)$ $\quad x < 1$ 0 $\quad x > 1$ $\mathrm{Re}\, z > -1, \ \mathrm{Re}\, \mu < 1$	$\pi^{\frac{1}{2}} 2^{\mu-z-1} a \Gamma(1+z)$ $\cdot [\Gamma(1-\frac{1}{2}\nu-\frac{1}{2}\mu+\frac{1}{2}z)\Gamma(\frac{3}{2}+\frac{1}{2}\nu-\frac{1}{2}\mu+\frac{1}{2}z)]^{-1}$ $\cdot {}_2F_3\left({\frac{1}{2}+\frac{1}{2}z, 1+\frac{1}{2}z; \atop \frac{3}{2}, 1-\frac{1}{2}\nu-\frac{1}{2}\mu+\frac{1}{2}z, \frac{3}{2}+\frac{1}{2}\nu-\frac{1}{2}\mu+\frac{1}{2}z;} -\frac{1}{4}a^2\right)$
8.41	$(1-x^2)^{-\frac{1}{2}\mu} \cos(ax)$ $P_\nu^\mu(x)$ $\quad x < 1$ 0 $\quad x > 1$ $\mathrm{Re}\, z > 0, \ \mathrm{Re}\, \mu < 1$	$\pi^{\frac{1}{2}} 2^{\mu-z} \Gamma(z) [\Gamma(1+\frac{1}{2}\nu-\frac{1}{2}\mu+\frac{1}{2}z)\Gamma(\frac{1}{2}-\frac{1}{2}\nu-\frac{1}{2}\mu+\frac{1}{2}z)]^{-1}$ $\cdot {}_2F_3\left({\frac{1}{2}z, \frac{1}{2}+\frac{1}{2}z; \atop \frac{1}{2}, \frac{1}{2}-\frac{1}{2}\nu-\frac{1}{2}\mu+\frac{1}{2}z, 1+\frac{1}{2}\nu-\frac{1}{2}\mu+\frac{1}{2}z;} -\frac{1}{4}a^2\right)$

	$\phi(x)$	$\Phi(z) = \int_0^\infty \phi(x) x^{z-1} dx$
8.42	$P_\nu(x-1)$ $\quad x < 2$ 0 $\quad x > 2$	$2^z \Gamma^2(z) [\Gamma(z-\nu)\Gamma(1+\nu+z)]^{-1}$ $\mathrm{Re}\, z > 0$
8.43	$(1-x)^{-\frac{1}{2}}$ $\cdot\{P_\nu^\mu[(1-x)^{\frac{1}{2}}]$ $+ P_\nu^\mu[-(1-x)^{\frac{1}{2}}]\}$ $\quad x < 1$ 0 $\quad x > 1$	$2^{\mu+1}\pi\Gamma(z+\frac{1}{2}\mu)\Gamma(z-\frac{1}{2}\mu)$ $\cdot[\Gamma(\frac{1}{2}+\frac{1}{2}\nu+z)\Gamma(z-\frac{1}{2}\nu)\Gamma(1-\frac{1}{2}\mu+\frac{1}{2}\nu)$ $\cdot\Gamma(\frac{1}{2}-\frac{1}{2}\mu-\frac{1}{2}\nu)]^{-1}$ $\mathrm{Re}\, z > \frac{1}{2}\lvert\mathrm{Re}\, \mu\rvert$
8.44	$(1-x)^{-\frac{1}{2}}(1-a^2+a^2x)^{\frac{1}{2}\mu}$ $\cdot\{P_\nu[a(1-x)^{\frac{1}{2}}]$ $+P_\nu[-a(1-x)^{\frac{1}{2}}]\}$ $\quad x < 1$ 0 $\quad x > 1$	$2^{\mu+1}\pi\Gamma(z)$ $\cdot[\Gamma(\frac{1}{2}+z)\Gamma(\frac{1}{2}-\frac{1}{2}\mu-\frac{1}{2}\nu)\Gamma(1-\frac{1}{2}\mu+\frac{1}{2}\nu)]^{-1}$ $\cdot {}_2F_1(-\frac{1}{2}\nu-\frac{1}{2}\mu, \frac{1}{2}+\frac{1}{2}\nu-\frac{1}{2}\mu; \frac{1}{2}+z; a^2)$ $-1 < a < 1,$ $\quad \mathrm{Re}\, z > 0$
8.45	$P_\nu^\mu(x)$ $\quad x < 1$ 0 $\quad x > 1$ $\mathrm{Re}\, \mu < 2,$ $\quad \mathrm{Re}\, z < 2$	$\pi^{\frac{1}{2}} 2^{2\mu-1}\Gamma(\frac{1}{2}z)[\Gamma(\frac{1}{2}-\frac{1}{2}\mu)\Gamma(1-\frac{1}{2}\mu+\frac{1}{2}z)]^{-1}$ $\cdot {}_3F_2\left(\begin{matrix}\frac{1}{2}+\frac{1}{2}\nu-\frac{1}{2}\mu, -\frac{1}{2}\nu-\frac{1}{2}\mu, 1-\frac{1}{2}\mu; \\ 1-\mu, 1-\frac{1}{2}\mu+\frac{1}{2}z;\end{matrix}\ 1\right)$
8.46	$[(b+x)^2-1]^{\frac{1}{2}\mu}$ $\cdot q_\nu^{-\mu}(b+x)$	$e^{i\pi z}\Gamma(z)(b^2-1)^{\frac{1}{2}\mu+\frac{1}{2}z} q_\nu^{-\mu-z}(b)$ $0 < \mathrm{Re}\, z < 1+\mathrm{Re}(\nu-\mu)$

	$\phi(x)$	$\Phi(z) = \int_0^\infty \phi(x)x^{z-1}dx$
8.47	$(1+x)^{-\frac{1}{2}}(a^2-1+a^2x)^{\frac{1}{2}\mu}$ $\cdot p_\nu^\mu[a(1+x)^{\frac{1}{2}}]$ $0<\mathrm{Re}\ z<\begin{cases}1+\frac{1}{2}\mathrm{Re}(\nu-\mu)\\ \frac{1}{2}-\frac{1}{2}\mathrm{Re}(\nu+\mu)\end{cases}$	$2^\mu a^{\mu-\nu-1}\Gamma(z)\Gamma(1+\frac{1}{2}\nu-\frac{1}{2}\mu-z)$ $\cdot\Gamma(\frac{1}{2}-\frac{1}{2}\nu-\frac{1}{2}\mu-z)[\Gamma(1-\frac{1}{2}\mu+\frac{1}{2}\nu)$ $\cdot\Gamma(\frac{1}{2}-\frac{1}{2}\nu-\frac{1}{2}\mu)\Gamma(1-\mu-z)]^{-1}$ $\cdot{}_2F_1(\frac{1}{2}+\frac{1}{2}\nu-\frac{1}{2}\mu,1+\frac{1}{2}\nu-\frac{1}{2}\mu-z;1-\mu-z;1-a^{-2})$
8.48	$(1+x)^{-\frac{1}{2}}(a^2-1+a^2x)^{-\frac{1}{2}\mu}$ $\cdot e^{-i\pi\mu}q_\nu^\mu[a(1+x)^{\frac{1}{2}}]$ $0<\mathrm{Re}\ z<1+\frac{1}{2}\mathrm{Re}(\nu+\mu)$	$2^{\mu-1}a^{-\mu-\nu-1}[\Gamma(3/2+\nu)]^{-1}\Gamma(z)$ $\cdot\Gamma(\frac{1}{2}+\frac{1}{2}\nu+\frac{1}{2}\mu)\Gamma(1+\frac{1}{2}\nu+\frac{1}{2}\mu-z)$ $\cdot{}_2F_1(\frac{1}{2}+\frac{1}{2}\nu+\frac{1}{2}\mu,1+\frac{1}{2}\nu+\frac{1}{2}\mu-z;3/2+\nu;a^{-2})$
8.49	$(a^2+x^2)^{\frac{1}{2}\nu}e^{-bx}$ $\cdot P_\nu^\mu[x(a^2+x^2)^{-\frac{1}{2}}]$ $\mathrm{Re}\ z>0$	$2^{-\nu-2}[\pi\Gamma(-\nu-\mu)]^{-1}a^{\nu+z}$ $\cdot G_{24}^{32}\left(\frac{1}{4}a^2b^2\middle\vert\begin{matrix}1-\frac{1}{2}z,\frac{1}{2}-\frac{1}{2}z\\ 0,\frac{1}{2},-\frac{1}{2}z-\frac{1}{2}\nu-\frac{1}{2}\mu,-\frac{1}{2}z+\frac{1}{2}\mu-\frac{1}{2}\nu\end{matrix}\right)$
8.50	$(b-x)^{\mu-1}(1-x)^{-\frac{1}{2}\lambda}$ $\cdot P_\nu^\lambda(1-2x)$ $x<b$ 0 $x>b$ $\mathrm{Re}\ \mu>0,\ b<1$	$[\Gamma(1-\lambda)\Gamma(\mu-\frac{1}{2}\lambda+z)]^{-1}\Gamma(\mu)$ $\cdot\Gamma(z-\frac{1}{2}\lambda)b^{\mu-1-\frac{1}{2}\lambda+z}\cdot{}_3F_2(-\nu,1+\nu,$ $z-\frac{1}{2}\lambda;1-\lambda,\mu-\frac{1}{2}\lambda+z;b)$ $\mathrm{Re}\ z>-\frac{1}{2}\mathrm{Re}\ \lambda$

	$\phi(x)$	$\Phi(z) = \int_0^\infty \phi(x) x^{z-1} dx$
8.51	$(a^2+x^2)^{-\frac{1}{2}-\frac{1}{2}\nu}$ $\cdot P_\nu^\mu[x(a^2+x^2)^{-\frac{1}{2}}]$	$\frac{1}{2}(\frac{1}{2}a)^{z-\nu-1}\Gamma(z)\Gamma(\frac{1}{2}-\frac{1}{2}\mu+\frac{1}{2}\nu-\frac{1}{2}z)$ $\cdot[\Gamma(\nu-\mu+1)\Gamma(\frac{1}{2}-\frac{1}{2}\mu-\frac{1}{2}\nu+\frac{1}{2}z)]^{-1}$ $0 < \mathrm{Re}\, z < 1+\mathrm{Re}(\nu+\mu)$
8.52	$p_\nu^\mu[(1+x^2/a^2)^{\frac{1}{2}}]$	$2^{\mu-1}a^z\Gamma(\frac{1}{2}z-\frac{1}{2}\mu)\Gamma(\frac{1}{2}+\frac{1}{2}\nu-\frac{1}{2}z)\Gamma(-\frac{1}{2}z-\frac{1}{2}\nu)$ $\cdot[\Gamma(1-\frac{1}{2}\mu-\frac{1}{2}z)\Gamma(\frac{1}{2}+\frac{1}{2}\nu-\frac{1}{2}\mu)\Gamma(-\frac{1}{2}\nu-\frac{1}{2}\mu)]^{-1}$ $\mathrm{Re}\,\mu < \mathrm{Re}\, z < \begin{cases} -\mathrm{Re}\,\nu \\ 1+\mathrm{Re}\,\nu \end{cases}$
8.53	$(a^2+x^2)^{-\frac{1}{2}}$ $\cdot p_\nu^\mu[(1+x^2/a^2)^{\frac{1}{2}}]$	$2^{\mu-1}a^{z-1}\Gamma(\frac{1}{2}z-\frac{1}{2}\mu)\Gamma(1-\frac{1}{2}z+\frac{1}{2}\nu)\Gamma(\frac{1}{2}-\frac{1}{2}z-\frac{1}{2}\nu)$ $\cdot[\Gamma(1-\frac{1}{2}\mu-\frac{1}{2}z)\Gamma(1-\frac{1}{2}\mu+\frac{1}{2}\nu)\Gamma(\frac{1}{2}-\frac{1}{2}\mu-\frac{1}{2}\nu)]^{-1}$ $\mathrm{Re}\,\mu < \mathrm{Re}\, z < \begin{cases} 1-\mathrm{Re}\,\nu \\ 2+\mathrm{Re}\,\nu \end{cases}$
8.54	$e^{-i\pi\mu}q_\nu^\mu[(1+x^2/a^2)^{\frac{1}{2}}]$ $\pm 2\mathrm{Re}\,\mu < \mathrm{Re}\, z < \frac{1}{2}+\mathrm{Re}\,\nu$	$2^{-\nu-2}\pi^{\frac{1}{2}}a^z\Gamma(\frac{1}{2}+\frac{1}{2}\nu-\frac{1}{2}z)\Gamma(1+\nu+\mu)$ $\cdot\Gamma(-\frac{1}{2}\mu-\frac{1}{2}z)\Gamma(\frac{1}{2}\mu-\frac{1}{2}z)\ [\Gamma(1+\frac{1}{2}\nu+\frac{1}{2}z)$ $\cdot\Gamma(\frac{1}{2}+\frac{1}{2}\nu-\frac{1}{2}\mu)\Gamma(\frac{1}{2}+\frac{1}{2}\nu+\frac{1}{2}\mu)]^{-1}$
8.55	$(a^2+x^2)^{-\frac{1}{2}}$ $\cdot e^{-i\pi\mu}q_\nu^\mu[(1+x^2/a^2)^{\frac{1}{2}}]$ $\pm\mathrm{Re}\,\mu < \mathrm{Re}\, z < \frac{3}{2}+\mathrm{Re}\,\nu$	$2^{\mu-2}a^{z-1}\Gamma(1+\frac{1}{2}\nu-\frac{1}{2}z)\Gamma(\frac{1}{2}+\frac{1}{2}\nu+\frac{1}{2}\mu)$ $\cdot\Gamma(\frac{1}{2}z+\frac{1}{2}\mu)\Gamma(\frac{1}{2}z-\frac{1}{2}\mu)$ $\cdot[\Gamma(\frac{1}{2}+\frac{1}{2}\nu+\frac{1}{2}z)\Gamma(1+\frac{1}{2}\nu-\frac{1}{2}\mu)]^{-1}$

	$\phi(x)$	$\Phi(z) = \int_0^\infty \phi(x) x^{z-1} dx$
8.56	$\{P_\nu^\mu[(1+x^2/a^2)^{\frac{1}{2}}]\}^2$ $2\mathrm{Re}\ \mu < \mathrm{Re}\ z < \begin{cases} -2\ \mathrm{Re}\ \nu \\ 2+2\ \mathrm{Re}\ \nu \end{cases}$	$-\frac{1}{2}\pi^{-\frac{1}{2}} a^z (\nu+\frac{1}{2}z)^{-1} \Gamma(\frac{1}{2}-\frac{1}{2}z)$ $\cdot\Gamma(\frac{1}{2}z-\mu)\Gamma(1+\nu-\frac{1}{2}z)$ $\cdot[\Gamma(1+\nu-\mu)\Gamma(-\nu-\mu)\Gamma(1-\frac{1}{2}z)]^{-1}$
8.57	$e^{i2\pi\mu}\{Q_\nu^\mu[(1+x^2/a^2)^{\frac{1}{2}}]\}^2$ $\pm 2\mathrm{Re}\ \mu < \mathrm{Re}\ z < 2+2\mathrm{Re}\ \nu$	$\frac{1}{4}\pi^{\frac{1}{2}} a^z \Gamma(1+\nu-\mu)\Gamma(1+\nu-\frac{1}{2}z)\Gamma(\frac{1}{2}z)$ $\cdot\Gamma(\frac{1}{2}z+\mu)\Gamma(\frac{1}{2}z-\mu)$ $\cdot[\Gamma(\frac{1}{2}+\frac{1}{2}z)\Gamma(1+\nu+\mu)\Gamma(1+\nu+\frac{1}{2}z)]^{-1}$
8.58	$(a^2+x^2)^{-\frac{1}{2}} P_\nu^\mu[(1+x^2/a^2)^{\frac{1}{2}}]$ $\cdot P_{-\nu}^\mu[(1+x^2/a^2)^{\frac{1}{2}}]$ $2\mathrm{Re}\ \mu < \mathrm{Re}\ z < 2\pm 2\ \mathrm{Re}\ \nu$	$\frac{1}{2}\pi^{-\frac{1}{2}} a^{z-1} \Gamma(\frac{1}{2}z-\mu)\Gamma(\frac{1}{2}-\frac{1}{2}z)$ $\cdot\Gamma(1+\nu-\frac{1}{2}z)\Gamma(1-\nu-\frac{1}{2}z)$ $\cdot[\Gamma(1-\frac{1}{2}z)\Gamma(1-\mu-\frac{1}{2}z)\Gamma(1-\mu+\nu)\Gamma(1-\mu-\nu)]^{-1}$
8.59	$P_\nu^\mu[(1+x^2/a^2)^{\frac{1}{2}}]$ $\cdot e^{i\pi\mu} Q_\nu^{-\mu}[(1+x^2/a^2)^{\frac{1}{2}}]$ $1+2\mathrm{Re}\ \mu < \mathrm{Re}\ z < 2+2\mathrm{Re}\ \nu$	$\frac{1}{4}\pi^{-\frac{1}{2}} a^z \Gamma(\frac{1}{2}-\frac{1}{2}z)\Gamma(\frac{1}{2}z)$ $\cdot\Gamma(\frac{1}{2}z-\mu)\Gamma(1+\nu-\frac{1}{2}z)$ $\cdot[\Gamma(1-\mu-\frac{1}{2}z)\Gamma(1+\nu+\frac{1}{2}z)]^{-1}$
8.60	$P_\nu^\mu[(1+x^2/a^2)^{\frac{1}{2}}]$ $\cdot e^{-i\pi\mu} Q_\nu^\mu[(1+x^2/a^2)^{\frac{1}{2}}]$ $\left.\begin{matrix} 0 \\ 2\mathrm{Re}\ \mu \end{matrix}\right\} < \mathrm{Re}\ z < \left\{\begin{matrix} 1 \\ 2+2\mathrm{Re}\ \nu \end{matrix}\right.$	$\frac{1}{2}\pi^{-\frac{1}{2}} a^z \Gamma(\frac{1}{2}z)\Gamma(\frac{1}{2}-\frac{1}{2}z)\Gamma(\frac{1}{2}z-\mu)$ $\cdot\ \Gamma(1+\nu-\frac{1}{2}z)\Gamma(1+\nu+\mu)$ $\cdot[\Gamma(1+\nu-\mu)\Gamma(1+\nu+\frac{1}{2}z)\Gamma(1-\frac{1}{2}z-\mu)]^{-1}$

1.9 Orthogonal Polynomials

	$\phi(x)$	$\Phi(z) = \int_0^\infty \phi(x) x^{z-1} dx$
9.1	$e^{-ax} He_n(x)$ Re $z > 0$ if n even Re $z > -1$ if n odd	$n! \sum_{m=0}^{[\frac{1}{2}n]} [m!(n-2m)!]^{-1}$ $\cdot \Gamma(z+n-2m)(-\frac{1}{2})^m a^{2m-n-z}$
9.2	$e^{-ax} He_n[(2x)^{\frac{1}{2}}]$ Re $z > 1$ if n even Re $z > \frac{1}{2}$ if n odd	$2^{-\frac{1}{2}n} n! a^{-z} \Gamma(z-\frac{1}{2}n)$ $\cdot C_n^{z-\frac{1}{2}n}(a^{-\frac{1}{2}})$
9.3	$e^{-ax}[He_n(x^{\frac{1}{2}})]^2$ Re $z > 0$	$h^z(a) n! \Gamma(z)$ where $(1-t^2)^{-\frac{1}{2}}[a-t(1+t)^{-1}]^{-z}$ $= \sum_{n=0}^{\infty} h_n^z(a) t^n$
9.4	$e^{-a^2x^2} He_k(\alpha x)$ $\cdot He_n(\beta x) He_n(\gamma x) \cdots$	See Appel, P. and M.J. Kampé de Fériet, 1926: Fonctions hypergéo-mitriques et hypersphériques. Polynomes d'Hermite. Gauthiers-Villars, p. 343. Erdélyi, A., 1936: Math. Z. 40, 693-702.

	$\phi(x)$	$\Phi(z) = \int_0^\infty \phi(x) x^{z-1} dx$
9.5	$(1-x^2)^{-\frac{1}{2}} T_n(x) \quad x < 1$ $0 \quad x > 1$	$\Pi 2^{-z} \Gamma(z) [\Gamma(\frac{1}{2}+\frac{1}{2}z+\frac{1}{2}n) \Gamma(\frac{1}{2}+\frac{1}{2}z-\frac{1}{2}n)]^{-1}$ $\mathrm{Re}\, z > 0$
9.6	$(1-x^2)^{-\frac{1}{2}} T_n(1/x) \quad x < 1$ $0 \quad x > 1$	$2^{z-2} [\Gamma(z)]^{-1} \Gamma(\frac{1}{2}n+\frac{1}{2}z) \Gamma(\frac{1}{2}z-\frac{1}{2}n)$ $\mathrm{Re}\, z > 0$
9.7	$(2-x)^{-\frac{1}{2}} T_n(x-1) \quad x < 2$ $0 \quad x > 2$ $\mathrm{Re}\, z > 0$	$\Pi 2^{\frac{1}{2}+2n-z} (n!)^2 \Gamma(2z)$ $\cdot [(2n)! \Gamma(\frac{1}{2}+z+n) \Gamma(\frac{1}{2}+z-n)]^{-1}$
9.8	$(2-x)^{\frac{1}{2}} U_n(x-1) \quad x < 2$ $0 \quad x > 2$ $\mathrm{Re}\, z > -\frac{1}{2}$	$\pi 2^{\frac{1}{2}+2n-z} [(n+1)!]^2 \Gamma(2z-1)$ $[(2n+2)! \Gamma(z+\frac{3}{2}+n) \Gamma(z-\frac{1}{2}-n)]^{-1}$
9.9	$(2-x)^{\nu} T_n(1-x) \quad x < 2$ $0 \quad x > 2$ $\mathrm{Re}\, z > 0,\ \mathrm{Re}\, \nu > -1$	$\dfrac{2^{2n+\nu+z} (n!)^2 \Gamma(1+\nu) \Gamma(z)}{(2n)! \Gamma(1+\nu+z)}$ $\cdot {}_3F_2(-n,n,z;\frac{1}{2},1+\nu+z;1)$
9.10	$(2-x)^{\nu} U_n(1-x) \quad x < 2$ $0 \quad x > 2$	$\dfrac{2^{1+2n+\nu+z} [(n+1)!]^2 \Gamma(1+\nu) \Gamma(z)}{(2n+2)! \Gamma(1+\nu+z)}$ $\cdot {}_3F_2(-n,n+1,z;\frac{3}{2},1+\nu+z;1)$

	$\phi(x)$	$\Phi(z) = \int_0^\infty \phi(x)x^{z-1}dx$
9.11	$(2-x)^{\nu}T_n(x-1)$ $x<2$ 0 $x>2$ $\mathrm{Re}\,\nu>-1$, $\mathrm{Re}\,z>0$	$\dfrac{2^{2n+\nu+z}(n!)^2\Gamma(1+\nu)\Gamma(z)}{(2n)!\Gamma(1+\nu+z)}$ $\cdot\ {}_3F_2(-n,n,1+\nu;\frac{1}{2},1+\nu+z;1)$
9.12	$(2-x)^{\nu}U_n(x-1)$ $x<2$ 0 $x>2$ $\mathrm{Re}\,\nu>-1$, $\mathrm{Re}\,z>0$	$\dfrac{2^{2n+1+\nu+z}[(n+1)!]^2\Gamma(1+\nu)\Gamma(z)}{(2n+2)!\Gamma(1+\nu+z)}$ $\cdot\ {}_3F_2(-n,n+1,1+\nu;\frac{3}{2},1+\nu+z;1)$
9.13	$(2-x)^{-\frac{1}{2}}$ $T_m(1-x)T_n(1-x)$ $\mathrm{Re}\,z>0$	$\pi^{\frac{1}{2}}2^{z-\frac{1}{2}}\Gamma(z)\Gamma(\frac{1}{2}+n-z)[\Gamma(\frac{1}{2}-z)\Gamma(\frac{1}{2}+n+z)]^{-1}$ $\cdot\ {}_4F_3(-m,m,z,\frac{1}{2}+z;\frac{1}{2},\frac{1}{2}+n+z,\frac{1}{2}-n+z;1)$
9.14	$P_{2n}(x)$ $x<1$ 0 $x>1$	$\dfrac{(-1)^n(\frac{1}{2}-\frac{1}{2}z)_n}{2(\frac{1}{2}z)_{n+1}}$ $\mathrm{Re}\,z>0$
9.15	$P_{2n+1}(x)$ $x<1$ 0 $x>1$	$\dfrac{(-1)^n(1-\frac{1}{2}z)_n}{2(\frac{1}{2}+\frac{1}{2}z)_{n+1}}$ $\mathrm{Re}\,z>-1$
9.16	$(2-x)^{\nu}P_n(x-1)$ $x<2$ 0 $x>2$ $\mathrm{Re}\,\nu>-1$, $\mathrm{Re}\,z>0$	$[\Gamma(1+\nu+z)]^{-1}2^{\nu+z}\Gamma(\nu+1)\Gamma(z)$ $\cdot\ {}_3F_2(-n,1+n,\nu+1;1;\nu+1+z;1)$

	$\phi(x)$	$\Phi(z) = \int_0^\infty \phi(x) x^{z-1} dx$
9.17	$(2-x)^{\nu} P_n(1-x)$ $x < 2$ 0 $x > 2$ $\mathrm{Re}\,\nu > -1$, $\mathrm{Re}\,z > 0$	$[\Gamma(1+\nu+z)]^{-1} 2^{\nu+z} \Gamma(\nu+1)\Gamma(z)$ $\cdot {}_3F_2(-n,1+n,z;1,1+\nu+z;1)$
9.18	$P_n(1-2x^2)$ $x < 1$ 0 $x > 1$	$\frac{1}{2}(-1)^n \Gamma^2(\frac{1}{2}z)[\Gamma(\frac{1}{2}z+n)\Gamma(\frac{1}{2}z-n)]^{-1}$ $\mathrm{Re}\,z > 0$
9.19	$P_n(1-x)$ $x < 2$ $\cdot P_m(1-z)$ 0 $x > 2$	$2^z \Gamma(z)\Gamma(1+n-z)[\Gamma(1-z)\Gamma(1+n+z)]^{-1}$ $\cdot {}_4F_3(-m,m+1,z,z;1,1+n+z,1-n+z;1)$ $\mathrm{Re}\,z > 0$
9.20	$(1-x^2)^{\nu-\frac{1}{2}}$ $x < 1$ $\cdot C_n^{\nu}(x)$ 0 $x > 1$	$2^{2\nu-1-z}\Gamma(n+2\nu)\Gamma(z)$ $\cdot [n!\Gamma(\nu)\Gamma(\frac{1}{2}+\frac{1}{2}n+\nu+\frac{1}{2}z)\Gamma(\frac{1}{2}+\frac{1}{2}z-\frac{1}{2}n)]^{-1}$ $\mathrm{Re}\,\nu > -\frac{1}{2}$, $\mathrm{Re}\,z > 0$
9.21	$(1-x^2)^{\nu-\frac{1}{2}}$ $x < 1$ $\cdot C_n^{\nu}(1/x)$ 0 $x > 1$	$2^{z-1}\Gamma(n+2\nu)\Gamma(\frac{1}{2}z-\frac{1}{2}n)\Gamma(\frac{1}{2}z+\nu+\frac{1}{2}n)$ $\cdot [n!\Gamma(\nu)\Gamma(z+2\nu)]^{-1}$ $\mathrm{Re}\,\nu > -\frac{1}{2}$, $\mathrm{Re}\,z > 0$

	$\phi(x)$	$\Phi(z) = \int_0^\infty \phi(x) x^{z-1} dx$
9.22	$P_n(1-\gamma b+\gamma x)$ $x < b$ 0 $x > b$	$n![\Gamma(n+1+z)]^{-1}\Gamma(z) b^z$ $P_n^{(z,-z)}(1-\gamma b)$ $\text{Re } z > 0$
9.23	$(b-x)^{\mu-1}$ $\cdot P_n(1-\gamma x)$ $x < b$ 0 $x > b$	$\Gamma(\mu)\Gamma(z)[\Gamma(\mu+z)]^{-1} b^{\mu-1+z}$ $\cdot {}_3F_2(-n,n+1,z;1,\mu+z;\tfrac{1}{2}\gamma b)$ $\text{Re}(\mu,z) > 0$
9.24	$(2-x)^{\nu-\frac{1}{2}}$ $\cdot C_n^\nu(x-1)$ $x < 2$ 0 $x > 2$	$\Gamma(\tfrac{1}{2}+\nu)\Gamma(n+2\nu) 2^{\nu-\frac{1}{2}+z}\Gamma(z)\Gamma(\tfrac{1}{2}-\nu+z)$ $\cdot [n!\Gamma(2\nu)\Gamma(\tfrac{1}{2}-\nu-n+z)\Gamma(\tfrac{1}{2}+\nu+n+z)]^{-1}$ $\text{Re } \nu > -\tfrac{1}{2}, \quad \text{Re } z > 0$
9.25	$(2-x)^\beta$ $\cdot C_n^\nu(1-x)$ $x < 2$ 0 $x > 2$ $\text{Re } z > 0, \quad \text{Re } \beta > -1$	$\Gamma(1+\beta) 2^{\beta+z}\Gamma(n+2\nu)\Gamma(z)$ $\cdot [n!\Gamma(2\nu)\Gamma(1+\beta+z)]^{-1}$ $\cdot {}_3F_2(-n,n+2\nu,z;\tfrac{1}{2}+\nu,1+\beta+z;1)$
9.26	$(2-x)^{\nu-\frac{1}{2}}$ $x < 2$ $\cdot C_n^\nu(1-x)$ $\cdot C_m^\mu(1-x)$ 0 $x > 2$	$\dfrac{2^{\nu+z-\frac{1}{2}}\Gamma(\tfrac{1}{2}+\nu)\Gamma(m+2\mu)\Gamma(n+2\nu)\Gamma(z)\Gamma(\tfrac{1}{2}+\nu+n-z)}{m!n!\Gamma(2\nu)\Gamma(2\mu)\Gamma(\tfrac{1}{2}+\nu-z)\Gamma(\tfrac{5}{2}+\nu+n-z)}$ $\cdot {}_4F_3\left(\begin{matrix} -m,m+2\mu,z,\tfrac{1}{2}-\nu+z; \\ \tfrac{1}{2}+\nu+n+z,\tfrac{1}{2}+\mu,\tfrac{1}{2}-\nu-n+z; \end{matrix} 1\right)$ $\text{Re } z > 0, \quad \text{Re } \nu > -\tfrac{1}{2}$

	$\phi(x)$	$\Phi(z) = \int_0^\infty \phi(x) x^{z-1} dx$
9.27	$(b-x)^{\mu-1}$ $\cdot C_n^\lambda(1-\gamma x) \quad x < b$ $0 \quad x > b$ $2\lambda \neq 0,-1,-2,\cdots$	$(2\lambda)_n \Gamma(\mu)[n!\Gamma(\mu+z)]^{-1}\Gamma(z) b^{\mu-1+z}$ $\cdot {}_3F_2(-n,n+2\lambda,z;\tfrac{1}{2}+\lambda,\mu+z;\tfrac{1}{2}\gamma b)$ $\mathrm{Re}(\mu,z) > 0$
9.28	$(b-x)^{\mu-1}$ $\cdot C_{2n}^\lambda(\gamma x^{\frac{1}{2}}) \quad x < b$ $0 \quad x > b$	$(-1)^n (\lambda)_n \Gamma(\mu)[n!\Gamma(\mu+z)]^{-1} b^{\mu-1+z}\Gamma(z)$ $\cdot {}_3F_2(-n,n+\lambda,z;\tfrac{1}{2},\mu+z;\gamma^2 b)$ $\mathrm{Re}(\mu,z) > 0$
9.29	$(b-x)^{\mu-1}$ $\cdot C_{2n+1}^\lambda(\gamma x^{\frac{1}{2}}) \quad x < b$ $0 \quad x > b$ $\mathrm{Re}\,\mu > 0$	$2(-1)^n \gamma (\lambda)_{n+1} \Gamma(\mu)[n!\Gamma(\tfrac{1}{2}+\mu+z)]^{-1}$ $\cdot \Gamma(\tfrac{1}{2}+z) b^{\mu-\frac{1}{2}+z}$ $\cdot {}_3F_2(-n,1+n+\lambda,\tfrac{1}{2}+z;\tfrac{3}{2},\tfrac{1}{2}+\mu+z;\gamma^2 b)$ $\mathrm{Re}\, z > -\tfrac{1}{2}$
9.30	$(b-x)^{\lambda-\frac{1}{2}}$ $\cdot C_n^\lambda(1-\gamma b+\gamma x)$ $x < b$ $0 \quad x > b$ $\mathrm{Re}\,\lambda > -1,\ 2\lambda \neq 0,-1,-2,\cdots$	$(2\lambda)_n \Gamma(\tfrac{1}{2}+\lambda)[\Gamma(\tfrac{1}{2}+\lambda+n+z)]^{-1}$ $\Gamma(z) b^{\lambda-\frac{1}{2}+z} P_n^{(\alpha,\beta)}(1-\gamma b)$ $\alpha = \lambda-\tfrac{1}{2}+z,\ \beta = \lambda-\tfrac{1}{2}-z, \quad \mathrm{Re}\, z > 0$

	$\phi(x)$	$\Phi(z) = \int_0^\infty \phi(x)x^{z-1}dx$
9.31	$e^{-x}L_n^{\alpha}(x)$	$[n!\Gamma(1+\alpha-z)]^{-1}\Gamma(\alpha+n+1-z)\Gamma(z)$ Re $z > 0$
9.32	$(b-x)^{\mu-1}$ $\cdot L_n^{\alpha}(\beta x)$ $x < b$ 0 $x > b$	$\Gamma(\mu)[n!\Gamma(1+\alpha)\Gamma(\mu+z)]^{-1}\Gamma(z)$ $\Gamma(\alpha+n+1)b^{z+\mu-1}{}_2F_2(-n,z;\alpha+1,$ $\mu+z;\beta b)$ Re$(\mu,z) > 0$
9.33	$(b-x)^{\lambda-1}$ $\cdot L_n^{\alpha}[\beta(b-x)]$ $x < b$ 0 $x > b$	$\Gamma(1+\alpha+n)[n!\Gamma(1+\alpha)\Gamma(\lambda+z)]^{-1}$ $\cdot\Gamma(\lambda)\Gamma(z)b^{\lambda+z-1}$ $\cdot{}_2F_2(-n,\lambda;1+\alpha,\lambda+z;\beta b)$ Re$(\lambda,z) > 0$
9.34	$(b-x)^{\alpha}$ $\cdot L_n^{\alpha}[\beta(b-x)]$ $x < b$ 0 $x > b$	$\Gamma(1+\alpha+n)[\Gamma(1+\alpha+n+z)]^{-1}b^{\alpha+z}$ $\cdot\Gamma(z)L_n^{\alpha+z}(\beta b)$ Re $\alpha > -1$, Re $z > 0$
9.35	$e^{-ax}L_n^{\nu}(bx)$ Re $z > 0$	$\Gamma(z+n)(n!)^{-1}(a-b)^n a^{-n-z}$ $\cdot{}_2F_1[-n,1+\nu-z;1-n-z;a(a-b)^{-1}]$

	$\phi(x)$	$\Phi(z) = \int_0^\infty \phi(x) x^{z-1} dx$
9.36	$(b-x)^{\mu-1} e^{-\beta x}$ $L_n^\alpha(\beta x) \quad x < b$ $0 \quad x > b$	$\Gamma(1+\alpha+n)[n!\Gamma(1+\alpha)\Gamma(\mu+z)]^{-1}$ $\cdot\Gamma(\mu)\Gamma(z) b^{z+\mu-1}$ $\cdot {}_2F_2(\alpha+n+1, z; \alpha+1, \mu+z; -\beta b)$ $\mathrm{Re}(\mu, z) > 0$
9.37	$(b-x)^{\lambda-1} e^{\beta x}$ $\cdot L_n^\alpha[\beta(b-x)] \quad x < b$ $0 \quad x > b$	$\Gamma(1+\alpha+n)[n!\Gamma(1+\alpha)\Gamma(\lambda+z)]^{-1}$ $\cdot\Gamma(\lambda)\Gamma(z) b^{\lambda+z-1} e^{-\beta b}$ $\cdot {}_2F_2(1+\alpha+n, \lambda; \alpha+1, \lambda+z; -\beta b)$ $\mathrm{Re}(\lambda, z) > 0$
9.38	$e^{-x} L_n^\alpha(\lambda x) L_n^\alpha(\mu x)$	Buchholz, H., 1953: Die konfluente hypergeometrische Funktion. Springer Verlag.
9.39	$e^{-x} L_{m_1}^{\alpha_1}(\lambda_1 x)$ $\cdots L_{m_n}^{\alpha_n}(\lambda_n x)$	Erdélyi, A., 1936: Math. Z. 40, 693-702.
9.40	$(2-x)^\beta$ $\cdot P_n^{(\alpha,\beta)}(1-x) \quad x < 2$ $0 \quad x > 2$	$[n!\Gamma(1+\alpha-z)\Gamma(1+\beta+n+z)]^{-1} 2^{\beta+z}$ $\cdot\Gamma(z)\Gamma(1+\beta+n)\Gamma(1+\alpha+n-z)$ $\mathrm{Re}\ \beta > -1,\ \mathrm{Re}\ z > 0$

	$\phi(x)$	$\Phi(z) = \int_0^\infty \phi(x) x^{z-1} dx$
9.41	$(b-x)^{\beta}$ $\cdot P_n^{(\alpha,\beta)}(\gamma b-1-\gamma x)$ $\quad x < b$ $0 \quad x > b$	$[\Gamma(\beta+n+1+z)]^{-1}\Gamma(\beta+n+1)\Gamma(z)$ $\cdot\, b^{\beta+z} P_n^{(\alpha-z,\beta+z)}(\gamma b-1)$ $\mathrm{Re}\,\beta > -1, \quad \mathrm{Re}\,z > 0$
9.42	$(b-x)^{\alpha}$ $\cdot P_n^{(\alpha,\beta)}(1-\gamma b+\gamma x)$ $\quad x < b$ $0 \quad x > b$	$[\Gamma(\alpha+n+1+z)]^{-1}\Gamma(\alpha+n+1)\Gamma(z)$ $\cdot b^{\alpha+z} P_n^{(\alpha+z,\beta-z)}(1-\gamma b)$ $\mathrm{Re}\,\alpha > -1, \quad \mathrm{Re}\,z > 0$
9.43	$(b-x)^{\mu-1}$ $\cdot P_n^{(\alpha,\beta)}(1-\gamma x) \quad x < b$ $0 \quad x > b$	$[n!\Gamma(1+\alpha)\Gamma(\mu+z)]^{-1}\Gamma(\alpha+n+1)$ $\cdot\Gamma(\mu)\Gamma(z) b^{z+\mu-1} {}_3F_2(-n, 1+n+\alpha+\beta,$ $z; 1+\alpha, \mu+z; \tfrac{1}{2}\gamma b)$ $\mathrm{Re}(\mu, z) > 0$
9.44	$(b-x)^{\mu-1}$ $\cdot P_n^{(\alpha,\beta)}(\gamma x-1) \quad x < b$ $0 \quad x > b$	$(-1)^n [n!\Gamma(1+\beta)\Gamma(\mu+z)]^{-1}\Gamma(\beta+n+1)$ $\cdot\Gamma(\mu)\Gamma(z) b^{z+\mu-1} {}_3F_2(-n, 1+n+\alpha+\beta,$ $z; 1+\beta, \mu+z; \tfrac{1}{2}\gamma b)$ $\mathrm{Re}(\mu, z) > 0$

	$\phi(x)$	$\Phi(z) = \int_0^\infty \phi(x)x^{z-1}dx$
9.45	$(b-x)^{\mu-1}(1-\frac{1}{2}\gamma x)^{\beta}$ $P_n^{(\alpha,\beta)}(1-\gamma x) \quad x < b$ $0 \quad x > b$	$[n!\Gamma(1+\alpha)\Gamma(\mu+z)]^{-1}\Gamma(1+\alpha+n)$ $\cdot\Gamma(\mu)\Gamma(z)b^{z+\mu-1}{}_3F_2(1+\alpha+n,-n-\beta,$ $z;1+\alpha,\mu+z;\frac{1}{2}\gamma b)$ $\mathrm{Re}(\mu,z) > 0$

1.10 Bessel Functions

	$\phi(x)$	$\Phi(z) = \int_0^\infty \phi(x) x^{z-1} dx$
10.1	$J_\nu(ax)$	$\frac{1}{2}(\frac{1}{2}a)^{-z}\Gamma(\frac{1}{2}\nu+\frac{1}{2}z)[\Gamma(1+\frac{1}{2}\nu-\frac{1}{2}z)]^{-1}$ $-\mathrm{Re}\ \nu < \mathrm{Re}\ z < \frac{3}{2}$
10.2	$Y_\nu(ax)$ $\pm\ \mathrm{Re}\ \nu < \mathrm{Re}\ z < \frac{3}{2}$	$-\pi^{-1}2^{z-1}a^{-z}\cos[\frac{1}{2}\pi(z-\nu)]$ $\cdot\Gamma(\frac{1}{2}z-\frac{1}{2}\nu)\Gamma(\frac{1}{2}\nu+\frac{1}{2}z)$
10.3	$(a+x)^{-1}J_\nu(bx)$ $-\mathrm{Re}_\nu < \mathrm{Re}\ z < \frac{5}{2}$	$(2a)^{z-1}[\frac{\Gamma(\frac{1}{2}z+\frac{1}{2}\nu)}{\Gamma(1+z+\frac{1}{2}\nu)} S_{1-z,\nu}(ab)$ $-2\frac{\Gamma(\frac{1}{2}+\frac{1}{2}z+\frac{1}{2}\nu)}{\Gamma(\frac{1}{2}+\frac{1}{2}\nu-\frac{1}{2}z)} S_{-z,\nu}(ab)]$
10.4	$(x^2-a^2)^{-1}J_\nu(bx)$ Principal value $\mathrm{Re}\ \nu < \mathrm{Re}\ z < \frac{7}{2}$	$-\frac{1}{2}\pi a^{z-2}Y_\nu(ab)$ $-2(2a)^{z-2}\frac{\Gamma(\frac{1}{2}\nu+\frac{1}{2}z)}{\Gamma(1+\frac{1}{2}\nu-\frac{1}{2}z)} S_{1-z,\nu}(ab)$
10.5	$(x^2-a^2)^{-1}Y_\nu(bx)$ Principal value	$\frac{1}{2}\pi a^{z-2}J_\nu(ab)+2^{z-1}\pi^{-1}\cos[\frac{1}{2}\pi(z-\nu)]$ $\cdot\Gamma(\frac{1}{2}z-\frac{1}{2}\nu)\Gamma(\frac{1}{2}z+\frac{1}{2}\nu)S_{1-z,\nu}(ab)$ $\frac{7}{2} < \mathrm{Re}\ z < \pm\ \mathrm{Re}\ \nu$
10.6	$J_\nu(ax)$ $\quad x < 1$ 0 $\quad x > 1$ $\mathrm{Re}\ z > -\mathrm{Re}\ \nu$	$(\frac{1}{2}a)^\nu[(z+\nu)\Gamma(1+\nu)]^{-1}$ $\cdot {}_1F_2(\frac{1}{2}\nu+\frac{1}{2}z;\nu+1,1+\frac{1}{2}\nu+\frac{1}{2}z;-\frac{1}{4}a^2)$

	$\phi(x)$	$\Phi(z) = \int_0^\infty \phi(x)\, x^{z-1} dx$
10.7	$e^{-ax} J_\nu(bx)$ $\mathrm{Re}\, z > -\mathrm{Re}\, \nu$	$(a^2+b^2)^{-\frac{1}{2}z}\Gamma(\nu+z)$ $\cdot P_{z-1}^{-\nu}[a(a^2+b^2)^{-\frac{1}{2}}]$ $= (\frac{1}{2}b/a)^\nu [\Gamma(1+\nu)]^{-1} a^{-z}\Gamma(z+\nu)$ $\cdot {}_2F_1(\frac{1}{2}z+\frac{1}{2}\nu, \frac{1}{2}+\frac{1}{2}z+\frac{1}{2}\nu; \nu+1; -b^2/a^2)$
10.8	$\exp(-b^2x^2) J_\nu(ax)$ $\mathrm{Re}\, z > -\mathrm{Re}\, \nu$	$[a\Gamma(1+\nu)]^{-1} b^{1-z}\Gamma(\frac{1}{2}\nu+\frac{1}{2}z)$ $\cdot \exp(-\frac{1}{8}a^2/b^2) M_{\frac{1}{2}z-\frac{1}{2}, \frac{1}{2}\nu}(\frac{1}{4}a^2/b^2)$
10.9	$\sin(ax) J_\nu(ax)$ $-1-\mathrm{Re}\, \nu < \mathrm{Re}\, z < \frac{1}{2}$	$\dfrac{2^{\nu-1} a^{-z}\Gamma(\frac{1}{2}-z)\Gamma(\frac{1}{2}+\frac{1}{2}\nu+\frac{1}{2}z)}{\Gamma(1+\nu-z)\Gamma(1-\frac{1}{2}\nu-\frac{1}{2}z)}$
10.10	$\cos(ax) J_\nu(ax)$ $-\mathrm{Re}\, \nu < \mathrm{Re}\, z < \frac{1}{2}$	$\dfrac{2^{\nu-1} a^{-z}\Gamma(\frac{1}{2}-z)\Gamma(\frac{1}{2}\nu+\frac{1}{2}z)}{\Gamma(\frac{1}{2}-\frac{1}{2}\nu-\frac{1}{2}z)\Gamma(1+\nu-z)}$
10.11	$\sin(ax) Y_\nu(ax)$ $-1\pm\mathrm{Re}\, \nu < \mathrm{Re}\, z < \frac{1}{2}$	$2^{z-1}\pi^{-\frac{1}{2}} a^{-z}\sin[\frac{1}{2}\pi(z-\nu)]$ $\cdot \dfrac{\Gamma(\frac{1}{2}+\frac{1}{2}z+\frac{1}{2}\nu)\Gamma(\frac{1}{2}+\frac{1}{2}z-\frac{1}{2}\nu)}{\Gamma(1-\frac{1}{2}z+\frac{1}{2}\nu)\Gamma(1-\frac{1}{2}z-\frac{1}{2}\nu)}$
10.12	$\cos(ax) Y_\nu(ax)$ $\pm\mathrm{Re}\, \nu < \mathrm{Re}\, z < \frac{1}{2}$	$2^{z-1}\pi^{-\frac{1}{2}} a^{-z}\cos[\frac{1}{2}\pi(z-\nu)]$ $\cdot \dfrac{\Gamma(\frac{1}{2}z+\frac{1}{2}\nu)\Gamma(\frac{1}{2}z-\frac{1}{2}\nu)}{\Gamma(\frac{1}{2}-\frac{1}{2}z+\frac{1}{2}\nu)\Gamma(\frac{1}{2}-\frac{1}{2}z-\frac{1}{2}\nu)}$

	$\phi(x)$	$\Phi(z) = \int_0^\infty \phi(x)x^{z-1}dx$
10.13	$\cos(ax)J_\nu(bx)$ $-\mathrm{Re}\,\nu < \mathrm{Re}\,z < 3/2$	$(\tfrac{1}{2}b/a)^\nu a^{-z}\cos[\tfrac{1}{2}\pi(\nu+z)]$ $\cdot\Gamma(\nu+z)[\Gamma(1+\nu)]^{-1}{}_2F_1(\tfrac{1}{2}\nu+\tfrac{1}{2}z,$ $\tfrac{1}{2}+\tfrac{1}{2}\nu+\tfrac{1}{2}z;\nu+1;b^2/a^2)$ $= (\tfrac{1}{2}\pi b)^{-\frac{1}{2}}\cos[\tfrac{1}{2}\pi(\nu+z)](a^2-b^2)^{\frac{1}{4}-\frac{1}{2}z}$ $\cdot e^{-i\pi(z-\frac{1}{2})}Q_{\nu-\frac{1}{2}}^{z-\frac{1}{2}}(a/b)$ $= \cos[\tfrac{1}{2}\pi(\nu+z)]\Gamma(\nu+z)$ $\cdot(a^2-b^2)^{-\frac{1}{2}z}P_{-z}^{-\nu}[a(a^2-b^2)^{-\frac{1}{2}}]$ $a > b$
10.14	$\cos(ax)J_\nu(bx)$ $-\mathrm{Re}\,\nu < \mathrm{Re}\,z < 3/2$	$2^{z-1}b^{-z}\Gamma(\tfrac{1}{2}\nu+\tfrac{1}{2}z)[\Gamma(1+\tfrac{1}{2}\nu-\tfrac{1}{2}z)]^{-1}$ $\cdot{}_2F_1(\tfrac{1}{2}\nu+\tfrac{1}{2}z,\tfrac{1}{2}z-\tfrac{1}{2}\nu;\tfrac{1}{2};a^2/b^2)$ $= b^{-\frac{1}{2}}2^{z-\nu-3/2}\Gamma(\nu+z)\Gamma(\tfrac{1}{2}+\tfrac{1}{2}z-\tfrac{1}{2}\nu)$ $\cdot[\Gamma(1+\tfrac{1}{2}\nu-\tfrac{1}{2}z)]^{-1}(b^2-a^2)^{\frac{1}{4}-\frac{1}{2}z}$ $\cdot[P_{\nu-\frac{1}{2}}^{\frac{1}{2}-z}(a/b)+P_{\nu-\frac{1}{2}}^{\frac{1}{2}-z}(-a/b)]$ $a < b$
10.15	$\cos(a-x)J_\nu(x)$ $\quad x < a$ $0 \quad x > a$ $\mathrm{Re}\,z > -\mathrm{Re}\,\nu$	$(\nu+z)^{-1}a^zJ_\nu(a)+2a^z$ $\cdot\sum_{n=1}^\infty(-1)^n[(\nu+z)_{2n+1}]^{-1}(\nu+1-z)_{2n-1}$ $\cdot(\nu+2n)J_{\nu+2n}(a)$

	$\phi(x)$	$\Phi(z) = \int_0^\infty \phi(x) x^{z-1} dx$
10.16	$\sin(ax) J_\nu(bx)$ $-1-\mathrm{Re}\ \nu < \mathrm{Re}\ z < 3/2$	$(\tfrac{1}{2}b/a)^\nu a^{-z} \sin[\tfrac{1}{2}\pi(\nu+z)]$ $\Gamma(\nu+z)[\Gamma(1+\nu)]^{-1} {}_2F_1(\tfrac{1}{2}+\tfrac{1}{2}\nu+\tfrac{1}{2}z,$ $\tfrac{1}{2}\nu+\tfrac{1}{2}z; 1+\nu; b^2/a^2)$ $=(\tfrac{1}{2}\pi b)^{-\frac{1}{2}} \sin[\tfrac{1}{2}\pi(\nu+z)](a^2-b^2)^{\frac{1}{4}-\frac{1}{2}z}$ $\cdot e^{-i\pi(z-\frac{1}{2})} q_{\nu-\frac{1}{2}}^{z-\frac{1}{2}}(a/b)$ $= \sin[\tfrac{1}{2}\pi(\nu+z)]\Gamma(\nu+z)$ $\cdot(a^2-b^2)^{-\frac{1}{2}z} P_{-z}^{-\nu}[a(a^2-b^2)^{-\frac{1}{2}}]$ $a > b$
10.17	$\sin(ax) J_\nu(bx)$ $-1-\mathrm{Re}\ \nu < \mathrm{Re}\ z < 3/2$	$a 2^z b^{-z-1} \Gamma(\tfrac{1}{2}+\tfrac{1}{2}\nu+\tfrac{1}{2}z)[\Gamma(\tfrac{1}{2}+\tfrac{1}{2}\nu-\tfrac{1}{2}z)]^{-1}$ $\cdot {}_2F_1(\tfrac{1}{2}+\tfrac{1}{2}\nu+\tfrac{1}{2}z, \tfrac{1}{2}-\tfrac{1}{2}\nu+\tfrac{1}{2}z; 3/2; a^2/b^2)$ $= -b^{-\frac{1}{2}} 2^{z-\nu-3/2} \Gamma(\nu+z)\Gamma(\tfrac{1}{2}z-\tfrac{1}{2}\nu)$ $\cdot[\Gamma(\tfrac{1}{2}+\tfrac{1}{2}\nu-\tfrac{1}{2}z)]^{-1}(b^2-a^2)^{\frac{1}{4}-\frac{1}{2}z}$ $\cdot[P_{\nu-\frac{1}{2}}^{\frac{1}{2}-z}(a/b) - P_{\nu-\frac{1}{2}}^{\frac{1}{2}-z}(-a/b)]$ $a < b$
10.18	$\sin(a-x) J_\nu(x) \quad x < a$ $0 \quad x > a$ $\mathrm{Re}\ z > -\mathrm{Re}\ \nu$	$2a^z \sum_{n=0}^\infty (-1)^n [(\nu+z)_{2n+2}]^{-1}$ $\cdot(\nu+2n+1)(\nu+1-z)_{2n} J_{\nu+2n+1}(a)$

	$\phi(x)$	$\Phi(z) = \int_0^\infty \phi(x) x^{z-1} dx$
10.19	$(b-x)^{\lambda} \cdot J_{\nu}[a(b-x)]$ $x < b$ 0 $x > b$ $\mathrm{Re}(\lambda+\nu) > -1$	$\Gamma(1+\lambda+\nu)(\frac{1}{2}a)^{\nu} b^{\lambda+\nu+z} \Gamma(z)$ $\cdot [\Gamma(1+\nu)\Gamma(1+\nu+\mu+z)]^{-1} {}_2F_3(\frac{1}{2}+\frac{1}{2}\lambda+\frac{1}{2}\nu,$ $1+\frac{1}{2}\lambda+\frac{1}{2}\nu; \nu+1, \frac{1}{2}+\frac{1}{2}\nu+\frac{1}{2}\lambda+\frac{1}{2}z,$ $1+\frac{1}{2}\lambda+\frac{1}{2}\nu+\frac{1}{2}z; -\frac{1}{4}a^2 b)$ $\mathrm{Re}\, z > 0$
10.20	$e^{\pm iaz}(b-x)^{\mu-1} \cdot J_{\nu}(ax)$ $x < b$ 0 $x > b$ $\mathrm{Re}\, \mu > 0$	$(\frac{1}{2}a)^{\nu} \Gamma(\mu)\Gamma(\nu+z) b^{\nu+\mu-1+z}$ $\cdot [\Gamma(1+\nu)\Gamma(\nu+\mu+z)]^{-1} {}_2F_2(\nu+z, \frac{1}{2}+\nu;$ $\nu+\mu+z, 2\nu+1; \pm i2ab)$ $\mathrm{Re}\, z > -\mathrm{Re}\, \nu$
10.21	$(b-x)^{\lambda} e^{\pm iax} \cdot J_{\nu}[a(b-x)]$ $x < b$ 0 $x > b$ $\mathrm{Re}(\nu+\lambda) > -1$	$(\frac{1}{2}a)^{\nu} b^{\nu+\lambda+z} \Gamma(z)\Gamma(1+\nu+\lambda) e^{\pm iab}$ $\cdot [\Gamma(1+\nu)\Gamma(1+\nu+\lambda+z)]^{-1} {}_2F_2(1+\nu+\lambda,$ $\frac{1}{2}+\nu; 1+\nu+\lambda+z, 2\nu+1; \pm 2iab)$ $\mathrm{Re}\, z > 0$
10.22	$(b-x)^{\lambda} \cdot J_{2\nu}[a(b-x)^{\frac{1}{2}}]$ $x < b$ 0 $x > b$ $\mathrm{Re}(\lambda+\nu) > -1$	$(\frac{1}{2}a)^{2\nu} b^{\lambda+\nu+\mu} \Gamma(1+\lambda+\nu)\Gamma(z)$ $\cdot [\Gamma(1+2\nu)\Gamma(1+\lambda+\nu+z)]^{-1}$ $\cdot {}_1F_2(\lambda+1+\nu; 1+2\nu, 1+\lambda+\nu+z; -\frac{1}{4}a^2 b)$ $\mathrm{Re}\, z > 0$

	$\phi(x)$	$\Phi(z) = \int_0^\infty \phi(x) x^{z-1} dx$
10.23	$J_\nu^2(ax)$ $-2\mathrm{Re}\,\nu < \mathrm{Re}\,z < 1$	$\tfrac{1}{2}(\tfrac{1}{2}a)^{-z}\Gamma(1-z)\Gamma(\nu+\tfrac{1}{2}z)$ $\cdot[\Gamma(1-\tfrac{1}{2}z)]^{-2}[\Gamma(1+\nu-\tfrac{1}{2}z)]^{-1}$
10.24	$J_\mu(ax)J_\nu(ax)$ $-\mathrm{Re}(\nu+\mu) < \mathrm{Re}\,z < 1$	$\tfrac{1}{2}(\tfrac{1}{2}a)^{-z}\Gamma(1-z)\Gamma(\tfrac{1}{2}\nu+\tfrac{1}{2}\mu+\tfrac{1}{2}z)$ $[\Gamma(1+\tfrac{1}{2}\nu-\tfrac{1}{2}\mu-\tfrac{1}{2}z)\Gamma(1+\tfrac{1}{2}\nu+\tfrac{1}{2}\mu-\tfrac{1}{2}z)$ $\cdot\Gamma(1+\tfrac{1}{2}\mu-\tfrac{1}{2}\nu-\tfrac{1}{2}z)]^{-1}$
10.25	$J_\nu(ax)J_\nu(bx)$ $-2\mathrm{Re}\,\nu < \mathrm{Re}\,z < 2$	$2^{z-1}[\Gamma(1-\tfrac{1}{2}z)]^{-1}\Gamma(\nu+\tfrac{1}{2}z)$ $\cdot(\lvert a^2-b^2\rvert)^{-\frac{1}{2}z}P_{-\frac{1}{2}z}^{-\nu}(\frac{a^2+b^2}{\lvert a^2-b^2\rvert})$
10.26	$J_\nu^2(ax)+Y_\nu^2(ax)$ $\pm 2\mathrm{Re}\,\nu < \mathrm{Re}\,z < 1$	$\pi^{-5/2}\cos(\pi\nu)\,a^{-z}\Gamma(\tfrac{1}{2}z)\Gamma(\tfrac{1}{2}-\tfrac{1}{2}z)$ $\cdot\Gamma(\nu+\tfrac{1}{2}z)$
10.27	$J_\nu(ax)Y_\mu(ax)$ $-Y_\nu(ax)J_\mu(ax)$ $\mathrm{Re}(\pm\mu\pm\nu) < \mathrm{Re}\,z < 1$	$\tfrac{1}{2}\sin[\pi(\nu-\mu)](\tfrac{1}{2}a)^{-z}\Gamma(1-z)$ $\cdot\Gamma(\tfrac{1}{2}z+\tfrac{1}{2}\nu+\tfrac{1}{2}\mu)\Gamma(\tfrac{1}{2}z-\tfrac{1}{2}\nu+\tfrac{1}{2}\mu)$ $\cdot\Gamma(\tfrac{1}{2}z+\tfrac{1}{2}\nu-\tfrac{1}{2}\mu)[\Gamma(1-\tfrac{1}{2}z+\tfrac{1}{2}\nu+\tfrac{1}{2}\mu)]^{-1}$

	$\phi(x)$	$\Phi(z) = \int_0^\infty \phi(x) x^{z-1} dx$
10.28	$J_\nu(ax) J_{-\nu}(ax)$ $0 < \operatorname{Re} z < 1$	$\frac{1}{2}\pi^{-\frac{1}{2}} a^{-z} \Gamma(\frac{1}{2}z) \Gamma(\frac{1}{2}-\frac{1}{2}z)$ $\cdot [\Gamma(1+\nu-\frac{1}{2}z) \Gamma(1-\nu-\frac{1}{2}z)]^{-1}$
10.29	$J_{-\nu}(ax) J_\nu(bx)$ $0 < \operatorname{Re} z < 2$	$2^{z-1} \Gamma(\frac{1}{2}z) [\Gamma(1+\nu-\frac{1}{2}z)]^{-1}$ $\cdot (b^2-a^2)^{-\frac{1}{2}z} P^{\nu}_{-\frac{1}{2}z}(\frac{b^2+a^2}{b^2-a^2})$, $b > a$ $2^{z-1} \Gamma(\frac{1}{2}z) [\Gamma(1-\nu-\frac{1}{2}z)]^{-1}$ $\cdot (a^2-b^2)^{-\frac{1}{2}z} P^{-\nu}_{-\frac{1}{2}z}(\frac{a^2+b^2}{a^2-b^2})$, $b < a$
10.30	$J_\nu(ax) Y_{-\nu}(ax)$ $\left.\begin{matrix} 0 \\ -2\operatorname{Re} \nu \end{matrix}\right\} < \operatorname{Re} z < 1$	$-\frac{1}{2}\pi^{-\frac{3}{2}} a^{-z} \cos[\pi(\nu+\frac{1}{2}z)]$ $\Gamma(\frac{1}{2}z) \Gamma(\frac{1}{2}-\frac{1}{2}z) \Gamma(\nu+\frac{1}{2}z)$ $\cdot [\Gamma(1+\nu-\frac{1}{2}z)]^{-1}$
10.31	$J_\nu(ax) Y_\mu(ax)$ $\operatorname{Re}(\pm\mu-\nu) < \operatorname{Re} z < 1$	$-\pi^{-1} 2^{z-1} a^{-z} \cos[\frac{1}{2}\pi(\nu-\mu+z)]$ $\cdot \Gamma(\frac{1}{2}z+\frac{1}{2}\mu+\frac{1}{2}\nu) \Gamma(\frac{1}{2}z-\frac{1}{2}\mu+\frac{1}{2}\nu) \Gamma(1-z)$ $\cdot [\Gamma(\frac{1}{2}\nu-\frac{1}{2}\mu-\frac{1}{2}z+1) \Gamma(\frac{1}{2}\nu+\frac{1}{2}\mu-\frac{1}{2}z+1)]^{-1}$
10.32	$J_\nu(bx) Y_{-\nu}(ax)$ $+J_{-\nu}(ax) Y_\nu(bx)$ $\pm 2 \operatorname{Re} \nu < \operatorname{Re} z < 2$	$-\pi^{-\frac{1}{2}} 2^{z-1} (ab)^{-\frac{1}{2}} \Gamma(\frac{1}{2}z+\nu) \Gamma(\frac{1}{2}z-\nu)$ $\cdot [\Gamma(1-\frac{1}{2}z)]^{-1} \lvert a^2-b^2 \rvert^{\frac{1}{2}-\frac{1}{2}z} P^{\frac{1}{2}-\frac{1}{2}z}_{\nu-\frac{1}{2}}(\frac{a^2+b^2}{2ab})$ $= -\pi^{-1} 2^z \Gamma(\frac{1}{2}z+\nu) [\Gamma(1-\frac{1}{2}z)]^{-1} \lvert a^2-b^2 \rvert^{-\frac{1}{2}z}$ $e^{i\pi\nu} q^{-\nu}_{\frac{1}{2}z-1}(\frac{a^2+b^2}{\lvert a^2-b^2 \rvert})$

	$\phi(x)$	$\Phi(z) = \int_0^\infty \phi(x) x^{z-1} dx$
10.33	$Y_\nu(ax) J_\nu(bx)$ $\left.\begin{matrix}0\\-2\ \mathrm{Re}\ \nu\end{matrix}\right\} < \mathrm{Re}\ z < 2$	$-2^{z-1}\pi^{-1}\cos(\tfrac{1}{2}\pi z)\Gamma(\tfrac{1}{2}z)\Gamma(\tfrac{1}{2}z+\nu)$ $\cdot(a^2-b^2)^{-\frac{1}{2}z} P^{-\nu}_{-\frac{1}{2}z}\left(\frac{a^2+b^2}{a^2-b^2}\right), \quad a > b$ $-2^{z-1}\pi^{-1}(b^2-a^2)^{-\frac{1}{2}z}\Gamma(\tfrac{1}{2}z)\Gamma(\nu+\tfrac{1}{2}z)$ $\cdot\{\cos(\tfrac{1}{2}\pi z) P^{-\nu}_{-\frac{1}{2}z}\left(\frac{b^2+a^2}{b^2-a^2}\right)$ $+ 2\pi^{-1} e^{-i\pi\nu}\sin[\pi(\nu+\tfrac{1}{2}z)]$ $\cdot q^{\nu}_{-\frac{1}{2}z}\left(\frac{b^2+a^2}{b^2-a^2}\right)\}, \quad a < b$
10.34	$J_\nu(bx) Y_{-\nu}(ax)$ $\left.\begin{matrix}0\\-2\ \mathrm{Re}\ \nu\end{matrix}\right\} < \mathrm{Re}\ z < 2$	$-\pi^{-1}2^{z-1}\cos[\pi(\nu+\tfrac{1}{2}z)]\Gamma(\tfrac{1}{2}z)\Gamma(\nu+\tfrac{1}{2}z)$ $\cdot(a^2-b^2)^{-\frac{1}{2}z} P^{-\nu}_{-\frac{1}{2}z}\left(\frac{a^2+b^2}{a^2-b^2}\right), \quad a > b$ $-\pi^{-1}2^{z-1}\Gamma(\tfrac{1}{2}z)\Gamma(\nu+\tfrac{1}{2}z)(b^2-a^2)^{-\frac{1}{2}z}$ $\cdot\{\cos[\pi(\nu+\tfrac{1}{2}z)] P^{-\nu}_{-\frac{1}{2}z}\left(\frac{b^2+a^2}{b^2-a^2}\right)$ $+2\pi^{-1} e^{-i\pi\nu}\cos(\pi\nu)\sin[\pi(\nu+\tfrac{1}{2}z)]$ $\cdot q^{\nu}_{-\frac{1}{2}z}\left(\frac{b^2+a^2}{b^2-a^2}\right), \quad a < b$
10.35	$\exp(-b^2x^2)$ $\cdot J_\mu(ax) J_\nu(ax)$ $\mathrm{Re}\ z > -\mathrm{Re}(\mu+\nu)$	$\tfrac{1}{2}(\tfrac{1}{2}a/b)^{\mu+\nu}[\Gamma(1+\nu)]^{-2} b^{-z}$ $\cdot\Gamma(\tfrac{1}{2}z+\tfrac{1}{2}\mu+\tfrac{1}{2}\nu)\ {}_3F_3(\tfrac{1}{2}+\tfrac{1}{2}\mu+\tfrac{1}{2}\nu, 1+\tfrac{1}{2}\mu+\tfrac{1}{2}\nu,$ $\tfrac{1}{2}z+\tfrac{1}{2}\mu+\tfrac{1}{2}\nu; \mu+1, \nu+1, \mu+\nu+1; -a^2/b^2)$

	$\phi(x)$	$\Phi(z) = \int_0^\infty \phi(x) x^{z-1} dx$
10.36	$J_\mu(ax) J_\nu(bx)$ $-\mathrm{Re}(\mu+\nu) < \mathrm{Re}\, z < 2$	$2^{z-1} a^\mu b^{-z-\mu} [\Gamma(1+\mu)]^{-1}$ $\cdot \Gamma(\tfrac{1}{2}\nu+\tfrac{1}{2}\mu+\tfrac{1}{2}z) [\Gamma(1+\tfrac{1}{2}\nu-\tfrac{1}{2}\mu-\tfrac{1}{2}z)]^{-1}$ $\cdot {}_2F_1(\tfrac{1}{2}\nu+\tfrac{1}{2}\mu+\tfrac{1}{2}z, \tfrac{1}{2}\mu-\tfrac{1}{2}\nu+\tfrac{1}{2}z; \mu+1; a^2/b^2)$ $a < b$ $b^\nu 2^{z-1} a^{-\nu-z} \Gamma(\tfrac{1}{2}\nu+\tfrac{1}{2}\mu+\tfrac{1}{2}z)$ $\cdot [\Gamma(1+\nu) \Gamma(1+\tfrac{1}{2}\mu-\tfrac{1}{2}\nu-\tfrac{1}{2}z)]^{-1}$ $\cdot {}_2F_1(\tfrac{1}{2}\nu+\tfrac{1}{2}\mu+\tfrac{1}{2}z, \tfrac{1}{2}\nu-\tfrac{1}{2}\mu+\tfrac{1}{2}z; \nu+1; b^2/a^2)$ $a > b$
10.37	$e^{-ax} J_\mu(bx) J_\nu(cx)$	Eason, G. Noble, B. and Sneddon, I.N., 1955: Phil. Trans. Roy. Soc. London (A), 247, 529-
10.38	$J_\mu(ax) J_\nu(bx) J_\zeta(cx)$	Bailey, W. N., 1936: Proc. London Math. Soc. 40, 37-49. J. London Math. Soc. 11, 16-20.

	$\phi(x)$	$\Phi(z) = \int_0^\infty \phi(x) z^{z-1} dx$
10.39	$J_\nu(bx) Y_\mu(ax)$ $\mathrm{Re}(-\nu\pm\mu) < \mathrm{Re}\, z < 2$	$-\frac{1}{2}\pi^{-1}[\Gamma(1+\nu)]^{-1} a^{-\nu}(\frac{1}{2}a)^{-z}$ $\cos[\frac{1}{2}\pi(\mu-\nu-z)]\Gamma(\frac{1}{2}z+\frac{1}{2}\mu+\frac{1}{2}\nu)$ $\cdot\Gamma(\frac{1}{2}z-\frac{1}{2}\mu+\frac{1}{2}\nu)\ {}_2F_1(\frac{1}{2}z+\frac{1}{2}\mu+\frac{1}{2}\nu,$ $\frac{1}{2}z-\frac{1}{2}\mu+\frac{1}{2}\nu;1+\nu;b^2/a^2), \quad b < a$ $-\int_0^\infty \{J_\mu(ax)Y_\nu(bx)+4\pi^{-2}$ $\cdot\sin[\frac{1}{2}\pi(z-\nu-\mu)]K_\nu(bx)K_\mu(ax)\}x^{-z}dx$ $a < b$
10.40	$Y_\mu(ax) Y_\nu(bx)$ $\lvert\mathrm{Re}(\mu\pm\nu)\rvert < \mathrm{Re}\, z < 2$	$\int_0^\infty \{J_\mu(ax)J_\nu(bx)+4\pi^{-2}\cos[\frac{1}{2}\pi(z-\nu-\mu)]$ $\cdot K_\mu(ax)K_\nu(bx)\}x^{z-1}dx$
10.41	$H_\mu^{(2)}(ax) H_\nu^{(2)}(bx)$ $\lvert\mathrm{Re}\,\mu\rvert+\lvert\mathrm{Re}\,\nu\rvert<\mathrm{Re}\, z<1$ For $H_\nu^{(1)} H_\mu^{(1)}$ change i into $-i$	$-2^{z-1}\exp[-\frac{1}{2}\pi i(\mu+\nu+z)]\pi^{-2}a^{-\nu-z}b^\nu$ $[\Gamma(z)]^{-1}\Gamma(\frac{1}{2}z+\frac{1}{2}\mu+\frac{1}{2}\nu)\Gamma(\frac{1}{2}-\frac{1}{2}\mu+\frac{1}{2}\nu)$ $\cdot\Gamma(\frac{1}{2}z+\frac{1}{2}\mu-\frac{1}{2}\nu)\Gamma(\frac{1}{2}z-\frac{1}{2}\mu-\frac{1}{2}\nu)$ $\cdot\ {}_2F_1(\frac{1}{2}z+\frac{1}{2}\mu+\frac{1}{2}\nu,\frac{1}{2}z-\frac{1}{2}\mu+\frac{1}{2}\nu;z;1-b^2/a^2)$

	$\phi(x)$	$\Phi(z) = \int_0^\infty \phi(x) x^{z-1} dx$
10.42	$(1-x^2)^\sigma J_\mu(ax) J_\nu(bx)$ $\quad x < 1$ $0 \quad x > 1$	Bailey, W. N., 1938: Quart. J. Math., Oxford Ser. 9, 141-147
10.43	$(1-x^2)^\sigma J_\mu(ax)$ $\cdot J_\nu[b(1-x^2)^{\frac{1}{2}}]$ $\quad x < 1$ $0 \quad x > 1$	as before
10.44	$(a^2+x^2)^{-\sigma}$ $\cdot J_\mu(x) J_\nu(x)$	Watson, G. N., 1922: A treatise on the theory of Bessel functions, Cambridge, p. 436
10.45	$(a^2+x^2)^{-\mu}$ $\cdot Y_\nu(x)$ $\lvert\mathrm{Re}\ \nu\rvert < \mathrm{Re}\ z < 2\ \mathrm{Re}\ \mu + \frac{3}{2}$	$[\Gamma(\mu)]^{-1} a^{z-1-2\mu}$ $\cdot G_{24}^{31}\left(\frac{1}{4}a^2 \middle\vert \begin{matrix} \frac{3}{2}, & -\frac{1}{2}\nu \\ \mu+\frac{1}{2}-\frac{1}{2}z, \frac{1}{2}+\frac{1}{2}\nu, & \frac{1}{2}-\frac{1}{2}\nu, -\frac{1}{2}\nu \end{matrix}\right)$

	$\phi(x)$	$\Phi(z) = \int_0^\infty \phi(x) x^{z-1} dx$
10.46	$J_\mu(x) J_\nu(a-x)$ $x < a$ 0 $x > a$ $\mathrm{Re}\, z > -\mathrm{Re}\, \mu$, $\mathrm{Re}\, \nu > -1$	$2^z \sum_{n=0}^{\infty} (-1)^n [n!\Gamma(\mu+n+1)]^{-1}$ $\cdot \Gamma(\mu+n+z)(z)_n J_{\mu+\nu+2n+z}(a)$
10.47	$(a-x)^{-1}$ $J_\nu(x) J_\nu(a-x)$ $x < a$ 0 $x > a$ $\mathrm{Re}\, \nu > 0$, $\mathrm{Re}\, z > -\mathrm{Re}\, \nu$	$(\nu a)^{-1} 2^z \sum_{n=0}^{\infty} (-1)^n [n!\Gamma(\mu+n+1)]^{-1}$ $\cdot (z)_n \Gamma(n+\mu+z)(z+\mu+\nu+2n)$ $\cdot J_{z+\mu+\nu+2n}(a)$
10.48	$(a-x)^{\sigma-1} J_\mu(x)$ $\cdot J_\nu(a-x)$ $x < a$ 0 $x > a$	Bailey, W. N., 1930: Proc. London Math. Soc. (2) 30, 422-421 and 31, 200-208.
10.49	$(a-x)^{\sigma-1} J_\lambda(bx)$ $\cdot J_\mu(cx) J_\nu(a-x)$ $x < a$ 0 $x > a$	as before

	$\phi(x)$	$\Phi(z) = \int_0^\infty \phi(x)x^{z-1}dx$
10.50	$(a^2+x^2)^{\frac{1}{2}\nu}$ $\cdot J_\nu[b(a^2+x^2)^{\frac{1}{2}}]$ $0 < \text{Re } z < \frac{3}{2} - \text{Re } \nu$	$\frac{1}{2}a^\nu(2a/b)^{\frac{1}{2}z}\Gamma(\frac{1}{2}z)$ $\cdot [J_{\nu+\frac{1}{2}z}(ab)\cos(\frac{1}{2}\pi z)$ $-Y_{\nu+\frac{1}{2}z}(ab)\sin(\frac{1}{2}\pi z)]$
10.51	$(a^2+x^2)^{-\frac{1}{2}\nu}$ $\cdot J_\nu[b(a^2+x^2)^{\frac{1}{2}}]$	$\frac{1}{2}a^{-\nu}(2a/b)^{\frac{1}{2}z}\Gamma(\frac{1}{2}z)J_{\nu-\frac{1}{2}z}(ab)$ $0 < \text{Re } z < \frac{3}{2} + \text{Re } \nu$
10.52	$(a^2+x^2)^{\frac{1}{2}\nu}$ $\cdot Y_\nu[b(a^2+x^2)^{\frac{1}{2}}]$ $0 < \text{Re } z < \frac{3}{2} - \text{Re } \nu$	$\frac{1}{2}a^\nu(2a/b)^{\frac{1}{2}z}\Gamma(\frac{1}{2}z)$ $\cdot [Y_{\nu+\frac{1}{2}z}(ab)\cos(\frac{1}{2}\pi z)$ $+ J_{\nu+\frac{1}{2}z}(ab)\sin(\frac{1}{2}\pi z)]$
10.53	$(a^2+x^2)^{-\frac{1}{2}\nu}$ $\cdot Y_\nu[b(a^2+x^2)^{\frac{1}{2}}]$	$\frac{1}{2}a^{-\nu}(2a/b)^{\frac{1}{2}z}\Gamma(\frac{1}{2}z)Y_{\nu-\frac{1}{2}z}(ab)$ $0 < \text{Re } z < \frac{3}{2} + \text{Re } \nu$
10.54	$(a^2-x^2)^{\frac{1}{2}\nu}$ $\cdot J_\nu[b(a^2-x^2)^{\frac{1}{2}}]$ $x < a$; 0 $x > a$; $\text{Re } \nu > -1$	$2^{\frac{1}{2}z-1}\Gamma(\frac{1}{2}z)a^{\nu+\frac{1}{2}z}b^{-\frac{1}{2}z}J_{\nu+\frac{1}{2}z}(ab)$ $\text{Re } z > 0$

	$\phi(x)$	$\Phi(z) = \int_0^\infty \phi(x) x^{z-1} dx$
10.55	$(a^2-x^2)^{-\frac{1}{2}\nu}$ $J_\nu[b(a^2-x^2)^{\frac{1}{2}}]$ $x < a$ 0 $x > a$	$2^{1-\nu}[\Gamma(\nu)]^{-1} a^{\frac{1}{2}z-\nu} b^{-\frac{1}{2}z}$ $\cdot s_{\nu-1+\frac{1}{2}z, \frac{1}{2}z-\nu}(ab)$ $\mathrm{Re}\, z > 0$
10.56	0 $x < a$ $(x^2-a^2)^{\frac{1}{2}\nu}$ $\cdot J_\nu[b(x^2-a^2)^{\frac{1}{2}}]$ $x > a$ $\mathrm{Re}\, \nu > -1$	$a^\nu (2a/b)^{\frac{1}{2}z} [\Gamma(1-\frac{1}{2}z)]^{-1}$ $\cdot K_{\nu+\frac{1}{2}z}(ab)$ $\mathrm{Re}\, z < 3/2 - \mathrm{Re}\, \nu$
10.57	0 $x < a$ $(x^2-a^2)^{-\frac{1}{2}\nu}$ $\cdot J_\nu[b(x^2-a^2)^{\frac{1}{2}}]$ $x > a$	$a^{\frac{1}{2}z-\nu} b^{-\frac{1}{2}z} [2^{\frac{1}{2}z-1} \frac{\Gamma(\frac{1}{2}z)}{\Gamma(1-\nu)} I_{\nu-\frac{1}{2}z}(ab)$ $-\frac{2^{1-\nu}}{\Gamma(\nu)} e^{-i\frac{1}{2}\pi(\frac{1}{2}z+\nu)} s_{\nu-1+\frac{1}{2}z, \frac{1}{2}z-\nu}(iab)]$ $\mathrm{Re}\, z < 3/2 + \mathrm{Re}\, \nu$
10.58	$(a^2-x^2)^{\frac{1}{2}\nu}$ $\cdot Y_\nu[b(a^2-x^2)^{\frac{1}{2}}]$ $x < a$ 0 $x > a$ $\mathrm{Re}\, \nu > -1$	$a^{\frac{1}{2}z+\nu} b^{-\frac{1}{2}z} [2^{\frac{1}{2}z-1} \Gamma(\frac{1}{2}z) Y_{\frac{1}{2}z+\nu}(ab)$ $+\pi^{-1} a^{\nu+1} \Gamma(1+\nu) S_{\frac{1}{2}z-1-\nu, \frac{1}{2}z+\nu}(ab)]$ $\mathrm{Re}\, z > 0$

	$\phi(x)$	$\Phi(z) = \int_0^\infty \phi(x) x^{z-1} dx$
10.59	$(a^2-x^2)^{-\frac{1}{2}\nu}$ $\cdot Y_\nu[b(a^2-x^2)^{\frac{1}{2}}]$ $x < a$ 0 $x > a$ $\mathrm{Re}\ \nu < 1$	$a^{\frac{1}{2}z-\nu} b^{-\frac{1}{2}z} [\pi^{-1} 2^{1-\nu} \cos(\pi\nu) \Gamma(1-\nu)$ $\cdot S_{\frac{1}{2}z+\nu-1,\frac{1}{2}z-\nu}(ab)$ $-2^{\frac{1}{2}z-1} \Gamma(\frac{1}{2}z) \csc(\pi\nu) J_{\frac{1}{2}z-\nu}(ab)]$ $\mathrm{Re}\ z > 0$
10.60	$J_\nu(ax) J_\mu[b(1-x^2)^{\frac{1}{2}}]$ $x < 1$ 0 $x > 1$	Bailey, W. N. 1938: Quart. Journal of Math. 9, 141-147.
10.61	$(a^2-x^2)^\mu J_\nu(bx)$ $x < 1$ 0 $x > 1$ $\mathrm{Re}\ \mu > -1$, $\mathrm{Re}\ z > -\mathrm{Re}\ \nu$	$b^\nu a^{z+\nu+2\mu} 2^{-\nu-1} [\Gamma(1+\nu)]^{-1}$ $\cdot B(\mu+1, \frac{1}{2}z+\frac{1}{2}\nu)$ $\cdot {}_1F_2(\frac{1}{2}\nu+\frac{1}{2}z; \nu+1, \mu+1+\frac{1}{2}\nu+\frac{1}{2}z; -\frac{1}{4}a^2b^2)$
10.62	$J_\nu(u) J_\nu(v)$ $+ Y_\nu(u) Y_\nu(v)$ $\begin{matrix} u \\ v \end{matrix} = b[(a^2+x^2)^{\frac{1}{2}} \pm a]$	$-\pi^{-3/2} \cos(\pi\nu) (a/b)^{\frac{1}{2}z}$ $\cdot \Gamma(\frac{1}{2}-\frac{1}{2}z) Y_{\frac{1}{2}z}(2ab)$ $\pm 2\ \mathrm{Re}\ \nu < \mathrm{Re}\ z < 1$
10.63	$Y_\nu(u) J_\nu(v)$ $-J_\nu(u) Y_\nu(v)$ $\begin{matrix} u \\ v \end{matrix} = b[(a^2+x^2)^{\frac{1}{2}} \pm a]$	$\pi^{-3/2} \cos(\pi\nu) (a/b)^{\frac{1}{2}z}$ $\cdot \Gamma(\frac{1}{2}-\frac{1}{2}z) J_{\frac{1}{2}z}(2ab)$ $\pm 2\ \mathrm{Re}\ \nu < \mathrm{Re}\ z < 1$

	$\phi(x)$	$\Phi(z) = \int_0^\infty \phi(x) x^{z-1} dx$
10.64	$J_\nu(v) J_{-\nu}(u)$ ${u \atop v} = b[(a^2+x^2)^{\frac{1}{2}} \pm a]$ $-2 \text{ Re } \nu < \text{Re } z < 1$	$\frac{1}{2}\pi^{\frac{1}{2}} \Gamma(\nu+\frac{1}{2}z) \Gamma(\frac{1}{2}-\frac{1}{2}z) [\Gamma(1+\nu-\frac{1}{2}z)]^{-1}$ $(a/b)^{\frac{1}{2}z} [J_{-\frac{1}{2}z}(2ab) \cos(\pi\nu)$ $- Y_{-\frac{1}{2}z}(2ab) \sin(\pi\nu)]$
10.65	$J_\nu(u) J_\nu(v)$ ${u \atop v} = b[(a^2+x^2)^{\frac{1}{2}} \pm a]$ $-2 \text{ Re } \nu < \text{Re } z < 1$	$\frac{1}{2}\pi^{-\frac{1}{2}} \Gamma(\nu+\frac{1}{2}z) \Gamma(\frac{1}{2}-\frac{1}{2}z) [\Gamma(1+\nu-\frac{1}{2}z)]^{-1}$ $\cdot (a/b)^{\frac{1}{2}z} J_{-\frac{1}{2}z}(2ab)$
10.66	$J_\nu(v) Y_{-\nu}(u)$ ${u \atop v} = b[(a^2+x^2)^{\frac{1}{2}} \pm a]$ $-2 \text{ Re } \nu < \text{Re } z < 1$	$\frac{1}{2}\pi^{\frac{1}{2}} \Gamma(\nu+\frac{1}{2}z) \Gamma(\frac{1}{2}-\frac{1}{2}z) [\Gamma(1+\nu-\frac{1}{2}z)]^{-1}$ $\cdot (a/b)^{\frac{1}{2}z} [J_{-\frac{1}{2}z}(2ab) \sin(\pi\nu)$ $+ Y_{-\frac{1}{2}z}(2ab) \cos(\pi\nu)]$
10.67	$J_\nu(v) Y_\nu(u)$ ${u \atop v} = b[(a^2+x^2)^{\frac{1}{2}} \pm a]$ $-2 \text{ Re } \nu < \text{Re } z < 1$	$\frac{1}{2}\pi^{-\frac{1}{2}} \Gamma(\nu+\frac{1}{2}z) \Gamma(\frac{1}{2}-\frac{1}{2}z) [\Gamma(1+\nu-\frac{1}{2}z)]^{-1}$ $\cdot (a/b)^{\frac{1}{2}z} Y_{-\frac{1}{2}z}(2ab)$

	$\phi(x)$	$\Phi(z) = \int_0^\infty \phi(x)x^{z-1}dx$
10.68	$(b-x)^{\mu-1}$ $\cdot J_\nu^2(ax^{\frac{1}{2}})$ $x < b$ 0 $x > b$ $\mathrm{Re}\ \mu > 0$	$\Gamma(\mu)(\frac{1}{2}a)^{2\nu}\Gamma(z+\nu)b^{z+\nu+\mu-1}$ $\cdot[\Gamma^2(1+\nu)\Gamma(z+\nu+\mu)]^{-1}$ $\cdot\ {}_2F_3(z+\nu,\frac{1}{2}+\nu;z+\nu+\mu,\nu+1,2\nu+1;-a^2b)$ $\mathrm{Re}\ z > -\mathrm{Re}\ \nu$
10.69	$(b-x)^{\lambda}$ $\cdot J_\nu^2[a(b-x)^{\frac{1}{2}}]$ $x < b$ 0 $x > b$ $\mathrm{Re}(\lambda+\nu) > -1$	$(\frac{1}{2}a)^{2\nu}\Gamma(1+\lambda+\nu)\Gamma(z)b^{\lambda+\nu+z}$ $\cdot[\Gamma^2(1+\nu)\Gamma(\lambda+\nu+1+z)]^{-1}$ $\cdot\ {}_2F_3(\lambda+\nu+1,\frac{1}{2}+\nu;\lambda+\nu+1+z,\nu+1,2\nu+1;-a^2b)$ $\mathrm{Re}\ z > 0$
10.70	$(b-x)^{\mu-1}$ $J_\nu(ax^{\frac{1}{2}})J_{-\nu}(ax^{\frac{1}{2}})$ $x < b$ 0 $x > b$	$\Gamma(z)\sin(\pi\nu)\Gamma(\mu)b^{z+\mu-1}$ $[\pi\nu\Gamma(z+\mu)]^{-1}\ {}_2F_3(\frac{1}{2},z;1+\nu,$ $1-\nu,\mu+z;-a^2b)$ $\mathrm{Re}(\mu,z) > 0$
10.71	$(b-x)^{\lambda-1}J_\nu[a(b-x)^{\frac{1}{2}}]$ $\cdot J_{-\nu}[a(b-x)^{\frac{1}{2}}]$ $x < b$ 0 $x > b$	$\Gamma(\lambda)\sin(\pi\nu)\Gamma(z)b^{\lambda-1+z}$ $[\pi\nu\Gamma(\nu+z)]^{-1}\ {}_2F_3(\frac{1}{2},\lambda;1+\nu,$ $1-\nu,\lambda+z;-a^2b)$ $\mathrm{Re}(\lambda,z) > 0$

	$\phi(x)$	$\Phi(z) = \int_0^\infty \phi(x) x^{z-1} dx$
10.72	$(b^2+x^2)^{-\rho} J_\nu(ax)$ $-\mathrm{Re}\ \nu < \mathrm{Re}\ z < 3/2 + 2\mathrm{Re}\ \rho$	$2^{z-3} a^{2-z} [\Gamma(\rho)]^{-1} b^{2-2\rho}$ $\cdot G^{21}_{13}\left(\tfrac{1}{4}a^2b^2 \middle\vert \begin{matrix} 0 \\ \rho-1, \tfrac{1}{2}\nu-1+\tfrac{1}{2}z, \tfrac{1}{2}z-1-\tfrac{1}{2}\nu \end{matrix}\right)$
10.73	$(b+x)^{-\nu} e^{iax}$ $\cdot J_\nu[a(b+x)]$	$\pi^{-\frac{1}{2}} e^{-iab} [\Gamma(1+2\nu-z)]^{-1} \Gamma(z) e^{i\frac{1}{2}\pi z}$ $\cdot (2a)^{\nu-z} \Gamma(\tfrac{1}{2}+\nu-z)\, {}_1F_1(\tfrac{1}{2}+\nu-z; 1+2\nu-z; 2iab)$ $0 < \mathrm{Re}\ z < \tfrac{1}{2} + \mathrm{Re}\ \nu$
10.74	$(b^2+x^2)^{\lambda}$ $\cdot J_\nu[a(b^2+x^2)^{\frac{1}{2}}]$	$2^{2\lambda-1} a^{-2\lambda} \Gamma(\tfrac{1}{2}z) b^z$ $\cdot G^{20}_{13}\left(\tfrac{1}{4}a^2b^2 \middle\vert \begin{matrix} 0 \\ -\tfrac{1}{2}z, \lambda+\tfrac{1}{2}\nu, \lambda-\tfrac{1}{2}\nu \end{matrix}\right)$ $0 < \mathrm{Re}\ z < \tfrac{1}{2} - 2\mathrm{Re}\ \lambda$
10.75	$(b^2+x^2)^{-1}$ $\cdot J_\mu(ax) J_\nu(ax)$	$\tfrac{1}{2}\pi^{-\frac{1}{2}} a^{-z-2}\, G^{23}_{35}\left(a^2b^2 \middle\vert \begin{matrix} 0, \tfrac{1}{2}z-1, \tfrac{1}{2}z-\tfrac{1}{2} \\ 0, p, q, r, s \end{matrix}\right)$ $p=\tfrac{1}{2}z+\tfrac{1}{2}\nu+\tfrac{1}{2}\mu-1,\ q=\tfrac{1}{2}z+\tfrac{1}{2}\mu-\tfrac{1}{2}\nu-1,$ $r=\tfrac{1}{2}z-\tfrac{1}{2}\mu+\tfrac{1}{2}\nu-1,\ s=\tfrac{1}{2}z-\tfrac{1}{2}\mu-\tfrac{1}{2}\nu-1,$ $-\tfrac{1}{2}\mathrm{Re}(\nu+\mu) < \mathrm{Re}\ z < 0$
10.76	$(b-x)^{\mu-1}$ $\quad x < b$ $\cdot J_\nu(ax)$ 0 $\quad x > b$ $\mathrm{Re}\ \mu > 0,\ \mathrm{Re}\ z > -\mathrm{Re}\ \nu$	$\Gamma(z+\nu)\Gamma(\mu)(\tfrac{1}{2}a)^\nu b^{\mu+\nu-1+z}$ $\cdot [\Gamma(1+\nu)\Gamma(\nu+\mu+z)]^{-1}\, {}_2F_3(\tfrac{1}{2}\nu+\tfrac{1}{2}z,$ $\tfrac{1}{2}+\tfrac{1}{2}\nu+\tfrac{1}{2}z; \nu+1, \tfrac{1}{2}\mu+\tfrac{1}{2}\nu+\tfrac{1}{2}z, \tfrac{1}{2}+\tfrac{1}{2}\nu+\tfrac{1}{2}\mu+\tfrac{1}{2}z;$ $-\tfrac{1}{4}a^2b)$

	$\phi(x)$	$\Phi(z) = \int_0^\infty \phi(x) x^{z-1} dx$
10.77	$J_\nu[a(x^{-1}-x)]$ $x < 1$ 0 $x > 1$	$K_{\frac{1}{2}\nu-\frac{1}{2}z}(a) I_{\frac{1}{2}\nu+\frac{1}{2}z}(a)$ $\mathrm{Re}\ \nu > -1, \quad \mathrm{Re}\ z > -\frac{3}{2}$
10.78	0 $x < 1$ $J_\nu[a(x-x^{-1})]$ $x > 1$	$K_{\frac{1}{2}\nu+\frac{1}{2}z}(a) I_{\frac{1}{2}\nu-\frac{1}{2}z}(a)$ $\mathrm{Re}\ \nu > -1, \quad \mathrm{Re}\ z < \frac{3}{2}$
10.79	$J_\nu[a\lvert x-x^{-1}\rvert]$ $\mathrm{Re}\ \nu > -1$ Pricipal value	$I_{\frac{1}{2}\nu-\frac{1}{2}z}(a) K_{\frac{1}{2}\nu+\frac{1}{2}z}(a)$ $+ I_{\frac{1}{2}\nu+\frac{1}{2}z}(a) K_{\frac{1}{2}\nu-\frac{1}{2}z}(a)$ $-\frac{3}{2} < \mathrm{Re}\ z < \frac{3}{2}$
10.80	$(x+x^{-1})^{-1}$ $\cdot J_\nu[a(x+x^{-1})]$	$a^{-1}[\Gamma(1+\nu)]^{-2}\Gamma(\frac{1}{2}+\frac{1}{2}\nu+\frac{1}{2}z)$ $\cdot\Gamma(\frac{1}{2}+\frac{1}{2}\nu-\frac{1}{2}z) M_{\frac{1}{2}z,\frac{1}{2}\nu}(\frac{1}{2}a) M_{-\frac{1}{2}\nu,\frac{1}{2}z}(\frac{1}{2}a)$ $-1-\mathrm{Re}\ \nu < \mathrm{Re}\ z < 1+\mathrm{Re}\ \nu$
10.81	$J_{2\nu}[a(x+x^{-1})]$ $-\frac{3}{2} < \mathrm{Re}\ z < \frac{3}{2}$	$-\frac{1}{2}\pi[J_{\nu-\frac{1}{2}z}(a) Y_{\nu+\frac{1}{2}z}(a)$ $+ J_{\nu+\frac{1}{2}z}(a) Y_{\nu-\frac{1}{2}z}(a)]$
10.82	$Y_{2\nu}[a(x+x^{-1})]$ $-\frac{3}{2} < \mathrm{Re}\ z < \frac{3}{2}$	$\frac{1}{2}\pi[J_{\nu-\frac{1}{2}z}(a) J_{\nu+\frac{1}{2}z}(a)$ $-Y_{\nu-\frac{1}{2}z}(a) Y_{\nu+\frac{1}{2}z}(a)]$

	$\phi(x)$	$\Phi(z) = \int_0^\infty \phi(x) x^{z-1} dx$
10.83	$Y_o[a\|x-x^{-1}\|]$ Principal value	$2\pi^{-1}\cos(\frac{1}{2}\pi z) K^2_{\frac{1}{2}z}(a)$ $-3/2 < \mathrm{Re}\, z < 3/2$
10.84	$(a+bx)^{\nu}(b+ax)^{-\nu}$ $\cdot J_{2\nu}\{[a^2+b^2+ab(x+x^{-1})]^{\frac{1}{2}}\}$	$-\pi[J_{\nu-z}(a) Y_{\nu+z}(b)$ $+J_{\nu+z}(b) Y_{\nu-z}(a)]$ $-3/2 < \mathrm{Re}\, z < 3/2$
10.85	$(a+bx)^{\nu}(b+ax)^{-\nu}$ $\cdot Y_{2\nu}\{[a^2+b^2+ab(x+x^{-1})]^{\frac{1}{2}}\}$	$\pi[J_{\nu-z}(a) J_{\nu+z}(b)$ $-Y_{\nu-z}(a) Y_{\nu+z}(b)]$ $-3/2 < \mathrm{Re}\, z < 3/2$
10.86	$(1+x^2)^{-1} J_{\nu}(\frac{2ax}{1+x^2})$ $\cdot\exp[-b(\frac{1-x^2}{1+x^2})]$ $-\mathrm{Re}\,\nu < \mathrm{Re}\, z < 2+\mathrm{Re}\,\nu$	$\frac{1}{2}a^{-1}[\Gamma(1+\nu)]^{-2}\Gamma(1+\frac{1}{2}\nu-\frac{1}{2}z)\Gamma(\frac{1}{2}\nu+\frac{1}{2}z)$ $\cdot M_{\frac{1}{2}-\frac{1}{2}z,\frac{1}{2}\nu}[b+(b^2-a^2)^{\frac{1}{2}}]$ $\cdot M_{\frac{1}{2}-\frac{1}{2}z,\frac{1}{2}\nu}(b-(b^2-a^2)^{\frac{1}{2}}]$
10.87	$(1+x^2)^{-1} J_{\nu}(\frac{2ax}{1+x^2})$ $\cdot\exp[-\frac{2b}{1+x^2})$ $-\mathrm{Re}\,\nu < \mathrm{Re}\, z < 2+\mathrm{Re}\,\nu$	$\frac{1}{2}a^{-1}[\Gamma(1+\nu)]^{-2}e^{-b}\Gamma(1+\frac{1}{2}\nu-\frac{1}{2}z)\Gamma(\frac{1}{2}\nu+\frac{1}{2}z)$ $\cdot M_{\frac{1}{2}-\frac{1}{2}z,\frac{1}{2}\nu}[b+(b^2-a^2)^{\frac{1}{2}}]$ $\cdot M_{\frac{1}{2}-\frac{1}{2}z,\frac{1}{2}\nu}[b-(b^2-a^2)^{\frac{1}{2}}]$

	$\phi(x)$	$\Phi(z) = \int_0^\infty \phi(x) x^{z-1} dx$
10.88	$(1+x^2)^{-1} J_\nu\left(\frac{2ax}{1+x^2}\right)$ $\cdot \exp\left[b\left(\frac{1-x^2}{1+x^2}\right)\right]$ $-\mathrm{Re}\ \nu < \mathrm{Re}\ z < 2+\mathrm{Re}\ \nu$	$\tfrac{1}{2} a^{-1} [\Gamma(1+\nu)]^{-2} \Gamma(\tfrac{1}{2}\nu+\tfrac{1}{2}z) \Gamma(1+\tfrac{1}{2}\nu-\tfrac{1}{2}z)$ $\cdot M_{\frac{1}{2}z-\frac{1}{2},\frac{1}{2}\nu}[b+(b^2-a^2)^{\frac{1}{2}}]$ $\cdot M_{\frac{1}{2}z-\frac{1}{2},\frac{1}{2}\nu}[b-(b^2-a^2)^{\frac{1}{2}}]$
10.89	$(1+x^2)^{-1} J_\nu\left(\frac{2ax}{1+x^2}\right)$ $\cdot \exp\left(-\frac{2bx^2}{1+x^2}\right)$ $-\mathrm{Re}\ \nu < \mathrm{Re}\ z < 2+\mathrm{Re}\ \nu$	$\tfrac{1}{2} a^{-1} [\Gamma(1+\nu)]^{-2} e^{-b} \Gamma(\tfrac{1}{2}\nu+\tfrac{1}{2}z) \Gamma(1+\tfrac{1}{2}\nu-\tfrac{1}{2}z)$ $\cdot M_{\frac{1}{2}z-\frac{1}{2},\frac{1}{2}\nu}[b+(b^2-a^2)^{\frac{1}{2}}]$ $\cdot M_{\frac{1}{2}z-\frac{1}{2},\frac{1}{2}\nu}[b-(b^2-a^2)^{\frac{1}{2}}]$
10.90	$(1-x^2)^{-1} J_\nu\left(\frac{2ax}{1-x^2}\right)$ $\cdot \exp\left[-b\left(\frac{1+x^2}{1-x^2}\right)\right]$, $x < 1$ 0 $x > 1$ $\mathrm{Re}\ z > -\mathrm{Re}\ \nu$	$\tfrac{1}{2} a^{-1} [\Gamma(1+\nu)]^{-1} \Gamma(\tfrac{1}{2}\nu+\tfrac{1}{2}z)$ $\cdot M_{\frac{1}{2}z-\frac{1}{2},\frac{1}{2}\nu}[(a^2+b^2)^{\frac{1}{2}}-b]$ $\cdot W_{\frac{1}{2}-\frac{1}{2}z,\frac{1}{2}\nu}[(a^2+b^2)^{\frac{1}{2}}+b]$
10.91	$(1-x^2)^{-1} J_\nu\left(\frac{2ax}{1-x^2}\right)$ $\cdot \exp\left(-\frac{2bx^2}{1-x^2}\right)$ $x < 1$ 0 $x > 1$ $\mathrm{Re}\ z > -\mathrm{Re}\ \nu$	$\tfrac{1}{2} a^{-1} [\Gamma(1+\nu)]^{-1} e^{b} \Gamma(\tfrac{1}{2}\nu+\tfrac{1}{2}z)$ $\cdot M_{\frac{1}{2}z-\frac{1}{2},\frac{1}{2}\nu}[(a^2+b^2)^{\frac{1}{2}}-b]$ $\cdot W_{\frac{1}{2}-\frac{1}{2}z,\frac{1}{2}\nu}[(a^2+b^2)^{\frac{1}{2}}+b]$

	$\phi(x)$	$\Phi(z) = \int_0^\infty \phi(x)x^{z-1}dx$
10.92	$0 \qquad x < 1$ $(x^2-1)^{-1}J_\nu(\frac{2ax}{x^2-1})$ $\cdot\exp[-b(\frac{x^2+1}{x^2-1})], \quad x > 1$ $\mathrm{Re}\ z < 2+\mathrm{Re}\ \nu$	$\frac{1}{2}a^{-1}[\Gamma(1+\nu)]^{-1}\Gamma(1+\frac{1}{2}\nu-\frac{1}{2}z)$ $\cdot M_{\frac{1}{2}-\frac{1}{2}z,\frac{1}{2}\nu}[(a^2+b^2)^{\frac{1}{2}}-b]$ $\cdot W_{\frac{1}{2}z-\frac{1}{2},\frac{1}{2}\nu}[(a^2+b^2)^{\frac{1}{2}}+b]$
10.93	$0 \qquad x < 1$ $(x^2-1)^{-1}J_\nu(\frac{2ax}{x^2-1})$ $\cdot\exp(-\frac{2b}{x^2-1}), \quad x > 1$ $\mathrm{Re}\ z < 2+\mathrm{Re}\ \nu$	$\frac{1}{2}a^{-1}[\Gamma(1+\nu)]^{-1}e^b\Gamma(1+\frac{1}{2}\nu-\frac{1}{2}z)$ $\cdot M_{\frac{1}{2}-\frac{1}{2}z,\frac{1}{2}\nu}[(a^2+b^2)^{\frac{1}{2}}-b]$ $\cdot W_{\frac{1}{2}z-\frac{1}{2},\frac{1}{2}\nu}[(a^2+b^2)^{\frac{1}{2}}+b]$
10.94	$J_\mu(ax)J_\nu(b/x)$	$\frac{1}{2}(\frac{1}{2}a)^{-z}G^{20}_{04}(\frac{1}{16}a^2b^2 \mid \frac{1}{2}\nu, \frac{1}{2}z+\frac{1}{2}\mu, \frac{1}{2}z-\frac{1}{2}\mu; -\frac{1}{2}\nu)$ $\mathrm{Re}\ z > \Big\}\begin{matrix} -3/2-\mathrm{Re}\ \mu \\ 3/2+\mathrm{Re}\ \nu \end{matrix}$

1.11 Modified Bessel Function

	$\phi(x)$	$\Phi(z) = \int_0^\infty \phi(x) x^{z-1} dx$
11.1	$K_\nu(ax)$	$2^{z-2} a^{-z} \Gamma(\tfrac{1}{2}z+\tfrac{1}{2}\nu)\Gamma(\tfrac{1}{2}z-\tfrac{1}{2}\nu)$ $\operatorname{Re} z > \pm \operatorname{Re} \nu$
11.2	$(b^2+x^2)^{-1} K_\nu(ax)$ $\operatorname{Re} z > \pm \operatorname{Re} \nu$	$(2b)^{z-2} \Gamma(\tfrac{1}{2}z-\tfrac{1}{2}\nu)\Gamma(\tfrac{1}{2}z+\tfrac{1}{2}\nu)$ $\cdot S_{1-z,\nu}(ab)$
11.3	$e^{-bx} I_\nu(ax)$ $b > a$	$\Gamma(\nu+z)(b^2-a^2)^{-\frac{1}{2}z} P_{z-1}^{-\nu}[b(b^2-a^2)^{-\frac{1}{2}}]$ $\operatorname{Re} z > -\operatorname{Re} \nu$
11.4	$e^{ax} K_\nu(ax)$ $\pm\operatorname{Re} \nu < \operatorname{Re} z < \tfrac{1}{2}$	$\pi^{-\frac{1}{2}} \cos(\pi\nu)(2a)^{-z}\Gamma(\tfrac{1}{2}-z)$ $\cdot\Gamma(z+\nu)\Gamma(z-\nu)$
11.5	$e^{-ax} K_\nu(ax)$ $\operatorname{Re} z > \pm \operatorname{Re} \nu$	$\pi^{\frac{1}{2}}(2a)^{-z}[\Gamma(\tfrac{1}{2}+z)]^{-1}$ $\cdot\Gamma(z-\nu)\Gamma(z+\nu)$
11.6	$e^{-bx} K_\nu(ax)$	$(\tfrac{1}{2}\pi/a)^{\frac{1}{2}}\Gamma(z-\nu)\Gamma(z+\nu)$ $(a^2-b^2)^{\frac{1}{4}-\frac{1}{2}z} P_{\nu-\frac{1}{2}}^{\frac{1}{2}-z}(b/a)$ $-a<b<a$ $(\tfrac{1}{2}\pi/a)^{\frac{1}{2}}\Gamma(z-\nu)\Gamma(z+\nu)$ $(b^2-a^2)^{\frac{1}{4}-\frac{1}{2}z} P_{\nu-\frac{1}{2}}^{\frac{1}{2}-z}(b/a)$ $b > a$

	$\phi(x)$	$\Phi(z) = \int_0^\infty \phi(x) x^{z-1} dx$
11.7	$e^{-bx^2} I_\nu(ax^2)$ Re $z > -$Re ν	$a^{-1} b^{\frac{1}{2}-\frac{1}{2}z} [\Gamma(1+\nu)]^{-1} \Gamma(\tfrac{1}{2}z+\tfrac{1}{2}\nu)$ $\cdot \exp(\tfrac{1}{8}a^2/b) M_{\frac{1}{2}-\frac{1}{2}z,\frac{1}{2}\nu}(\tfrac{1}{4}a^2/b)$
11.8	$e^{-bx^2} K_\nu(ax^2)$ Re $z > \pm$ Re ν	$\tfrac{1}{2} a^{-1} b^{\frac{1}{2}-\frac{1}{2}z} \Gamma(\tfrac{1}{2}z+\tfrac{1}{2}\nu) \Gamma(\tfrac{1}{2}z-\tfrac{1}{2}\nu)$ $\exp(\tfrac{1}{8}a^2/b) W_{\frac{1}{2}-\frac{1}{2}z,\frac{1}{2}\nu}(\tfrac{1}{4}a^2/b)$
11.9	$(1+x^2)^{-1} I_\nu(\frac{ax}{1+x^2})$ $-$Re $\nu<$Re $z<2+$Re ν	$a^{-1} [\Gamma(1+\nu)]^{-2} \Gamma(\tfrac{1}{2}z+\tfrac{1}{2}\nu) \Gamma(1-\tfrac{1}{2}z+\tfrac{1}{2}\nu)$ $\cdot M_{\frac{1}{2}z-\frac{1}{2},\frac{1}{2}\nu}(a) M_{\frac{1}{2}-\frac{1}{2}z,\frac{1}{2}\nu}(a)$
11.10	$\sin(ax) K_\nu(ax)$ Re $z > -1$	$2^{z-2} a^{-z} \Gamma(\tfrac{1}{2}+\tfrac{1}{2}z+\tfrac{1}{2}\nu) \Gamma(\tfrac{1}{2}+\tfrac{1}{2}z-\tfrac{1}{2}\nu)$ $\cdot {}_2F_1(\tfrac{1}{2}+\tfrac{1}{2}z+\tfrac{1}{2}\nu, \tfrac{1}{2}+\tfrac{1}{2}z-\tfrac{1}{2}\nu; \tfrac{3}{2}; -1)$
11.11	$\cos(ax) K_\nu(ax)$ Re $z > 0$	$2^{z-2} a^{-z} \Gamma(\tfrac{1}{2}z+\tfrac{1}{2}\nu) \Gamma(\tfrac{1}{2}z-\tfrac{1}{2}\nu)$ $\cdot {}_2F_1(\tfrac{1}{2}z+\tfrac{1}{2}\nu, \tfrac{1}{2}z-\tfrac{1}{2}\nu; \tfrac{1}{2}; -1)$
11.12	$\sin(bx) K_\nu(ax)$ Re $z>-1 \pm$ Re ν	$-\tfrac{1}{4}\pi \sec[\tfrac{1}{2}\pi(z-\nu)] \Gamma(\nu+z) (a^2+b^2)^{-\frac{1}{2}z}$ $\cdot \{P_{z-1}^{-\nu}[b(a^2+b^2)^{-\frac{1}{2}}] - P_{z-1}^{-\nu}[-b(a^2+b^2)^{-\frac{1}{2}}]\}$
11.13	$\cos(bx) K_\nu(ax)$	$\tfrac{1}{4}\pi \csc[\tfrac{1}{2}\pi(z-\nu)] \Gamma(\nu+z) (a^2+b^2)^{-\frac{1}{2}z}$ $\cdot \{P_{z-1}^{-\nu}[b(a^2+b^2)^{-\frac{1}{2}}] + P_{z-1}^{-\nu}[-b(a^2+b^2)^{-\frac{1}{2}}]\}$

	$\phi(x)$	$\Phi(z) = \int_0^\infty \phi(x)x^{z-1}dx$
11.14	$(a+x)^{-\nu}e^{-bx}$ $\cdot I_\nu[b(a+x)]$	$(2\pi ab)^{-\frac{1}{2}}a^{-\nu}(2b/a)^{-\frac{1}{2}z}\Gamma(z)$ $\cdot[\Gamma(1+2\nu-z)]^{-1}M_{-\frac{1}{2}z,\nu-\frac{1}{2}z}(2ab)$ $0 < \mathrm{Re}\, z < \frac{1}{2} + \mathrm{Re}\,\nu$
11.15	$(a+x)^{-\nu}e^{-bx}$ $\cdot K_\nu[b(a+x)]$	$(2ab/\pi)^{-\frac{1}{2}}a^{-\nu}(2b/a)^{-\frac{1}{2}z}\Gamma(z)$ $\cdot W_{-\frac{1}{2}z,\nu-\frac{1}{2}z}(2ab)$ $\mathrm{Re}\, z > 0$
11.16	$(a+x)^{-\nu}e^{bx}$ $\cdot K_\nu[b(a+x)]$	$(2ab/\pi)^{-\frac{1}{2}}a^{-\nu}\Gamma(z)[\Gamma(\frac{1}{2}+\nu)]^{-1}(2b/a)^{-\frac{1}{2}z}$ $\cdot\Gamma(\frac{1}{2}+\nu-z)W_{\frac{1}{2}z,\nu-\frac{1}{2}z}(2ab)$ $0 < \mathrm{Re}\, z < \frac{1}{2} + \mathrm{Re}\,\nu$
11.17	$e^{-x}\sin(ax^{\frac{1}{2}})K_\nu(x)$ $\mathrm{Re}\, z > \lvert\mathrm{Re}\,\nu\rvert - \frac{1}{2}$	$(\frac{1}{2}\pi)^{\frac{1}{2}}a2^{-z}\Gamma(\frac{1}{2}+\nu+z)\Gamma(\frac{1}{2}-\nu+z)[\Gamma(1+z)]^{-1}$ $\cdot{}_2F_2(\frac{1}{2}+\nu+z,\frac{1}{2}-\nu+z;3/2,1+z;-1/8\,a^2)$
11.18	$e^{-x}\cos(ax^{\frac{1}{2}})K_\nu(x)$ $\mathrm{Re}\, z > \lvert\mathrm{Re}\,\nu\rvert$	$\pi^{\frac{1}{2}}2^{-z}[\Gamma(\frac{1}{2}+z)]^{-1}\Gamma(z+\nu)\Gamma(z-\nu)$ $\cdot{}_2F_2(\nu+z,z-\nu;\frac{1}{2},\frac{1}{2}+z;-1/8\,a^2)$

	$\phi(x)$	$\Phi(z) = \int_0^\infty \phi(x)x^{z-1}dx$
11.19	$(b+x)^{-\rho}$ $e^{-ax}I_\nu(ax)$	$\pi^{-\frac{1}{2}}[\Gamma(\rho)]^{-1}b^{z-\rho}G^{22}_{23}\left(2ab\middle\vert\begin{matrix}1-z,\frac{1}{2}\\ \rho-z,\nu,-\nu\end{matrix}\right)$ $-\text{Re }\nu < \text{Re }z < \frac{1}{2}+\text{Re }\rho$
11.20	$(b+x)^{-\rho}e^{-ax}$ $\cdot K_\nu(ax)$	$\pi^{\frac{1}{2}}[\Gamma(\rho)]^{-1}b^{z-\rho}G^{31}_{23}\left(ab\middle\vert\begin{matrix}1-z,\frac{1}{2}\\ \rho-z,\nu,-\nu\end{matrix}\right)$ $\text{Re }z > \pm\text{ Re }\nu$
11.21	$(b+x)^{-\rho}$ $e^{ax}K_\nu(ax)$	$\pi^{-\frac{1}{2}}[\Gamma(\rho)]^{-1}\cos(\pi\nu)b^{z-\rho}G^{32}_{23}\left(2ab\middle\vert\begin{matrix}1-z,\frac{1}{2}\\ \rho-z,\nu,-\nu\end{matrix}\right)$ $\pm\text{ Re }\nu < \text{Re }z < \frac{1}{2}+\text{Re }\rho$
11.22	$(b+x)^{-\lambda}e^{-ax}$ $\cdot I_\nu[a(b+x)]$	$\pi^{-\frac{1}{2}}(2a)^\lambda\Gamma(z)b^ze^{ab}$ $\cdot G^{21}_{23}\left(2ab\middle\vert\begin{matrix}\frac{1}{2}-\lambda,0\\ -z,\nu-\lambda,-\nu-\lambda\end{matrix}\right)$ $0 < \text{Re }z < \frac{1}{2}+\text{Re }\lambda$
11.23	$(b+x)^{-\lambda}e^{-ax}$ $K_\nu[a(b+x)]$	$\pi^{\frac{1}{2}}(2a)^\lambda\Gamma(z)b^ze^{ab}$ $\cdot G^{30}_{23}\left(2ab\middle\vert\begin{matrix}0,\frac{1}{2}-\lambda\\ -z,\nu-\lambda,-\nu-\lambda\end{matrix}\right)$ $\text{Re }z > 0$

	$\phi(x)$	$\Phi(z) = \int_0^\infty \phi(x) x^{z-1} dx$
11.24	$(b+x)^{-\lambda} e^{ax}$ $K_\nu[a(b+x)]$ $0 < \text{Re } z <$	$\pi^{-\frac{1}{2}} (2a)^{\lambda} \cos(\pi\nu) b^z \Gamma(z) e^{-ab}$ $\cdot G^{31}_{23}\left(2ab \middle\vert \begin{matrix} \frac{1}{2}-\lambda, 0 \\ -z, \nu-\lambda, -\nu-\lambda \end{matrix}\right)$ $0 < \text{Re } z < \frac{1}{2} + \text{Re } \lambda$
11.25	$(b^2+x^2)^{\lambda}$ $\cdot K_\nu[a(b^2+x^2)^{\frac{1}{2}}]$ $\text{Re } z > 0$	$2^{-2\lambda-2} \Gamma(\frac{1}{2}z) a^{2\lambda} b^z$ $\cdot G^{30}_{13}\left(\frac{1}{4}a^2b^2 \middle\vert \begin{matrix} 0 \\ -z, \frac{1}{2}\nu-\lambda, -\frac{1}{2}\nu-\lambda \end{matrix}\right)$
11.26	$(b^2-x^2)^{\mu-1}$ $x < b$ $\cdot I_\nu(ax)$ 0 $x > b$ $\text{Re } \mu>0, \text{Re } z>-\text{Re } \nu$	$\frac{1}{2}(\frac{1}{2}a)^{\nu} \Gamma(\mu) b^{z+\nu+2\mu-2} \Gamma(\frac{1}{2}z+\frac{1}{2}\nu)$ $\cdot [\Gamma(1+\nu) \Gamma(\frac{1}{2}z+\frac{1}{2}\nu+\mu)]^{-1}$ $\cdot {}_1F_2(\frac{1}{2}z+\frac{1}{2}\nu; 1+\nu, \frac{1}{2}z+\frac{1}{2}\nu+\mu; \frac{1}{4}a^2b^2)$
11.27	$(b^2-x^2)^{\mu-1}$ $x < b$ $\cdot K_\nu(ax)$ 0 $x > b$ $\text{Re } \mu>0, \text{ Re } z>\pm\text{Re } \nu$	$\frac{1}{4}\Gamma(\mu) b^{z+2\mu-2} [f(\nu)+f(-\nu)]$ $f(\nu) = (\frac{1}{2}ab)^{-\nu} \Gamma(\nu) \Gamma(\frac{1}{2}z-\frac{1}{2}\nu) [\Gamma(\frac{1}{2}z+\mu-\frac{1}{2}\nu)]^{-1}$ $\cdot {}_1F_2(\frac{1}{2}z-\frac{1}{2}\nu; 1-\nu, \frac{1}{2}z-\frac{1}{2}\nu+\mu; \frac{1}{4}a^2b^2)$

	$\phi(x)$	$\Phi(z) = \int_0^\infty \phi(x) x^{z-1} dx$
11.28	$(b^2-x^2)^\lambda$ $\quad x < b$ $\cdot I_\nu[a(b^2-x^2)^{\frac{1}{2}}$ $0 \quad x > b$ $\mathrm{Re}\ \lambda > -1-\frac{1}{2}\mathrm{Re}\ \nu; \mathrm{Re}\ z > 0$	$\frac{1}{2}(\frac{1}{2}a)^\nu [\Gamma(1+\nu)\Gamma(1+\lambda+\frac{1}{2}\nu+\frac{1}{2}z)]^{-1}$ $\cdot \Gamma(\frac{1}{2}z)\Gamma(1+\lambda+\frac{1}{2}\nu) b^{2\lambda+z+\nu}$ $\cdot {}_1F_2(1+\lambda+\frac{1}{2}\nu; 1+\nu, 1+\lambda+\frac{1}{2}\nu+\frac{1}{2}z; \frac{1}{4}a^2b^2)$
11.29	$(b-x)^{\mu-1}$ $\quad x < b$ $\cdot e^{\pm ax} I_\nu(ax)$ $0 \quad x > b$	$(\frac{1}{2}a)^\nu [\Gamma(1+\nu)\Gamma(z+\mu+\nu)]^{-1} \Gamma(\mu)\Gamma(z+\mu)$ $\cdot b^{z+\mu+\nu-1} {}_2F_2(\frac{1}{2}+\nu, z+\nu; 2\nu+1, \mu+\nu+z; \pm 2ab)$ $\mathrm{Re}\ \mu > 0,\ \mathrm{Re}\ z > -\mathrm{Re}\ \nu$
11.30	$0 \quad x < b$ $(x-b)^\nu e^{-ax}$ $\cdot I_\nu[a(x-b)] \quad x > b$ $\mathrm{Re}\ \nu > -\frac{1}{2}$	$\pi^{-\frac{1}{2}} [\Gamma(1-z)]^{-1} \Gamma(\frac{1}{2}+\nu)\Gamma(\frac{1}{2}-\nu-z)$ $\cdot (2a)^{-\frac{1}{2}-\frac{1}{2}z} b^{-\frac{1}{2}+\nu+\frac{1}{2}z} W_{\frac{1}{2}z, \nu+\frac{1}{2}z}(2ab)$ $\mathrm{Re}\ z < \frac{1}{2}-\mathrm{Re}\ \nu$
11.31	$(b-x)^{\lambda-1} e^{\pm ax}$ $\cdot I_\nu[a(b-x)] \quad x < b$ $0 \quad x > b$ $\mathrm{Re}(\lambda+\nu) > 0$	$(\frac{1}{2}a)^\nu \Gamma(\lambda+\nu) [\Gamma(1+\nu)\Gamma(\lambda+\nu+z)]^{-1}$ $\cdot b^{\lambda+\nu-1+z} {}_2F_2(\frac{1}{2}+\nu, \lambda+\nu; 2\nu+1, \lambda+\nu+z, \pm 2ab)$ $\mathrm{Re}\ z > 0$

	$\phi(x)$	$\Phi(z) = \int_0^\infty \phi(x)x^{z-1}dx$
11.32	$0 \qquad x < b$ $(x-b)^{\mu-1} \cdot e^{-ax} I_\nu(ax) \qquad x > b$ $\mathrm{Re}\ \mu > 0$	$\pi^{-\frac{1}{2}} b^\mu \Gamma(\mu)(2a)^{1-z} \cdot G^{21}_{23}\left(2ab \middle\vert \begin{matrix} z-\frac{1}{2}, 0 \\ -\mu, \nu-1+z, -\nu-1+z \end{matrix}\right)$ $\mathrm{Re}\ z > \frac{3}{2} - \mathrm{Re}\ \mu$
11.33	$0 \qquad x < b$ $(x-b)^{\mu-1} \cdot e^{-ax} K_\nu(ax) \qquad x > b$ $\mathrm{Re}\ \mu > 0$	$\pi^{\frac{1}{2}} b^\mu \Gamma(\mu)(2a)^{1-z} \cdot G^{30}_{23}\left(2ab \middle\vert \begin{matrix} 0, z-\frac{1}{2} \\ -\mu, \nu-1+z, -\nu-1+z \end{matrix}\right)$
11.34	$0 \qquad x < b$ $(x-b)^{\mu-1} \cdot e^{ax} K_\nu(ax)$ $\mathrm{Re}\ \mu > 0$	$\pi^{-\frac{1}{2}} b^\mu \cos(\pi\nu)\Gamma(\mu)(2a)^{1-z} \cdot G^{31}_{23}\left(2ab \middle\vert \begin{matrix} -\frac{1}{2}+z, 0 \\ -\mu, \nu-1+z, -\nu-1+z \end{matrix}\right)$ $\mathrm{Re}\ z < \frac{3}{2} - \mathrm{Re}\ \mu$
11.35	$0 \qquad x < b$ $(x-b)^{\mu-1} \cdot K_\nu(ax^{\frac{1}{2}}) \qquad x > b$ $\mathrm{Re}\ \mu > 0$	$\frac{1}{2}\Gamma(\mu) b^\mu (\frac{1}{2}a)^{2-2z} \cdot G^{30}_{13}\left(\frac{1}{4}a^2 b \middle\vert \begin{matrix} 0 \\ -\mu, \frac{1}{2}\nu-1+z, z-1-\frac{1}{2}\nu \end{matrix}\right)$

	$\phi(x)$	$\Phi(z) = \int_0^\infty \phi(x) x^{z-1} dx$
11.36	$0 \quad x < b$ $(x-b)^\lambda e^{-ax} \cdot I_\nu[a(x-b)] \quad x > b$ $\mathrm{Re}(\lambda+\nu) > -1$	$\pi^{-\frac{1}{2}}[\Gamma(1-z)]^{-1} b^{\lambda+z} e^{-ab} \cdot G^{22}_{23}\left(2ab \middle\vert \begin{matrix} -\lambda, \frac{1}{2} \\ -z-\lambda, \nu, -\nu \end{matrix}\right)$ $\mathrm{Re}\, z < \frac{1}{2} - \mathrm{Re}\, \lambda$
11.37	$0 \quad x < b$ $(x-b)^\lambda e^{-ax} K_\nu[a(x-b)] \quad x > b$ $\mathrm{Re}(\lambda\pm\nu) > -1$	$\pi^{\frac{1}{2}}[\Gamma(1-z)]^{-1} b^{\lambda+z} e^{-ab} \cdot G^{31}_{23}\left(2ab \middle\vert \begin{matrix} -\lambda, \frac{1}{2} \\ -z-\lambda, \nu, -\nu \end{matrix}\right)$
11.38	$0 \quad x < b$ $(x-b)^\lambda e^{ax} \cdot K_\nu[a(x-b)] \quad x > b$ $\mathrm{Re}(\lambda\pm\nu) > -1$	$\pi^{-\frac{1}{2}}[\Gamma(1-z)]^{-1} \cos(\pi\nu) e^{ab} b^{\lambda+z} \cdot G^{22}_{23}\left(2ab \middle\vert \begin{matrix} -\lambda, \frac{1}{2} \\ -z-\lambda, \nu, -\nu \end{matrix}\right)$ $\mathrm{Re}\, z > \frac{1}{2} - \mathrm{Re}\, \lambda$
11.39	$0 \quad x < b$ $(x-b)^\lambda \cdot K_\nu[a(x-b)^{\frac{1}{2}}] \quad x > b$ $\mathrm{Re}(\lambda\pm\nu) > -1$	$\frac{1}{2}[\Gamma(1-z)]^{-1} b^{\lambda+z} \cdot G^{31}_{13}\left(\frac{1}{4}a^2 b \middle\vert \begin{matrix} -\lambda \\ -z-\lambda, \frac{1}{2}\nu, -\frac{1}{2}\nu \end{matrix}\right)$

	$\phi(x)$	$\Phi(z) = \int_0^\infty \phi(x) x^{z-1} dx$
11.40	$J_\nu(ax)K_\nu(ax)$	$2^{z-3}a^{-z}\Gamma(\frac{1}{2}z)\Gamma(\frac{1}{2}\nu+\frac{1}{4}z)[\Gamma(1-\frac{1}{4}z+\frac{1}{2}\nu)]^{-1}$ Re z $\begin{cases}0\\-2\ \mathrm{Re}\ \nu\end{cases}$
11.41	$J_\nu(bx)K_\nu(ax)$ Re $z > \begin{cases}0\\-2\ \mathrm{Re}\ \nu\end{cases}$	$2^{z-2}\Gamma(\frac{1}{2}z)\Gamma(\nu+\frac{1}{2}z)(a^2+b^2)^{-\frac{1}{2}z}$ $\cdot P_{-\frac{1}{2}z}^{-\nu}(\frac{a^2-b^2}{a^2+b^2})$
11.42	$Y_\nu(ax)K_\nu(ax)$ Re $z > \begin{cases}0\\\pm 2\ \mathrm{Re}\ \nu\end{cases}$	$-\pi^{-1}2^{z-3}a^{-z}\Gamma(\frac{1}{2}z)\Gamma(\frac{1}{4}z+\frac{1}{2}\nu)$ $\cdot\Gamma(\frac{1}{4}z-\frac{1}{2}\nu)\cos[\frac{1}{2}\pi(\frac{1}{2}z-\nu)]$
11.43	$I_\nu(ax)K_\nu(ax)$ $\left.\begin{matrix}0\\-2\ \mathrm{Re}\ \nu\end{matrix}\right\} <\mathrm{Re}\ z<1$	$\frac{1}{4}\pi^{-\frac{1}{2}}a^{-z}\Gamma(\frac{1}{2}z)\Gamma(\frac{1}{2}-\frac{1}{2}z)\Gamma(\nu+\frac{1}{2}z)$ $\cdot[\Gamma(1-\frac{1}{2}z+\nu)]^{-1}$
11.44	$K_\mu(ax)I_\nu(ax)$ $\mathrm{Re}(-\nu\pm\mu) < \mathrm{Re}\ z < 1$	$2^{z-2}a^{-z}\Gamma(\frac{1}{2}z+\frac{1}{2}\mu+\frac{1}{2}\nu)\ [\Gamma(1-\frac{1}{2}z+\frac{1}{2}\nu+\frac{1}{2}\mu)]^{-1}$ $\cdot B(1-z,\frac{1}{2}z-\frac{1}{2}\mu+\frac{1}{2}\nu)$
11.45	$K_\nu^2(ax)$ Re $z > \begin{cases}0\\\pm 2\ \mathrm{Re}\ \nu\end{cases}$	$\frac{1}{4}\pi^{\frac{1}{2}}a^{-z}\Gamma(\frac{1}{2}z+\nu)\Gamma(\frac{1}{2}z-\nu)\Gamma(\frac{1}{2}z)$ $\cdot[\Gamma(\frac{1}{2}+\frac{1}{2}z)]^{-1}$
11.46	$K_\mu(ax)K_\nu(ax)$ Re $z > \mathrm{Re}(\pm\mu\pm\nu)$	$a^{-z}2^{z-3}[\Gamma(z)]^{-1}\Gamma(\frac{1}{2}\nu+\frac{1}{2}\mu+\frac{1}{2}z)$ $\cdot\Gamma(\frac{1}{2}\nu-\frac{1}{2}\mu+\frac{1}{2}z)\Gamma(\frac{1}{2}\mu-\frac{1}{2}\nu+\frac{1}{2}z)$ $\cdot\Gamma(\frac{1}{2}z-\frac{1}{2}\nu-\frac{1}{2}\mu)$

	$\phi(x)$	$\Phi(z) = \int_0^\infty \phi(x) x^{z-1} dx$
11.47	$I_\nu(bx) K_\nu(ax)$ $a>b$, $\operatorname{Re} z > \begin{cases} 0 \\ -2\operatorname{Re}\nu \end{cases}$	$2^{z-2}\Gamma(\nu+\frac{1}{2}z)\Gamma(\frac{1}{2}z)(a^2-b^2)^{-\frac{1}{2}z}$ $\cdot P_{-\frac{1}{2}z}^{-\nu}(\frac{a^2+b^2}{a^2-b^2})$ $=(\pi ab)^{-\frac{1}{2}} 2^{z-2}(a^2-b^2)^{\frac{1}{2}-\frac{1}{2}z}\Gamma(\frac{1}{2}z)$ $\cdot e^{-i\frac{1}{2}\pi(z-1)} q_{\nu-\frac{1}{2}}^{\frac{1}{2}z-\frac{1}{2}}(\frac{a^2+b^2}{2ab})$
11.48	$K_\nu(ax) K_\nu(bx)$ $\operatorname{Re} z > \begin{cases} 0 \\ \pm 2 \operatorname{Re}\nu \end{cases}$	$\pi^{\frac{1}{2}} 2^{z-3}(ab)^{-\frac{1}{2}}\Gamma(\frac{1}{2}z)\Gamma(\nu+\frac{1}{2}z)$ $\cdot\Gamma(-\nu+\frac{1}{2}z)(\|a^2-b^2\|)^{\frac{1}{2}-\frac{1}{2}z} P_{\nu-\frac{1}{2}}^{\frac{1}{2}-\frac{1}{2}z}(\frac{a^2+b^2}{2ab})$ $= 2^{z-2}\Gamma(\frac{1}{2}z)\Gamma(\nu+\frac{1}{2}z)$ $(\|a^2-b^2\|)^{-\frac{1}{2}z} e^{i\pi\nu} q_{\frac{1}{2}z-1}^{-\nu}(\frac{a^2+b^2}{\|a^2-b^2\|})$
11.49	$K_\mu(ax) J_\nu(bx)$ $\operatorname{Re}(a\pm ib) > 0$ $\operatorname{Re} z > \operatorname{Re}(-\nu\pm\mu)$	$[\Gamma(1+\nu)]^{-1} 2^{z-2} a^{-\nu-z} b^\nu \Gamma(\frac{1}{2}\nu+\frac{1}{2}\mu+\frac{1}{2}z)$ $\cdot\Gamma(\frac{1}{2}\nu-\frac{1}{2}\mu+\frac{1}{2}z)\, {}_2F_1(\frac{1}{2}\nu+\frac{1}{2}\mu+\frac{1}{2}z,$ $\frac{1}{2}\nu-\frac{1}{2}\mu+\frac{1}{2}z;\nu+1;-b^2/a^2)$
11.50	$K_\mu(ax) I_\nu(bx)$ $b < a$, $\operatorname{Re} z > \operatorname{Re}(-\nu\pm\mu)$	$[\Gamma(1+\nu)]^{-1} 2^{z-2} a^{-\nu-z} b^\nu \Gamma(\frac{1}{2}\nu+\frac{1}{2}\mu+\frac{1}{2}z)$ $\cdot\Gamma(\frac{1}{2}\nu-\frac{1}{2}\mu+\frac{1}{2}z)\, {}_2F_1(\frac{1}{2}\nu+\frac{1}{2}\mu+\frac{1}{2}z,$ $\frac{1}{2}\nu-\frac{1}{2}\mu+\frac{1}{2}z;\nu+1;b^2/a^2)$

	$\phi(x)$	$\Phi(z) = \int_0^\infty \phi(x) x^{z-1} dx$
11.51	$K_\mu(ax) K_\nu(bx)$ $\mathrm{Re}(a+b) > 0$ $\mathrm{Re}\, z > \mathrm{Re}(\pm\mu\pm\nu)$	$b^\nu [\Gamma(z)]^{-1} 2^{z-3} a^{-\nu-z}$ $\cdot\Gamma(\tfrac{1}{2}\nu+\tfrac{1}{2}\mu+\tfrac{1}{2}z)\Gamma(\tfrac{1}{2}\nu-\tfrac{1}{2}\mu+\tfrac{1}{2}z)$ $\cdot\Gamma(\tfrac{1}{2}\mu-\tfrac{1}{2}\nu+\tfrac{1}{2}z)\Gamma(-\tfrac{1}{2}\mu-\tfrac{1}{2}\nu+\tfrac{1}{2}z)$ $\cdot {}_2F_1(\tfrac{1}{2}\nu+\tfrac{1}{2}\mu+\tfrac{1}{2}z, \tfrac{1}{2}\nu-\tfrac{1}{2}\mu+\tfrac{1}{2}z; z; 1-b^2/a^2)$
11.52	$[I_\nu(ax) + I_{-\nu}(ax)] K_\nu(ax)$ $\pm 2\, \mathrm{Re}\, \nu < \mathrm{Re}\, z < 1$	$2^{z-2} a^{-z} \cos(\pi\nu) \Gamma(1-z)$ $\cdot\Gamma(\tfrac{1}{2}z+\nu)\Gamma(\tfrac{1}{2}z-\nu)$ $\cdot[\Gamma(1-\nu-\tfrac{1}{2}z)\Gamma(1+\nu-\tfrac{1}{2}z)]^{-1}$
11.53	$J_\nu(ax) K_\mu(bx) K_\rho(cx)$	Bailey, W. N., 1936: Proc. London Math. Soc. 40, 37-48. Journal London Math. Soc. 11, 16-20.

	$\phi(x)$	$\Phi(z) = \int_0^\infty \phi(x) x^{z-1} dx$
11.54	$(a^2-x^2)^{\frac{1}{2}\nu}$ $\cdot I_\nu[b(a^2-x^2)^{\frac{1}{2}}]$ $x < a$ 0 $x > a$ $\mathrm{Re}\ \nu > -1$	$2^{\frac{1}{2}z-1}\Gamma(\frac{1}{2}z) a^{\nu+\frac{1}{2}z} b^{-\frac{1}{2}z} I_{\nu+\frac{1}{2}z}(ab)$ $\mathrm{Re}\ z > 0$
11.55	$(b^2+x^2)^{-\frac{1}{2}\nu}$ $\cdot K_\nu[a(b^2+x^2)^{\frac{1}{2}}]$	$\frac{1}{2}b^{-\nu}(2b/a)^{\frac{1}{2}z}\Gamma(\frac{1}{2}z) K_{\nu-\frac{1}{2}z}(ab)$ $\mathrm{Re}\ z > 0$
11.56	$(b^2+x^2)^{\frac{1}{2}\nu}$ $\cdot K_\nu[a(b^2+x^2)^{\frac{1}{2}}]$	$\frac{1}{2}b^{\nu}(2b/a)^{\frac{1}{2}z}\Gamma(\frac{1}{2}z) K_{\nu+\frac{1}{2}z}(ab)$ $\mathrm{Re}\ z > 0$
11.57	$(a^2-x^2)^{-\frac{1}{2}\nu}$ $\cdot K_\nu[b(a^2-x^2)^{\frac{1}{2}}]$ $x < a$ 0 $x > a$ $\mathrm{Re}\ z > \mathrm{Max}(0, \mathrm{Re}\ \nu)$	$2^{-\nu-1} a^z b^{\nu+1} z^{-1} \Gamma(-\nu)$ $\cdot {}_1F_2(1;\nu+1,1+\frac{1}{2}z;\frac{1}{4}a^2b^2)$ $+\pi 2^{\frac{1}{2}z-2} a^{\frac{1}{2}z-\nu} b^{-\frac{1}{2}z} \csc(\pi\nu)$ $\cdot\Gamma(\frac{1}{2}z) I_{\frac{1}{2}z-\nu}(ab)$
11.58	0 $x < a$ $(x^2-a^2)^{\frac{1}{2}\nu} K_\nu[b(x^2-a^2)^{\frac{1}{2}}]$ $x > a$	$(2b)^{\nu}\Gamma(1+\nu)(b/a)^{\frac{1}{2}z} S_{\frac{1}{2}z-1-\nu,\frac{1}{2}z+\nu}(ab)$ $\mathrm{Re}\ \nu > -1$

	$\phi(x)$	$\Phi(z) = \int_0^\infty \phi(x) x^{z-1} dx$
11.59	$J_\nu(v) K_\nu(u)$ $\begin{matrix} u \\ v \end{matrix} = (2b)^{\frac{1}{2}}[(a^2+x^2)^{\frac{1}{2}} \pm a]^{\frac{1}{2}}$	$2^{z-1}\Gamma(\frac{1}{2}\nu+\frac{1}{2}z)[\Gamma(1+\frac{1}{2}\nu-\frac{1}{2}z)]^{-1}$ $\cdot (a/b)^{\frac{1}{2}z} K_z[2(ab)^{\frac{1}{2}}]$ $\operatorname{Re} z > -\operatorname{Re} \nu$
11.60	$Y_\nu(v) K_\nu(u)$ $\begin{matrix} u \\ v \end{matrix} = (2b)^{\frac{1}{2}}[(a^2+x^2)^{\frac{1}{2}} \pm a]^{\frac{1}{2}}$	$-\pi^{-1} 2^{z-1} \cos[\frac{1}{2}\pi(\nu-z)]\Gamma(\frac{1}{2}z+\frac{1}{2}\nu)$ $\cdot \Gamma(\frac{1}{2}z-\frac{1}{2}\nu)(a/b)^{\frac{1}{2}z} K_z[2(ab)^{\frac{1}{2}}]$ $\operatorname{Re} z > \pm \operatorname{Re} \nu$
11.61	$I_\nu(v) K_\nu(u)$ $\begin{matrix} u \\ v \end{matrix} = b[(a^2+x^2)^{\frac{1}{2}} \pm a]$	$\frac{1}{2}\pi^{-\frac{1}{2}}\Gamma(\nu+\frac{1}{2}z)\Gamma(\frac{1}{2}-\frac{1}{2}z)[\Gamma(1+\nu-\frac{1}{2}z)]^{-1}$ $\cdot (a/b)^{\frac{1}{2}z} K_{\frac{1}{2}z}(2ab)$ $-\operatorname{Re} \nu < \operatorname{Re} z < 1$
11.62	$K_\nu(v) K_\nu(u)$ $\begin{matrix} u \\ v \end{matrix} = b[(a^2+x^2)^{\frac{1}{2}} \pm a]$	$\frac{1}{2}\pi^{\frac{1}{2}}\Gamma(\frac{1}{2}z-\nu)\Gamma(\frac{1}{2}z+\nu)[\Gamma(\frac{1}{2}+\frac{1}{2}z)]^{-1}$ $\cdot (a/b)^{\frac{1}{2}z} K_{\frac{1}{2}z}(2ab)$ $\operatorname{Re} z > \pm 2 \operatorname{Re} \nu$

	$\phi(x)$	$\Phi(z) = \int_0^\infty \phi(x) x^{z-1} dx$
11.63	$K_o(a\|x-x^{-1}\|)$ Principal value	$\frac{1}{4}\pi^2 [J^2_{\frac{1}{2}z}(a)+Y^2_{\frac{1}{2}z}(a)]$
11.64	$K_\nu[a(x^{-1}-x)]$ $x<1$ 0 $x>1$ $-1 < \operatorname{Re}\nu < 1$	$\frac{1}{4}\pi^2 \csc(\pi\nu)[J_{\frac{1}{2}\nu+\frac{1}{2}z}(a)Y_{-\frac{1}{2}\nu+\frac{1}{2}z}(a)$ $-J_{-\frac{1}{2}\nu+\frac{1}{2}z}(a)Y_{\frac{1}{2}\nu+\frac{1}{2}z}(a)]$
11.65	0 $x<1$ $K_\nu[a(x-x^{-1})]$ $x>1$ $-1 < \operatorname{Re}\nu < 1$	$\frac{1}{4}\pi^2 \csc(\pi\nu)[J_{\frac{1}{2}\nu-\frac{1}{2}z}(a)Y_{-\frac{1}{2}\nu-\frac{1}{2}z}(a)$ $-J_{-\frac{1}{2}\nu-\frac{1}{2}z}(a)Y_{\frac{1}{2}\nu-\frac{1}{2}z}(a)]$
11.66	$K_{2\nu}[a(x+x^{-1})]$	$K_{\nu+\frac{1}{2}z}(a)K_{\nu-\frac{1}{2}z}(a)$
11.67	$(a+bx)^\nu (b+ax)^{-\nu}$ $\cdot K_{2\nu}\{[a^2+b^2+ab(x+x^{-1})]^{\frac{1}{2}}\}$	$2K_{\nu+z}(a)K_{\nu-z}(b)$
11.68	$(1+x^2)^{-1} I_\nu(\frac{2ax}{1+x^2})$ $\cdot\exp[-b(\frac{1-x^2}{1+x^2})]$ $-\operatorname{Re}\nu < \operatorname{Re} z < 2+\operatorname{Re}\nu$	$\frac{1}{2}a^{-1}[\Gamma(1+\nu)]^{-2}\Gamma(1+\frac{1}{2}\nu-\frac{1}{2}z)\Gamma(\frac{1}{2}\nu+\frac{1}{2}z)$ $\cdot M_{\frac{1}{2}-\frac{1}{2}z,\frac{1}{2}\nu}[(b^2+a^2)^{\frac{1}{2}}+b]$ $\cdot M_{\frac{1}{2}z-\frac{1}{2},\frac{1}{2}\nu}[(b^2+a^2)^{\frac{1}{2}}-b]$

	$\phi(x)$	$\Phi(z) = \int_0^\infty \phi(x)x^{z-1}dx$
11.69	$(1+x^2)^{-1}I_\nu(\frac{2ax}{1+x^2})$ $\cdot\exp[b(\frac{1-x^2}{1+x^2})]$ $-\mathrm{Re}\ \nu<\mathrm{Re}\ z<2+\mathrm{Re}\ \nu$	$\frac{1}{2}a^{-1}[\Gamma(1+\nu)]^{-2}\Gamma(\frac{1}{2}\nu+\frac{1}{2}z)\Gamma(1+\frac{1}{2}\nu-\frac{1}{2}z)$ $\cdot M_{\frac{1}{2}z-\frac{1}{2},\frac{1}{2}\nu}[(a^2+b^2)^{\frac{1}{2}}+b]$ $\cdot M_{\frac{1}{2}-\frac{1}{2}z,\frac{1}{2}\nu}[(a^2+b^2)^{\frac{1}{2}}-b]$
11.70	$(1+x^2)^{-1}I_\nu(\frac{2ax}{1+x^2})$ $\cdot\exp(-\frac{2bx^2}{1+x^2})$ $-\mathrm{Re}\ \nu<\mathrm{Re}\ z<2+\mathrm{Re}\ \nu$	$\frac{1}{2}a^{-1}[\Gamma(1+\nu)]^{-2}e^{-b}\Gamma(\frac{1}{2}\nu+\frac{1}{2}z)\Gamma(1+\frac{1}{2}\nu-\frac{1}{2}z)$ $\cdot M_{\frac{1}{2}z-\frac{1}{2},\frac{1}{2}\nu}[(a^2+b^2)^{\frac{1}{2}}+b]$ $\cdot M_{\frac{1}{2}-\frac{1}{2}z,\frac{1}{2}\nu}[(a^2+b^2)^{\frac{1}{2}}-b]$
11.71	$(1-x^2)^{-1}I_\nu(\frac{2ax}{1-x^2})$ $\cdot\exp[-b(\frac{1+x^2}{1-x^2})]$, $x<1$ $0 \quad x>1$ $b>a$, $\mathrm{Re}\ z>-\mathrm{Re}\ \nu$	$\frac{1}{2}a^{-1}[\Gamma(1+\nu)]^{-1}\Gamma(\frac{1}{2}\nu+\frac{1}{2}z)$ $\cdot M_{\frac{1}{2}-\frac{1}{2}z,\frac{1}{2}\nu}[b-(b^2-a^2)^{\frac{1}{2}}]$ $\cdot W_{\frac{1}{2}-\frac{1}{2}z,\frac{1}{2}\nu}[b+(b^2-a^2)^{\frac{1}{2}}]$
11.72	$(1-x^2)^{-1}I_\nu(\frac{2ax}{1-x^2})$ $\cdot\exp(-\frac{2bx^2}{1-x^2})$, $x<1$ $0 \quad x>1$ $b>a$, $\mathrm{Re}\ z>-\mathrm{Re}\ \nu$	$\frac{1}{2}a^{-1}[\Gamma(1+\nu)]^{-1}e^{b}\Gamma(\frac{1}{2}\nu+\frac{1}{2}z)$ $\cdot M_{\frac{1}{2}-\frac{1}{2}z,\frac{1}{2}\nu}[b-(b^2-a^2)^{\frac{1}{2}}]$ $\cdot W_{\frac{1}{2}-\frac{1}{2}z,\frac{1}{2}\nu}[b+(b^2-a^2)^{\frac{1}{2}}]$

	$\phi(x)$	$\Phi(z) = \int_0^\infty \phi(x) x^{z-1} dx$
11.73	$(1+x^2)^{-1} I_\nu(\frac{2ax}{1+x^2})$ $\cdot \exp(-\frac{2b}{1+x^2})$ $-\text{Re}\ \nu < \text{Re}\ z < 2+\text{Re}\ \nu$	$\frac{1}{2}a^{-1}[\Gamma(1+\nu)]^{-2} e^{-b} \Gamma(1+\frac{1}{2}\nu-\frac{1}{2}z)\Gamma(\frac{1}{2}\nu+\frac{1}{2}z)$ $\cdot M_{\frac{1}{2}-\frac{1}{2}z,\frac{1}{2}\nu}[(b^2+a^2)^{\frac{1}{2}}+b]$ $\cdot M_{\frac{1}{2}z-\frac{1}{2},\frac{1}{2}\nu}[(b^2+a^2)^{\frac{1}{2}}-b]$
11.74	$0 \qquad x < 1$ $(x^2-1)^{-1} I_\nu(\frac{2ax}{x^2-1}$ $\cdot \exp[-b(\frac{x^2+1}{x^2-1})], \quad x > 1$ $b > a,\ \text{Re}\ z < 2+\text{Re}\ \nu$	$\frac{1}{2}a^{-1}[\Gamma(1+\nu)]^{-1}\Gamma(1+\frac{1}{2}\nu-\frac{1}{2}z)$ $\cdot M_{\frac{1}{2}z-\frac{1}{2},\frac{1}{2}\nu}[b-(b^2-a^2)^{\frac{1}{2}}]$ $\cdot W_{\frac{1}{2}z-\frac{1}{2},\frac{1}{2}\nu}[b+(b^2-a^2)^{\frac{1}{2}}]$
11.75	$0 \qquad x < 1$ $(x^2-1)^{-1} I_\nu(\frac{2ax}{x^2-1})$ $\cdot \exp(-\frac{2b}{x^2-1}), \qquad x > 1$ $\text{Re}\ z < 2+\text{Re}\ \nu, \qquad b > a$	$\frac{1}{2}a^{-1}[\Gamma(1+\nu)]^{-1}\Gamma(1+\frac{1}{2}\nu-\frac{1}{2}z)e^{b}$ $\cdot M_{\frac{1}{2}z-\frac{1}{2},\frac{1}{2}\nu}[b-(b^2-a^2)^{\frac{1}{2}}]$ $\cdot W_{\frac{1}{2}z-\frac{1}{2},\frac{1}{2}\nu}[b+(b^2-a^2)^{\frac{1}{2}}]$
11.76	$0 \quad x < 1$ $(x^2-1)^{-1} K_\nu(\frac{2ax}{x^2-1})$ $\cdot \exp[-b(\frac{x^2+1}{x^2-1})], \quad x > 1$ $\text{Re}\ z < 2 \pm \text{Re}\ \nu$	$\frac{1}{4}a^{-1}\Gamma(1-\frac{1}{2}\nu-\frac{1}{2}z)\Gamma(1+\frac{1}{2}\nu-\frac{1}{2}z)$ $\cdot W_{\frac{1}{2}z-\frac{1}{2},\frac{1}{2}\nu}[b-(b^2-a^2)^{\frac{1}{2}}]$ $\cdot W_{\frac{1}{2}z-\frac{1}{2},\frac{1}{2}\nu}[b+(b^2-a^2)^{\frac{1}{2}}]$

	$\phi(x)$	$\Phi(z) = \int_0^\infty \phi(x) x^{z-1} dx$
11.77	$(1-x^2)^{-1} K_\nu\left(\frac{2ax}{1-x^2}\right)$ $\cdot \exp\left[-b\left(\frac{1+x^2}{1-x^2}\right)\right]$, $x < 1$ 0 $x > 1$ $\text{Re } z > \pm \text{Re } \nu$	$\frac{1}{4} a^{-1} \Gamma(\frac{1}{2}z-\frac{1}{2}\nu) \Gamma(\frac{1}{2}z+\frac{1}{2}\nu)$ $\cdot W_{\frac{1}{2}-\frac{1}{2}z, \frac{1}{2}\nu}[b-(b^2-a^2)^{\frac{1}{2}}]$ $\cdot W_{\frac{1}{2}-\frac{1}{2}z, \frac{1}{2}\nu}[b+(b^2-a^2)^{\frac{1}{2}}]$
11.78	$(1-x^2)^{-1} K_\nu\left(\frac{2ax}{1-x^2}\right)$ $\cdot \exp\left(-\frac{2bx^2}{1-x^2}\right)$, $x < 1$ 0 $x > 1$	$\frac{1}{4} a^{-1} e^b \Gamma(\frac{1}{2}z-\frac{1}{2}\nu) \Gamma(\frac{1}{2}z+\frac{1}{2}\nu)$ $\cdot W_{\frac{1}{2}-\frac{1}{2}z, \frac{1}{2}\nu}[b-(b^2-a^2)^{\frac{1}{2}}]$ $\cdot W_{\frac{1}{2}-\frac{1}{2}z, \frac{1}{2}\nu}[b+(b^2-a^2)^{\frac{1}{2}}]$ $\text{Re } z > \pm \text{Re } \nu$

	$\phi(x)$	$\Phi(z) = \int_0^\infty \phi(x) x^{z-1} dx$
11.79	$0 \qquad x < 1$ $(x^2-1)^{-1} K_\nu\left(\frac{2ax}{x^2-1}\right)$ $\cdot \exp\left(-\frac{2b}{x^2-1}\right), \quad x > 1$ $\mathrm{Re}\, z < 2 \pm \mathrm{Re}\, \nu$	$\frac{1}{4} a^{-1} e^{b} \Gamma(1-\frac{1}{2}\nu-\frac{1}{2}z) \Gamma(1+\frac{1}{2}\nu-\frac{1}{2}z)$ $\cdot W_{\frac{1}{2}z-\frac{1}{2}, \frac{1}{2}\nu}[b-(b^2-a^2)^{\frac{1}{2}}]$ $\cdot W_{\frac{1}{2}z-\frac{1}{2}, \frac{1}{2}\nu}[b+(b^2-a^2)^{\frac{1}{2}}]$
11.80	$K_\mu(ax) J_\nu(b/x)$	$\frac{1}{4}(\frac{1}{2}a)^{-z} G^{30}_{04}\left(\frac{1}{16} a^2 b^2 \middle\vert \frac{1}{2}\nu, \frac{1}{2}z+\frac{1}{2}\mu, \frac{1}{2}z-\frac{1}{2}\mu, -\frac{1}{2}\nu\right)$ $\mathrm{Re}\, z > \lvert \mathrm{Re}\, \mu \rvert - 3/2$
11.81	$K_\mu(ax) Y_\nu(b/x)$ m integer,	$\frac{1}{4}(-1)^{m+1} (\frac{1}{2}a)^{-z}$ $\cdot G^{40}_{15}\left(\frac{1}{16} a^2 b^2 \middle\vert \begin{matrix} \frac{1}{2}-\frac{1}{2}\nu-m \\ \frac{1}{2}\nu, -\frac{1}{2}\nu, \frac{1}{2}z+\frac{1}{2}\mu, \frac{1}{2}z-\frac{1}{2}\mu, \frac{1}{2}-\frac{1}{2}\nu, m \end{matrix}\right)$ $\mathrm{Re}\, z > -3/2 + \lvert \mathrm{Re}\, \mu \rvert$
11.82	$K_\mu(ax) K_\nu(b/x)$	$\frac{1}{8}(\frac{1}{2}a)^{-z} G^{40}_{04}\left(\frac{1}{16} a^2 b^2 \middle\vert \frac{1}{2}\nu, -\frac{1}{2}\nu, \frac{1}{2}z+\frac{1}{2}\mu, \frac{1}{2}z-\frac{1}{2}\mu\right)$

1.12 Functions Related to Bessel Function

	$\phi(x)$	$\Phi(z) = \int_0^\infty \phi(x)x^{z-1}dx$
12.1	$\mathbf{H}_\nu(ax)$ $-1-\mathrm{Re}\,\nu < \mathrm{Re}\,z < \begin{cases} 3/2 \\ 1-\mathrm{Re}\,\nu \end{cases}$	$2^{z-1}a^{-z}\Gamma(\tfrac{1}{2}z+\tfrac{1}{2}\nu)\,[\Gamma(1+\tfrac{1}{2}\nu-\tfrac{1}{2}z)]^{-1}$ $\cdot\tan[\tfrac{1}{2}\pi(z+\nu)]$
12.2	$\mathbf{H}_\nu(ax)-Y_\nu(ax)$ $\pm\mathrm{Re}\,\nu < \mathrm{Re}\,z < 1-\mathrm{Re}\,\nu$	$2^{z}a^{-z}\Gamma(\tfrac{1}{2}z-\tfrac{1}{2}\nu)\,[\Gamma(1-\tfrac{1}{2}z-\tfrac{1}{2}\nu)]^{-1}$ $\cos(\pi\nu)\csc[\pi(\nu+z)]$
12.3	$I_\nu(ax)-\mathbf{L}_\nu(ax)$ $-\mathrm{Re}\,\nu < \mathrm{Re}\,z < 1-\mathrm{Re}\,\nu$	$2^{z-1}a^{-z}\sec[\tfrac{1}{2}\pi(\nu+z)]$ $\cdot\Gamma(\tfrac{1}{2}z+\tfrac{1}{2}\nu)\,[\Gamma(1+\tfrac{1}{2}\nu-\tfrac{1}{2}z)]^{-1}$
12.4	$I_\nu(ax)-\mathbf{L}_{-\nu}(ax)$ $-\mathrm{Re}\,\nu < \mathrm{Re}\,z < 1+\mathrm{Re}\,\nu$	$2^{z-1}a^{-z}\cos(\pi\nu)\sec[\tfrac{1}{2}\pi(\nu-z)]$ $\cdot\Gamma(\tfrac{1}{2}z+\tfrac{1}{2}\nu)\,[\Gamma(1+\tfrac{1}{2}\nu-\tfrac{1}{2}z)]^{-1}$
12.5	$e^{-a^2x^2}\mathbf{H}_\nu(bx)$ $\mathrm{Re}\,z > -1-\mathrm{Re}\,\nu$	$\pi^{-\frac{1}{2}}2^{-\nu-1}a^{-\frac{1}{2}\nu-\frac{1}{2}-\frac{1}{2}z}[\Gamma(\tfrac{3}{2}+\nu)]^{-1}$ $\cdot b^{\nu+1}\Gamma(\tfrac{1}{2}+\tfrac{1}{2}\nu+\tfrac{1}{2}z)$ $\cdot\,{}_2F_2(1,\tfrac{1}{2}+\tfrac{1}{2}\nu+\tfrac{1}{2}z;\tfrac{3}{2},\tfrac{3}{2}+\nu;-\tfrac{1}{4}b^2/a^2)$
12.6	$\mathbf{H}_0[a(b^2+x^2)^{\frac{1}{2}}]$ $0 < \mathrm{Re}\,z < 1$	$\tfrac{1}{2}(2b/a)^{\frac{1}{2}z}\Gamma(\tfrac{1}{2}z)$ $\cdot[\mathbf{H}_{\frac{1}{2}z}(ab)\sec(\tfrac{1}{2}\pi z)$ $+J_{-\frac{1}{2}z}(ab)\tan(\tfrac{1}{2}\pi z)$

	$\phi(x)$	$\Phi(z) = \int_0^\infty \phi(x) x^{z-1} dx$
12.7	$(b^2+x^2)^{-\frac{1}{2}\nu}$ $\cdot \mathbf{H}_\nu [a(b^2+x^2)^{\frac{1}{2}}]$	$(2b)^{-\nu} \Gamma(\frac{1}{2}z)(b/a)^{\frac{1}{2}z} [\pi^{-1} \frac{\Gamma(\frac{1}{2}-\frac{1}{2}z)}{\Gamma(\frac{1}{2}+\nu)}$ $\cdot S_{\frac{1}{2}z+\nu, \frac{1}{2}z-\nu}(ab) + 2^{\nu-1-\frac{1}{2}z} Y_{\nu-\frac{1}{2}z}(ab)]$ $0 < \mathrm{Re}\, z < \begin{cases} 1 \\ 3/2 + \mathrm{Re}\,\nu \end{cases}$
12.8	$(b^2+x^2)^{\frac{1}{2}\nu}$ $\cdot \{\mathbf{H}_\nu [a(b^2+x^2)^{\frac{1}{2}}]$ $-Y_\nu [a(b^2+x^2)^{\frac{1}{2}}]\}$	$\frac{1}{2}\pi^{-1} b^\nu \cos(\pi\nu)(2b/a)^{\frac{1}{2}z}$ $\cdot \Gamma(\frac{1}{2}z)\Gamma(\frac{1}{2}-\nu-\frac{1}{2}z)\Gamma(\frac{1}{2}+\nu+\frac{1}{2}z)$ $[\mathbf{H}_{\frac{1}{2}z+\nu}(ab) - Y_{\frac{1}{2}z+\nu}(ab)]$ $0 < \mathrm{Re}\, z < 1-2\, \mathrm{Re}\, \nu$
12.9	$(b^2+x^2)^{-\frac{1}{2}\nu}$ $\cdot \{\mathbf{H}_\nu [a(b^2+x^2)^{\frac{1}{2}}]$ $-Y_\nu [a(b^2+x^2)^{\frac{1}{2}}]\}$	$\pi^{-1}(2b)^{-\nu}(b/a)^{\frac{1}{2}z}\Gamma(\frac{1}{2}z)\Gamma(\frac{1}{2}-\frac{1}{2}z)$ $[\Gamma(\frac{1}{2}+\nu)]^{-1} S_{\frac{1}{2}z+\nu, \frac{1}{2}z-\nu}(ab)$ $0 < \mathrm{Re}\, z < 1$
12.10	$(b^2+x^2)^{\frac{1}{2}\nu}$ $\cdot \{I_{-\nu}[a(b^2+x^2)^{\frac{1}{2}}]$ $-\mathbf{L}_\nu [a(b^2+x^2)^{\frac{1}{2}}]\}$ $\mathrm{Re}\,\nu < \frac{1}{2}$	$\frac{1}{2}\cos(\pi\nu) b^\nu (2b/a)^{\frac{1}{2}z}\Gamma(\frac{1}{2}z)$ $\cdot \sec[\pi(\nu+\frac{1}{2}z)]$ $\cdot [I_{-\nu-\frac{1}{2}z}(ab) - \mathbf{L}_{\nu+\frac{1}{2}z}(ab)]$ $\mathrm{Re}\, z > 0$

	$\phi(x)$	$\Phi(z) = \int_0^\infty \phi(x) x^{z-1} dx$
12.11	$(b^2+x^2)^{\frac{1}{2}\nu}$ $\cdot\mathbf{H}_\nu[a(b^2+x^2)^{\frac{1}{2}}]$ $0 < \mathrm{Re}\ z < \begin{cases} 1-2\mathrm{Re}\ \nu \\ \frac{3}{2} - \mathrm{Re}\ \nu \end{cases}$	$\frac{1}{2}b^\nu \sec[\pi(\nu+\frac{1}{2}z)](2b/a)^{\frac{1}{2}z}\Gamma(\frac{1}{2}z)$ $\cdot[\cos(\pi\nu)\mathbf{H}_{\nu+\frac{1}{2}z}(ab)+\sin(\pi z)\ J_{-\nu-\frac{1}{2}z}(ab)]$
12.12	$(b^2-x^2)^{\frac{1}{2}\nu}$ $\cdot\mathbf{H}_\nu[a(b^2-x^2)^{\frac{1}{2}}]$ $x < b$; 0 $x > b$	$\frac{1}{2}(\frac{1}{2}a)^{-\frac{1}{2}z}\Gamma(\frac{1}{2}z)b^{\nu+\frac{1}{2}z}\mathbf{H}_{\nu+\frac{1}{2}z}(ab)$ $\mathrm{Re}\ \nu > -3/2, \quad \mathrm{Re}\ z > 0$
12.13	$(b^2-x^2)^{-\frac{1}{2}\nu}$ $\cdot\mathbf{H}_\nu[a(b^2-x^2)^{\frac{1}{2}}]$ $x < b$; 0 $x > b$	$(2b)^{-\nu}\Gamma(\frac{1}{2}+\nu)[\Gamma(\frac{1}{2}+\frac{1}{2}z)]^{-1}\Gamma(\frac{1}{2}z)$ $(b/a)^{\frac{1}{2}z}s_{\nu+\frac{1}{2}z,\nu-\frac{1}{2}z}(ab)$ $\mathrm{Re}\ z > 0$
12.14	$(b^2-x^2)^{\frac{1}{2}\nu}$ $\cdot\mathbf{L}_\nu[a(b^2-x^2)^{\frac{1}{2}}]$ $x < b$; 0 $x > b$	$\frac{1}{2}(\frac{1}{2}a)^{-\frac{1}{2}z}\Gamma(\frac{1}{2}z)b^{\nu+\frac{1}{2}z}\mathbf{L}_{\nu+\frac{1}{2}z}(ab)$ $\mathrm{Re}\ \nu > -3/2, \quad \mathrm{Re}\ z > 0$
12.15	$s_{\mu,\nu}(ax)$ $-1-\mathrm{Re}\ \mu < \mathrm{Re}\ z < 1-\mathrm{Re}\ \mu$	$2^{z+\mu-1}a^{-z}\Gamma(\frac{1}{2}-\frac{1}{2}\mu-\frac{1}{2}z)$ $\cdot\Gamma(\frac{1}{2}+\frac{1}{2}\mu+\frac{1}{2}z)\Gamma(\frac{1}{2}+\frac{1}{2}\mu+\frac{1}{2}\nu)$ $\cdot\Gamma(\frac{1}{2}+\frac{1}{2}\mu-\frac{1}{2}\nu)$ $\cdot[\Gamma(1-\frac{1}{2}\nu-\frac{1}{2}z)\Gamma(1+\frac{1}{2}\nu-\frac{1}{2}z)]^{-1}$

	$\phi(x)$	$\Phi(z) = \int_0^\infty \phi(x)x^{z-1}dx$
12.16	$(b^2+x^2)^{\frac{1}{2}\nu}$ $\cdot s_{\mu,\nu}[a(b^2+x^2)^{\frac{1}{2}}]$ $0<\mathrm{Re}\ z<\begin{cases}3-\mathrm{Re}\ \nu\\ -\frac{3}{2}-\mathrm{Re}(\nu+\mu)\end{cases}$	$\frac{1}{2}\Gamma(\frac{1}{2}z)b^{\nu}(b/a)^{\frac{1}{2}z}$ $\cdot\{\frac{\Gamma(\frac{1}{2}-\frac{1}{2}\nu-\frac{1}{2}\mu-\frac{1}{2}z)}{\Gamma(\frac{1}{2}-\frac{1}{2}\nu-\frac{1}{2}\mu)}S_{\frac{1}{2}z+\mu,\frac{1}{2}z+\nu}(ab)$ $-2^{\mu-1+\frac{1}{2}z}\Gamma(\frac{1}{2}+\frac{1}{2}\mu+\frac{1}{2}\nu)\Gamma(\frac{1}{2}+\frac{1}{2}\mu-\frac{1}{2}\nu)$ $(\sin[\frac{1}{2}\pi(\mu+\nu)]J_{-\nu-\frac{1}{2}}(ab)$ $-\cos[\frac{1}{2}\pi(\mu+\nu)]Y_{-\nu-\frac{1}{2}}(ab))\}$
12.17	$(b^2+x^2)^{\frac{1}{2}\nu}$ $\cdot S_{\mu,\nu}[a(b^2+x^2)^{\frac{1}{2}}]$ $0<\mathrm{Re}\ z<1-\mathrm{Re}(\nu+\mu)$	$\frac{1}{2}b^{\nu}(b/a)^{\frac{1}{2}z}\Gamma(\frac{1}{2}z)\frac{\Gamma(\frac{1}{2}-\frac{1}{2}z-\frac{1}{2}\mu-\frac{1}{2}\nu)}{\Gamma(\frac{1}{2}-\frac{1}{2}\nu-\frac{1}{2}\mu)}$ $\cdot S_{\frac{1}{2}z+\mu,\frac{1}{2}z+\nu}(ab)$
12.18	$\mathbf{J}_{\nu}(ax)-J_{\nu}(ax)$ $0<\mathrm{Re}\ z<1$	$\frac{1}{2}\sin(\pi\nu)(\frac{1}{2}a)^{-z}\csc(\pi z)$ $\cdot\Gamma(\frac{1}{2}\nu+\frac{1}{2}z)[\Gamma(1+\frac{1}{2}\nu-\frac{1}{2}z)]^{-1}$
12.19	$\mathbf{J}_{\nu}(ax)+\mathbf{J}_{-\nu}(ax)$ $0<\mathrm{Re}\ z<2$	$\pi\cos(\frac{1}{2}\pi\nu)(\frac{1}{2}a)^{-z}\csc(\frac{1}{2}\pi z)$ $\cdot[\Gamma(1-\frac{1}{2}z+\frac{1}{2}\nu)\Gamma(1-\frac{1}{2}z-\frac{1}{2}\nu)]^{-1}$
12.20	$\mathbf{J}_{\nu}(ax)-\mathbf{J}_{-\nu}(ax)$ $-1<\mathrm{Re}\ z<1$	$\pi\sin(\frac{1}{2}\pi\nu)(\frac{1}{2}a)^{-z}\sec(\frac{1}{2}\pi z)$ $\cdot[\Gamma(1-\frac{1}{2}\nu-\frac{1}{2}z)\Gamma(1+\frac{1}{2}\nu-\frac{1}{2}z)]^{-1}$

	$\phi(x)$	$\Phi(z) = \int_0^\infty \phi(x)x^{z-1}dx$
12.21	$(a^2+x^2)^{\frac{1}{2}\nu}$ $\{\mathbf{J}_\nu[b(a^2+x^2)^{\frac{1}{2}}]$ $-J_\nu[b(a^2+x^2)^{\frac{1}{2}}]\}$ $0 < \mathrm{Re}\, z < 1-\mathrm{Re}\,\nu$	$\pi^{-1}\sin(\pi\nu)(\frac{1}{2}a)^\nu(a/b)^{\frac{1}{2}z}\Gamma(\frac{1}{2}z)$ $\cdot[\frac{\Gamma(\frac{1}{2}-\frac{1}{2}\nu-\frac{1}{2}z)}{\Gamma(\frac{1}{2}-\frac{1}{2}\nu)} S_{\frac{1}{2}z,\nu+\frac{1}{2}z}(ab)$ $+2\frac{\Gamma(1-\frac{1}{2}\nu-\frac{1}{2}z)}{\Gamma(-\frac{1}{2}\nu)} S_{\frac{1}{2}z-1,\nu+\frac{1}{2}z}(ab)]$
12.22	$(a^2+x^2)^{\frac{1}{2}\nu}$ $\cdot\{\mathbf{J}_\nu[b(a^2+x^2)^{\frac{1}{2}}]$ $+\mathbf{J}_{-\nu}[b(a^2+x^2)^{\frac{1}{2}}]\}$ $0 < \mathrm{Re}\, z < 3/2-\mathrm{Re}\,\nu$	$\pi^{-1}\sin(\pi\nu)a^\nu\Gamma(\frac{1}{2}z)(a/b)^{\frac{1}{2}z}$ $\cdot\{\frac{\Gamma(1-\frac{1}{2}\nu-\frac{1}{2}z)}{\Gamma(-\nu)} S_{\frac{1}{2}z-1,\nu+\frac{1}{2}z}(ab)$ $-\frac{1}{2}\pi\csc(\frac{1}{2}\pi\nu)2^{\frac{1}{2}z}[\cos(\frac{1}{2}\pi\nu)J_{-\nu-\frac{1}{2}}(ab)$ $+\sin(\frac{1}{2}\pi\nu)Y_{-\nu-\frac{1}{2}z}(ab)]\}$
12.23	$(a^2+x^2)^{\frac{1}{2}\nu}$ $\cdot\{\mathbf{J}_\nu[b(a^2+x^2)^{\frac{1}{2}}$ $-\mathbf{J}_{-\nu}[b(a^2+x^2)^{\frac{1}{2}}\}$ $0 < \mathrm{Re}\, z < 1-\mathrm{Re}\,\nu$	$\pi^{-1}\sin(\pi\nu)a^\nu\Gamma(\frac{1}{2}z)(a/b)^{\frac{1}{2}z}$ $\cdot\{\frac{\Gamma(\frac{1}{2}-\frac{1}{2}\nu-\frac{1}{2}z)}{\Gamma(\frac{1}{2}-\frac{1}{2}\nu)} S_{\frac{1}{2}z,\nu+\frac{1}{2}z}(ab)$ $-\frac{1}{2}\pi 2^{\frac{1}{2}z}\sec(\frac{1}{2}\pi\nu)[\sin(\frac{1}{2}\pi\nu)J_{-\nu-\frac{1}{2}}(ab)$ $-\cos(\frac{1}{2}\pi\nu)Y_{-\frac{1}{2}-\nu}(ab)]\}$
12.24	$\mathrm{ker}_\nu^2(ax)+\mathrm{kei}_\nu^2(ax)$	$2^{z-4}a^{-z}\Gamma(\frac{1}{2}z)\Gamma(\frac{1}{4}z+\frac{1}{2}\nu)\Gamma(\frac{1}{4}z-\frac{1}{2}\nu)$ $\mathrm{Re}\, z > \pm 2\,\mathrm{Re}\,\nu$

1.13 Whittaker Functions and Special Cases*

	$\phi(x)$	$\Phi(z) = \int_0^\infty \phi(x) x^{z-1} dx$
13.1	$\mathrm{Erf}(ax)$	$-\pi^{-\frac{1}{2}} z^{-1} a^{-z} \Gamma(\frac{1}{2}+\frac{1}{2}z)$ $-1 < \mathrm{Re}\, z < 0$
13.2	$\exp(-a^2x^2)\mathrm{Erf}(iax)$	$\frac{1}{2} i \pi^2 a^{-z} [\Gamma(1-\frac{1}{2}z)]^{-1} \sec(\frac{1}{2}\pi z)$ $-1 < \mathrm{Re}\, z < 1$
13.3	$\mathrm{Erfc}(ax)$	$\pi^{-\frac{1}{2}} z^{-1} a^{-z} \Gamma(\frac{1}{2}+\frac{1}{2}z)$ $\mathrm{Re}\, z > 0$
13.4	$\exp(a^2x^2)\mathrm{Erfc}(ax)$	$\frac{1}{2} a^{-z} \sec(\frac{1}{2}\pi z) \Gamma(\frac{1}{2}z)$ $0 < \mathrm{Re}\, z < 1$
13.5	$\exp(b^2x^2)\mathrm{Erfc}(ax)$ $b < a,\ \mathrm{Re}\, z > 0$	$\pi^{-\frac{1}{2}} z^{-1} a^{-z} \Gamma(\frac{1}{2}+\frac{1}{2}z)$ $\cdot {}_2F_1(\frac{1}{2}z, \frac{1}{2}z+\frac{1}{2}; 1+\frac{1}{2}z; b^2/a^2)$
13.6	$\sin(bx)\mathrm{Erfc}(ax)$ $\mathrm{Re}\, z > -1$	$\pi^{-\frac{1}{2}} (z+1)^{-1} b a^{-z-1} \Gamma(1+\frac{1}{2}z)$ $\cdot {}_2F_2(\frac{1}{2}+\frac{1}{2}z, 1+\frac{1}{2}z; \frac{3}{2}; \frac{1}{2}z+\frac{3}{2}; -\frac{1}{4}b^2/a^2)$
13.7	$\cos(bx)\mathrm{Erfc}(ax)$ $\mathrm{Re}\, z > 0$	$\pi^{-\frac{1}{2}} z^{-1} a^{-z} \Gamma(\frac{1}{2}+\frac{1}{2}z)$ $\cdot {}_2F_2(\frac{1}{2}z, \frac{1}{2}+\frac{1}{2}z; \frac{1}{2}; 1+\frac{1}{2}z; -\frac{1}{4}b^2/a^2)$

* The Fresnel, exponential, sine, cosine and error-integrals; incomplete gamma and parabolic cylinder functions.

	$\phi(x)$	$\Phi(z) = \int_0^\infty \phi(x) x^{z-1} dx$
13.8	$\exp(a^2x^2)\operatorname{Erfc}(ax+b)$	$\pi^{-\frac{1}{2}}(2a)^{-z}\Gamma(z)\Gamma(\frac{1}{2}-\frac{1}{2}z,b^2)$ $0 < \operatorname{Re} z < 1$
13.9	$\operatorname{Ei}(-ax)$	$-z^{-1}a^{-z}\Gamma(z)$ $\quad \operatorname{Re} z > 0$
13.10	$e^{ax}\operatorname{Ei}(-ax)$	$-\pi a^{-z}\Gamma(z)\csc(\pi z)$ $0 < \operatorname{Re} z < 1$
13.11	$\operatorname{Ei}[-b(a+x)]$	$-a^{z}\Gamma(z)\Gamma(-z,ab)$ $\quad \operatorname{Re} z > 0$
13.12	$e^{-ax}\overline{\operatorname{Ei}}(ax)$	$-\pi a^{-z}\Gamma(z)\cot(\pi z)$ $0 < \operatorname{Re} z < 1$
13.13	$e^{ax}[\operatorname{Ei}(-2ax)$ $-\operatorname{Ei}(-ax)]$	$\frac{1}{2}a^{-z}\Gamma(z)[\psi(1-\frac{1}{2}z)-\psi(\frac{1}{2}-\frac{1}{2}z)]$ $0 < \operatorname{Re} z < 1$
13.14	$e^{ax}[\operatorname{Ei}(-ax-bx)$ $-\operatorname{Ei}(-ax)]$	$(a/b)^{-1}b^{1-z}\Gamma(z)\Psi(-\frac{b}{a},1,1-z)$ $0 < \operatorname{Re} z < 1$
13.15	$e^{ax}[\operatorname{Ei}(-ax)]^2$ $\operatorname{Re} z > 0$	$\Gamma(z)[2\pi\csc(\pi z)\cot(\pi z)$ $-\frac{1}{2}\psi'(1-\frac{1}{2}z)+\frac{1}{2}\psi'(\frac{1}{2}-\frac{1}{2}z)]$

	$\phi(x)$	$\Phi(z) = \int_0^\infty \phi(x)x^{z-1}dx$
13.16	$Ei(-ax)\overline{E}i(ax)$	$\pi z^{-1}a^{-z}\Gamma(z)\cot(\frac{1}{2}\pi z)$ $0 < Re\ z < 2$
13.17	$Si(ax)$	$-z^{-1}a^{-z}\sin(\frac{1}{2}\pi z)\Gamma(z)$ $-1 < Re\ z < 0$
13.18	$si(ax)$	$-z^{-1}a^{-z}\sin(\frac{1}{2}\pi z)\Gamma(z)$ $0 < Re\ z < 2$
13.19	$Ci(ax)$	$-z^{-1}a^{-z}\cos(\frac{1}{2}\pi z)\Gamma(z)$ $0 < Re\ z < -2$
13.20	$Ci(ax)\sin(ax)$ $-si(ax)\cos(ax)$	$\frac{1}{2}\pi a^{-z}\sec(\frac{1}{2}\pi z)\Gamma(z)$ $-1 < Re\ z < 1$
13.21	$Ci(ax)\cos(ax)$ $+si(ax)\sin(ax)$	$-\frac{1}{2}\pi a^{-z}\csc(\frac{1}{2}\pi z)\Gamma(z)$ $-1 < Re\ z < 1$
13.22	$[Ci(ax)]^2+[si(ax)]^2$	$\pi z^{-1}a^{-z}\Gamma(z)\csc(\frac{1}{2}\pi z)$ $0 < Re\ z < 2$
13.23	$Erf(ix^{\frac{1}{2}})Erfc(x^{\frac{1}{2}})$	$i\pi^{-\frac{1}{2}}z^{-1}\tan(\frac{1}{2}\pi z)\Gamma(\frac{1}{2}+z), -\frac{1}{2}<Re\ z<1$

	$\phi(x)$	$\Phi(z) = \int_0^\infty \phi(x)x^{z-1}dx$
13.24	$\sin[b(a^2+x^2)^{\frac{1}{2}}]$ $\cdot \mathrm{si}[b(a^2+x^2)^{\frac{1}{2}}]$ $+\cos[b(a^2+x^2)^{\frac{1}{2}}]$ $\cdot \mathrm{Ci}[b(a^2+x^2)^{\frac{1}{2}}]$	$-\pi^{\frac{1}{2}}(ab)^{\frac{1}{2}}(a/b)^{\frac{1}{2}z}\csc(\frac{1}{2}\pi z)$ $\cdot S_{\frac{1}{2}z-\frac{3}{2},\frac{1}{2}z+\frac{1}{2}}(ab)$ $0 < \mathrm{Re}\, z < 2$
13.25	$\sin[b(a^2+x^2)^{\frac{1}{2}}]$ $\cdot \mathrm{Ci}[b(a^2+x^2)^{\frac{1}{2}}]$ $-\cos[b(a^2+x^2)^{\frac{1}{2}}]$ $\cdot \mathrm{si}[b(a^2+x^2)^{\frac{1}{2}}]$	$\frac{1}{2}\pi^{-\frac{1}{2}}(ab)^{\frac{1}{2}}(a/b)^{\frac{1}{2}z}\Gamma(\frac{1}{2}z)\Gamma(\frac{1}{2}-\frac{1}{2}z)$ $\cdot S_{\frac{1}{2}z-\frac{1}{2},\frac{1}{2}z+\frac{1}{2}}(ab)$ $0 < \mathrm{Re}\, z < 1$
13.26	$(a^2+x^2)^{-\frac{1}{2}}$ $\cdot\{\sin[b(a^2+x^2)^{\frac{1}{2}}]$ $\cdot \mathrm{si}[b(a^2+x^2)^{\frac{1}{2}}]$ $+\cos[b(a^2+x^2)^{\frac{1}{2}}]$ $\cdot \mathrm{Ci}[b(a^2+x^2)^{\frac{1}{2}}]\}$	$-\pi^{-\frac{1}{2}}(b/a)^{\frac{1}{2}}(a/b)^{\frac{1}{2}z}\Gamma(\frac{3}{2}-\frac{1}{2}z)$ $\cdot\Gamma(\frac{1}{2}z)S_{\frac{1}{2}z-\frac{3}{2},\frac{1}{2}z-\frac{1}{2}}(ab)$ $0 < \mathrm{Re}\, z < 3$
13.27	$(a^2+x^2)^{-\frac{1}{2}}$ $\cdot\{\sin[b(a^2+x^2)^{\frac{1}{2}}]$ $\cdot \mathrm{Ci}[b(a^2+x^2)^{\frac{1}{2}}]$ $-\cos[b(a^2+x^2)^{\frac{1}{2}}]$ $\cdot \mathrm{si}[b(a^2+x^2)^{\frac{1}{2}}]\}$	$\frac{1}{2}(\frac{1}{2}\pi)^{\frac{3}{2}}(b/a)^{\frac{1}{2}}(2a/b)^{\frac{1}{2}z}\Gamma(\frac{1}{2}z)$ $\cdot\csc(\frac{1}{2}\pi z)[\mathbf{H}_{\frac{1}{2}z-\frac{1}{2}}-Y_{\frac{1}{2}z-\frac{1}{2}}(ab)]$ $0 < \mathrm{Re}\, z < 2$

	$\phi(x)$	$\Phi(z) = \int_0^\infty \phi(x) x^{z-1} dx$
13.28	$S(ax)$ $-\frac{3}{2} < \mathrm{Re}\, z < 0$	$-(2\pi)^{-\frac{1}{2}} z^{-1} a^{-z} \Gamma(\tfrac{1}{2}+z)$ $\cdot \sin[\tfrac{1}{2}\pi(\tfrac{1}{2}+z)]$
13.29	$C(ax)$ $-\frac{1}{2} < \mathrm{Re}\, z < 0$	$-(2\pi)^{-\frac{1}{2}} z^{-1} a^{-z} \Gamma(\tfrac{1}{2}+z)$ $\cdot \cos[\tfrac{1}{2}\pi(\tfrac{1}{2}+z)]$
13.30	$\cos(ax) C(ax)$ $+\sin(ax) S(ax)$	$2^{-\frac{3}{2}} \pi^2 a^{-z} [\Gamma(1-z)]^{-1} \sec[\tfrac{1}{2}\pi(\tfrac{1}{2}-z)]$ $-\frac{1}{2} < \mathrm{Re}\, z < 1$
13.31	$\sin(ax) C(ax)$ $-\cos(ax) S(ax)$	$2^{-\frac{3}{2}} \pi^2 a^{-z}\ \Gamma(1-z)]^{-1} \csc[\tfrac{1}{2}\pi(\tfrac{1}{2}-z)]$ $-\frac{3}{2} < \mathrm{Re}\, z < 1$
13.32	$[\tfrac{1}{2}-S(ax)]\cos(ax)$ $-[\tfrac{1}{2}-C(ax)]\sin(ax)$	$2^{-\frac{3}{2}} a^{-z} \Gamma(z) \csc[\tfrac{1}{2}\pi(\tfrac{1}{2}-z)]$ $0 < \mathrm{Re}\, z < 3$
13.33	$[\tfrac{1}{2}-C(ax)]\cos(ax)$ $+[\tfrac{1}{2}-S(ax)]\sin(ax)$	$2^{-\frac{3}{2}} a^{-z} \Gamma(z) \sec[\tfrac{1}{2}\pi(\tfrac{1}{2}-z)]$ $0 < \mathrm{Re}\, z < 3$
13.34	$[\tfrac{1}{2}-C(ax)]^2$ $+[\tfrac{1}{2}-S(ax)]^2$	$\tfrac{1}{2}\pi^{-\frac{1}{2}} z^{-1} a^{-z} \sec(\tfrac{1}{2}\pi z) \Gamma(\tfrac{1}{2}+z)$ $0 < \mathrm{Re}\, z < 1$

	$\phi(x)$	$\Phi(z) = \int_0^\infty \phi(x)x^{z-1}dx$
13.35	$\cos[b(a^2+x^2)^{\frac{1}{2}}]$ $\cdot\{\frac{1}{2}-C[b(a^2+x^2)^{\frac{1}{2}}]\}$ $+\sin[b(a^2+x^2)^{\frac{1}{2}}]$ $\cdot\{\frac{1}{2}-S[b(a^2+x^2)^{\frac{1}{2}}]\}$	$\frac{1}{4}(\frac{1}{2}ab/\pi)^{\frac{1}{2}}(a/b)^{\frac{1}{2}z}\Gamma(\frac{1}{2}z)\Gamma(\frac{3}{4}-\frac{1}{2}z)$ $\cdot[\Gamma(\frac{3}{4})]^{-1}S_{\frac{1}{2}z-1,\frac{1}{2}z+\frac{1}{2}}(ab)$ $0 < \mathrm{Re}\ z < \frac{3}{2}$
13.36	$\cos[b(a^2+x^2)^{\frac{1}{2}}]$ $\cdot\{\frac{1}{2}-S[b(a^2+x^2)^{\frac{1}{2}}]\}$ $-\sin[b(a^2+x^2)^{\frac{1}{2}}]$ $\cdot\{\frac{1}{2}-C[b(a^2+x^2)^{\frac{1}{2}}]\}$	$\frac{1}{2}(\frac{1}{2}ab/\pi)^{\frac{3}{2}}(a/b)^{\frac{1}{2}z}\Gamma(\frac{1}{2}z)\Gamma(\frac{1}{4}-\frac{1}{2}z)$ $\cdot[\Gamma(\frac{1}{4})]^{-1}S_{\frac{1}{2}z,\frac{1}{2}+\frac{1}{2}z}(ab)$ $0 < \mathrm{Re}\ z < \frac{1}{2}$
13.37	$(a^2+x^2)^{-\frac{1}{2}}\cos[b(a^2+x^2)^{\frac{1}{2}}]$ $\cdot\{\frac{1}{2}-C[b(a^2+x^2)^{\frac{1}{2}}]\}$ $+(a^2+x^2)^{-\frac{1}{2}}\sin[b(a^2+x^2)^{\frac{1}{2}}]$ $\cdot\{\frac{1}{2}-S[b(a^2+x^2)^{\frac{1}{2}}]\}$	$(2\pi a/b)^{-\frac{1}{2}}(a/b)^{\frac{1}{2}z}\Gamma(\frac{5}{4}-\frac{1}{2}z)\Gamma(\frac{1}{2}z)$ $\cdot[\Gamma(\frac{1}{4})]^{-1}S_{\frac{1}{2}z-1,\frac{1}{2}z-\frac{1}{2}}(ab)$ $0 < \mathrm{Re}\ z < \frac{5}{2}$
13.38	$(a^2+x^2)^{-\frac{1}{2}}\cos[b(a^2+x^2)^{\frac{1}{2}}]$ $\cdot\{\frac{1}{2}-S[b(a^2+x^2)^{\frac{1}{2}}]\}$ $-(a^2+x^2)^{-\frac{1}{2}}\sin[b(a^2+x^2)^{\frac{1}{2}}]$ $\cdot\{\frac{1}{2}-C[b(a^2+x^2)^{\frac{1}{2}}]\}$	$\frac{1}{2}(2\pi a/b)^{-\frac{1}{2}}(a/b)^{\frac{1}{2}z}\Gamma(\frac{1}{2}z)\Gamma(\frac{3}{4}-\frac{1}{2}z)$ $\cdot[\Gamma(\frac{3}{4})]^{-1}S_{\frac{1}{2}z,\frac{1}{2}z-\frac{1}{2}}(ab)$ $0 < \mathrm{Re}\ z < \frac{3}{2}$

	$\phi(x)$	$\Phi(z) = \int_0^\infty \phi(x) x^{z-1} dx$
13.39	$\Gamma(\nu,ax)$	$z^{-1}a^{-z}\Gamma(\nu+z)$ Re $z > 0$
13.40	$e^{ax}\Gamma(\nu,ax)$ $0 < \mathrm{Re}\ z < 1-\mathrm{Re}\ \nu$	$a^{-z}[\Gamma(1-\nu)]^{-1}\Gamma(z)\Gamma(\nu+z)$ $\cdot\Gamma(1-\nu-z)$ $=[\Gamma(1-\nu)]^{-1}\pi a^{-z}\Gamma(z)\csc[\pi(\nu+z)]$
13.41	$e^{-bx}\Gamma(\nu,ax)$ Re $z > 0$	$z^{-1}a^{-z}(1+b/a)^{-z-\nu}\Gamma(z+\nu)$ $\cdot {}_2F_1[1,\nu+z,1+z;b/(a+b)]$
13.42	$e^{-bx}\gamma(\nu,ax)$ Re $z > -$Re ν	$\nu^{-1}(1+b/a)^{-\nu-z}\Gamma(z+\nu)$ $\cdot {}_2F_1[1,\nu+z;\nu+1;(1+b/a)^{-1}]$
13.43	$e^{-2bu}\overline{\mathrm{Ei}}(u-2ba)$ $+e^{2bu}\mathrm{Ei}(-u-2ba)$	$-\pi^{\frac{1}{2}}b(a/b)^{\frac{1}{2}+\frac{1}{2}z}\Gamma(\frac{1}{2}z)\cot(\frac{1}{2}\pi z)K_{\frac{1}{2}+\frac{1}{2}z}(2ab)$ $u = 2b(a^2+x^2)^{\frac{1}{2}}$, $0 < \mathrm{Re}\ z < 1$
13.44	$e^{-2ab}\mathrm{Ei}(-u+2ba)$ $+e^{2ab}\mathrm{Ei}(-u-2ba)$	$-\pi^{-\frac{1}{2}}bz^{-1}(a/b)^{\frac{1}{2}+\frac{1}{2}z}\Gamma(\frac{1}{2}z)K_{\frac{1}{2}+\frac{1}{2}z}(2ab)$ $u = 2b(a^2+x^2)^{\frac{1}{2}}$, Re $z > 0$
13.45	$D_\nu[(ix)^{\frac{1}{2}}]$ $\cdot D_\nu[(-ix)^{\frac{1}{2}}]$	$\frac{1}{2}[\Gamma(-\nu)]^{-1}\Gamma(z)B(-\frac{1}{2}\nu-\frac{1}{2}z,\frac{1}{2}+z)$ $0 < \mathrm{Re}\ z < -\mathrm{Re}\ \nu$

	$\phi(x)$	$\Phi(z) = \int_0^\infty \phi(x) x^{z-1} dx$
13.46	$e^{\frac{1}{2}ax} D_\nu[(2ax)^{\frac{1}{2}}]$	$2^{-2z-\frac{1}{2}\nu} a^{-z} [\Gamma(-\nu)]^{-1} \Gamma(2z)\Gamma(-\frac{1}{2}\nu-z)$ $0 < \mathrm{Re}\, z < -\frac{1}{2}\mathrm{Re}\,\nu$
13.47	$e^{-\frac{1}{2}ax} D_\nu[(2ax)^{\frac{1}{2}}]$	$\pi^{\frac{1}{2}} 2^{1-2z+\frac{1}{2}\nu} a^{-z} [\Gamma(\frac{1}{2}+z-\frac{1}{2}\nu)]^{-1}$ $\cdot\Gamma(2z) \qquad \mathrm{Re}\, z > 0$
13.48	$e^{-bx} D_\nu(ax^{\frac{1}{2}})$	$\pi^{\frac{1}{2}} 2^{\frac{1}{2}\nu+1-z} a^{-2z} [\Gamma(\frac{1}{2}+z-\frac{1}{2}\nu)]^{-1}$ $\cdot {}_2F_1(\frac{1}{2}+z, z; \frac{1}{2}+z-\frac{1}{2}\nu; \frac{1}{2}-2b/a^2)$ $= \pi^{\frac{1}{2}} 2^{\frac{1}{2}\nu+1} \Gamma(2z)(4b+a^2)^{-z}$ $\cdot {}_2F_1(-\frac{1}{2}\nu, z; \frac{1}{2}+z-\frac{1}{2}\nu; \frac{4b-a^2}{4b+a^2})$ $= \pi^{\frac{1}{2}} 2^{\frac{1}{2}+z} \Gamma(2z)(4b+a^2)^{-\frac{1}{2}(\frac{1}{2}+z+\frac{1}{2}\nu)}$ $\cdot(4b-a^2)^{-\frac{1}{2}(z-\frac{1}{2}-\frac{1}{2}\nu)} P^{\frac{1}{2}\nu+\frac{1}{2}-z}_{\frac{1}{2}\nu+z-\frac{1}{2}}[(\frac{1}{2}+\frac{2b}{a^2})^{-\frac{1}{2}}]$ $\mathrm{Re}\, z > 0$
13.49	$e^{-\frac{1}{2}ax} M_{K,\mu}(ax)$ $-\frac{1}{2}-\mathrm{Re}\,\mu < \mathrm{Re}\, z < \mathrm{Re}\, K$	$\Gamma(1+2\mu)[\Gamma(\frac{1}{2}+\mu+K)\Gamma(\frac{1}{2}+\mu-z)]^{-1}$ $\cdot a^{-z}\Gamma(K-z)\Gamma(\frac{1}{2}+\mu+z)$ $\mathrm{Re}\, z > \pm\frac{1}{2}\mathrm{Re}\, a$

	$\phi(x)$	$\Phi(z) = \int_0^\infty \phi(x)x^{z-1}dx$
13.50	$e^{-bx}M_{K,\mu}(ax)$ $\cdot\mathrm{Re}\, z > -\tfrac{1}{2}-\mathrm{Re}\,\mu$ $b > a$	$a^{\frac{1}{2}+\mu}\Gamma(\tfrac{1}{2}+\mu+z)(b-\tfrac{1}{2}a)^{-\mu-\frac{1}{2}-z}$ $\cdot\,{}_2F_1[\tfrac{1}{2}+\mu+z,\mu+K+\tfrac{1}{2};1+2\mu;(\tfrac{1}{2}-b/a)^{-1}]$ $= a^{\frac{1}{2}+\mu}\Gamma(\tfrac{1}{2}+\mu+z)(b+\tfrac{1}{2}a)^{-\frac{1}{2}-\mu-z}$ $\cdot\,{}_2F_1[\tfrac{1}{2}+\mu+z,\mu-K+\tfrac{1}{2};1+2\mu;(\tfrac{1}{2}+b/a)^{-1}]$
13.51	$e^{\frac{1}{2}ax}W_{K,\mu}(ax)$ $-\tfrac{1}{2}\pm\mathrm{Re}\,\mu < \mathrm{Re}\, z < -\mathrm{Re}\, K$	$[\Gamma(\tfrac{1}{2}+\mu-K)]^{-1}a^{-z}$ $\cdot\Gamma(\tfrac{1}{2}+\mu+z)B(\tfrac{1}{2}-\mu+z,-K-z)$
13.52	$e^{-\frac{1}{2}ax}W_{K,\mu}(ax)$	$a^{-z}[\Gamma(1-K+z)]^{-1}\Gamma(\tfrac{1}{2}+\mu+z)\Gamma(\tfrac{1}{2}-\mu+z)$ $\mathrm{Re}\, z > -\tfrac{1}{2} \pm \mathrm{Re}\,\mu$
13.53	$e^{-bx}W_{K,\mu}(ax)$ $\mathrm{Re}\, z > -\tfrac{1}{2} \pm \mathrm{Re}\,\mu$ $b > a$	$a^{-z}[\Gamma(1+z-K)]^{-1}\Gamma(\tfrac{1}{2}+z+\mu)$ $\cdot\Gamma(\tfrac{1}{2}+z-\mu)\,{}_2F_1(\tfrac{1}{2}+\mu+z,\tfrac{1}{2}-\mu+z;$ $1-K+z;\ \tfrac{1}{2}-b/a)$ $= a^{\frac{1}{2}+\mu}\Gamma(\tfrac{1}{2}+\mu+z)\Gamma(\tfrac{1}{2}-\mu+z)$ $\cdot[\Gamma(1-K+z)]^{-1}(b+\tfrac{1}{2}a)^{-\mu-z-\frac{1}{2}}$ $\cdot\,{}_2F_1(\tfrac{1}{2}+\mu+z,\tfrac{1}{2}+\mu-K;1-K+z;\tfrac{2b-a}{2b+a})$

	$\phi(x)$	$\Phi(z) = \int_0^\infty \phi(x) x^{z-1} dx$
13.54	$(b+x)^{-\rho} e^{-\frac{1}{2}ax}$ $M_{K,\mu}(ax)$	$\Gamma(1+2\mu)[\Gamma(\rho)\Gamma(\frac{1}{2}+K+\mu)]^{-1} b^{z-\rho}$ $\cdot G^{22}_{23}\left(ab \middle\vert \begin{matrix} 1-z, 1-K \\ \rho-z, \frac{1}{2}+\mu, \frac{1}{2}-\mu \end{matrix}\right)$ $-\frac{1}{2}-\mathrm{Re}\ \mu < \mathrm{Re}\ z < \mathrm{Re}(\rho+K)$
13.55	$(b+x)^{-\rho} e^{-\frac{1}{2}ax}$ $\cdot W_{K,\mu}(ax)$	$[\Gamma(\rho)]^{-1} b^{z-\rho} G^{31}_{23}\left(ab \middle\vert \begin{matrix} 1-z, 1-K \\ \rho-z, \frac{1}{2}+\mu, \frac{1}{2}-\mu \end{matrix}\right)$ $\mathrm{Re}\ z > -\frac{1}{2} \pm \mathrm{Re}\ \mu$
13.56	$(b+x)^{-\rho} e^{\frac{1}{2}ax}$ $W_{K,\mu}(ax)$ $-\frac{1}{2} \pm \mathrm{Re}\ \mu < \mathrm{Re}\ z < \mathrm{Re}(\rho-K)$	$[\Gamma(\rho)\Gamma(\frac{1}{2}-K+\mu)\Gamma(\frac{1}{2}-K-\mu)]^{-1}$ $\cdot b^{z-\rho} G^{32}_{23}\left(ab \middle\vert \begin{matrix} 1-z, 1+K \\ \rho-z, \frac{1}{2}+\mu, \frac{1}{2}-\mu \end{matrix}\right)$

	$\phi(x)$	$\Phi(z) = \int_0^\infty \phi(x)x^{z-1}dx$
13.57	$(b+x)^{\lambda-\frac{1}{2}}e^{\frac{1}{2}ax}$ $\cdot W_{K,\lambda}[a(b+x)]$ $0 < \mathrm{Re}\ z < \frac{1}{2}-\mathrm{Re}(K+\lambda)$	$[\Gamma(\frac{1}{2}-K-\lambda)]^{-1}a^{-\frac{1}{2}z}b^{\lambda-\frac{1}{2}+\frac{1}{2}z}\Gamma(z)$ $\cdot\Gamma(\frac{1}{2}-K-\lambda-z)W_{K+\frac{1}{2}z,\lambda+\frac{1}{2}z}(ab)$
13.58	$(x+b)^{-\sigma}e^{\frac{1}{2}ax}$ $\cdot W_{\mu,\nu}[a(b+x)]$ $0 < \mathrm{Re}\ z < \mathrm{Re}(\sigma-\mu)$	$b^{z-\sigma}\Gamma(z)e^{-\frac{1}{2}ab}[\Gamma(\frac{1}{2}+\nu-\mu)\Gamma(\frac{1}{2}-\nu-\mu)]^{-1}$ $\cdot G^{31}_{23}\left(ab \middle\vert \begin{matrix} 1+\mu,\sigma \\ \sigma-z,\frac{1}{2}+\nu,\frac{1}{2}-\nu \end{matrix}\right)$
13.59	$(x+b)^{-\sigma}e^{-\frac{1}{2}ax}$ $\cdot W_{\mu,\nu}[a(b+x)]$	$b^{z-\sigma}e^{\frac{1}{2}ab}\Gamma(z)G^{30}_{23}\left(ab \middle\vert \begin{matrix} \sigma,1-\mu \\ \sigma-z,\frac{1}{2}+\nu,\frac{1}{2}-\nu \end{matrix}\right)$ $\mathrm{Re}\ z > 0$
13.60	$0 \qquad x < b$ $(x-b)^{\mu-1}e^{\frac{1}{2}ax}$ $\cdot W_{K,\lambda}(ax) \qquad x > b$ $\mathrm{Re}\ \mu > 0$	$b^{\mu-1+z}\Gamma(\mu)[\Gamma(\frac{1}{2}+\lambda-K)\Gamma(\frac{1}{2}-\lambda-K)]^{-1}$ $\cdot G^{31}_{23}\left(ab \middle\vert \begin{matrix} 1+K,1-z \\ 1-\mu-z,\frac{1}{2}+\lambda,\frac{1}{2}-\lambda \end{matrix}\right)$ $\mathrm{Re}\ z < 1-\mathrm{Re}(K+\mu)$
13.61	$0 \qquad x < b$ $(x-b)^{\mu-1}e^{-\frac{1}{2}ax}$ $\cdot W_{K,\lambda}(ax) \qquad x > b$	$b^{\mu-1+z}\Gamma(\mu)G^{30}_{23}\left(ab \middle\vert \begin{matrix} 1-z,1-K \\ 1-\mu-z,\frac{1}{2}+\lambda,\frac{1}{2}-\lambda \end{matrix}\right)$ $\mathrm{Re}\ \mu > 0$

	$\phi(x)$	$\Phi(z) = \int_0^\infty \phi(x) x^{z-1} dx$
13.62	$0 \quad x < b$ $(x-b)^{\mu-\frac{1}{2}} e^{-\frac{1}{2}ax}$ $\cdot M_{K,\mu}[a(x-b)] \quad x > b$	$\Gamma(1+2\mu)\Gamma(\frac{1}{2}+K-\mu-z)[\Gamma(1-z)]^{-1}$ $\cdot a^{-\frac{1}{2}z} b^{\mu+\frac{1}{2}z-\frac{1}{2}} W_{\frac{1}{2}z-K,\frac{1}{2}z+\mu}(ab)$ $\mathrm{Re}\ z < \frac{1}{2}+\mathrm{Re}(K-\mu)$
13.63	$0 \quad x < b$ $(x-b)^{\lambda} e^{-\frac{1}{2}ax}$ $\cdot M_{K,\mu}[a(x-b)] \quad x > b$	$\Gamma(1+2\mu)[\Gamma(1-z)\Gamma(\frac{1}{2}+K+\mu)]^{-1}$ $\cdot b^{\lambda+z} e^{-\frac{1}{2}ab} G^{22}_{23}\left(ab \middle\vert \begin{matrix} -\lambda, 1-K \\ -\lambda-z, \frac{1}{2}+\mu, \frac{1}{2}-\mu \end{matrix}\right)$ $\mathrm{Re}\ z < \mathrm{Re}(K-\lambda)$
13.64	$0 \quad x < b$ $(x-b)^{\lambda} e^{-\frac{1}{2}ax}$ $\cdot W_{K,\mu}[a(x-b)] \quad x > b$ $\mathrm{Re}\ \lambda > -\frac{3}{2} \pm \mathrm{Re}\ \mu$	$[\Gamma(1-z)]^{-1} b^{\lambda+z} e^{-\frac{1}{2}ab}$ $\cdot G^{31}_{23}\left(ab \middle\vert \begin{matrix} -\lambda, 1-K \\ -z-\lambda, \frac{1}{2}+\mu, \frac{1}{2}-\mu \end{matrix}\right)$
13.65	$0 \quad x < b$ $(x-b)^{\lambda} e^{\frac{1}{2}ax}$ $\cdot W_{K,\mu}[a(x-b)] \quad x > b$ $\mathrm{Re}\ \lambda > -\frac{3}{2} \pm \mathrm{Re}\ \mu$	$[\Gamma(1-z)\Gamma(\frac{1}{2}-K-\mu)\Gamma(\frac{1}{2}-K+\mu)]^{-1}$ $\cdot b^{\lambda+z} e^{\frac{1}{2}ab} G^{32}_{23}\left(ab \middle\vert \begin{matrix} -\lambda, 1+K \\ -z-\lambda, \frac{1}{2}+\mu, \frac{1}{2}-\mu \end{matrix}\right)$ $\mathrm{Re}\ z < -\mathrm{Re}(K+\lambda)$

	$\phi(x)$	$\Phi(z) = \int_0^\infty \phi(x) x^{z-1} dx$
13.66	$(1+x)^{-\nu-\frac{1}{2}} e^{-\frac{1}{2}ax}$ $\cdot M_{\mu,\nu}[a(1+x)]$ $0 < \text{Re } z < \frac{1}{2} + \text{Re}(\nu+\mu)$	$\Gamma(1+2\nu)\Gamma(z) a^{-\frac{1}{2}z}$ $\cdot [\Gamma(2\nu+1-z)\Gamma(\mu+\nu+\frac{1}{2})]^{-1}$ $\cdot M_{\mu-\frac{1}{2}z,\nu-\frac{1}{2}z}(a)$
13.67	$(1+x)^{\nu-\frac{1}{2}} e^{-\frac{1}{2}ax}$ $\cdot W_{\mu,\nu}[a(1+x)]$	$a^{-\frac{1}{2}z}\Gamma(z) W_{\mu-\frac{1}{2}z,\nu+\frac{1}{2}z}(a)$ $\text{Re } z > 0$
13.68	$D_{2\nu-\frac{1}{2}}[(ax)^{\frac{1}{2}}]$ $\cdot D_{-2\nu-\frac{1}{2}}[(ax)^{\frac{1}{2}}]$	$(2\pi)^{\frac{1}{2}} (\frac{1}{4}a)^{-z} \Gamma(2z)$ $\cdot [\Gamma(\frac{3}{4}+\frac{1}{2}z+\nu)\Gamma(\frac{3}{4}+\frac{1}{2}z-\nu)]^{-1}$ $\text{Re } z > 0$
13.69	$M_{-\mu,\nu}(ax) W_{\mu,\nu}(ax)$ $-1-2\text{Re } \nu < \text{Re } z < -2\text{Re } \mu$	$\frac{1}{2}\Gamma(1+2\nu) a^{-z} \Gamma(\nu+\frac{1}{2}+\frac{1}{2}z)$ $\cdot [\Gamma(\nu+\frac{1}{2}-\frac{1}{2}z)\Gamma(\nu+\frac{1}{2}-\mu)]^{-1}$ $\cdot B(-\mu-\frac{1}{2}z, 1+z)$
13.70	$W_{\mu,\nu}(ax) W_{-\mu,\nu}(ax)$ $\text{Re } z > \begin{cases} -1 \\ -1 \pm 2 \text{ Re } \nu \end{cases}$	$\frac{1}{2}a^{-z}\Gamma(1+z)\Gamma(\frac{1}{2}+\frac{1}{2}z+\nu)\Gamma(\frac{1}{2}+\frac{1}{2}z-\nu)$ $\cdot [\Gamma(1+\frac{1}{2}z+\mu)\Gamma(1+\frac{1}{2}z-\mu)]^{-1}$

	$\phi(x)$	$\Phi(z) = \int_0^\infty \phi(x) x^{z-1} dx$
13.71	$\sin(cx^{\frac{1}{2}})$ $\cdot e^{-\frac{1}{2}x} W_{\mu,\nu}(x)$	$c[\Gamma(3/2-\mu+z)]^{-1}\Gamma(1+\nu+z)\Gamma(1-\nu+z)$ $\cdot {}_2F_2(1+\nu+z, 1-\nu+z; 3/2, 3/2-\mu+z; -\frac{c^2}{4})$ $\operatorname{Re} z > -1 \pm \operatorname{Re} \nu$
13.72	$[x^{\frac{1}{2}}+(a+x)^{\frac{1}{2}}]^{2\sigma}$ $\cdot e^{-\frac{1}{2}x} W_{\mu,\nu}(x)$	$-\pi^{-\frac{1}{2}}\sigma a^{\sigma} G^{32}_{34}\left(a \middle\vert \begin{matrix} \frac{1}{2}, 1, 1-\mu+z \\ \frac{1}{2}+\nu+z, \frac{1}{2}-\nu+z, -\sigma, \sigma \end{matrix}\right)$ $\vert \arg a \vert < \pi, \quad \operatorname{Re} z > -\frac{1}{2} \pm \operatorname{Re} \nu$
13.73	$[x^{\frac{1}{2}}+(a+x)^{\frac{1}{2}}]^{2\sigma}$ $\cdot e^{\frac{1}{2}x} W_{\mu,\nu}(x)$ $\vert \arg a \vert < \pi$	$-\pi^{-\frac{1}{2}}[\Gamma(\frac{1}{2}-\mu+\nu)\Gamma(\frac{1}{2}-\mu-\nu)]^{-1}\sigma a^{\sigma}$ $\cdot G^{33}_{34}\left(a \middle\vert \begin{matrix} \frac{1}{2}, 1, 1+\mu+z \\ \frac{1}{2}+\nu+z, \frac{1}{2}-\nu+z, -\sigma, \sigma \end{matrix}\right)$ $-\frac{1}{2} \pm \operatorname{Re} \nu < \operatorname{Re} z < -\operatorname{Re}(\mu+\sigma)$
13.74	$(a+x)^{-\frac{1}{2}} e^{\frac{1}{2}x}$ $\cdot [x^{\frac{1}{2}}+(a+x)^{\frac{1}{2}}]^{2\sigma} W_{\mu,\nu}(x)$ $\vert \arg a \vert < \pi$	$\pi^{-\frac{1}{2}}[\Gamma(\frac{1}{2}-\mu+\nu)\Gamma(\frac{1}{2}-\mu-\nu)]^{-1} a^{\sigma}$ $\cdot G^{33}_{34}\left(a \middle\vert \begin{matrix} 0, \frac{1}{2}, \frac{1}{2}+\mu+z \\ -\sigma, \mu+z, z-\mu, \sigma \end{matrix}\right)$ $-\frac{1}{2} \pm \operatorname{Re} \nu < \operatorname{Re} z < \frac{1}{2} - \operatorname{Re}(\mu+\sigma)$
13.75	$(a+x)^{-\frac{1}{2}} e^{-\frac{1}{2}x}$ $\cdot [x^{\frac{1}{2}}+(a+x)^{\frac{1}{2}}]^{2\sigma} W_{\mu,\nu}(x)$ $\vert \arg a \vert < \pi$	$\pi^{-\frac{1}{2}} a^{\sigma} G^{32}_{34}\left(a \middle\vert \begin{matrix} 0, \frac{1}{2}, \frac{1}{2}-\mu+z \\ -\sigma, z+\nu, z-\nu, \sigma \end{matrix}\right)$ $\operatorname{Re} z > -\frac{1}{2} \pm \operatorname{Re} \nu$

	$\phi(x)$	$\Phi(z) = \int_0^\infty \phi(x)x^{z-1}dx$
13.76	$\sin(cx^{\frac{1}{2}})$ $e^{\frac{1}{2}x}W_{\mu,\nu}(x)$	$\pi^{\frac{1}{2}}[\Gamma(\frac{1}{2}-\mu+\nu)\Gamma(\frac{1}{2}-\mu-\nu)]^{-1}$ $\cdot G^{22}_{23}\left(\frac{1}{4}c^2\middle\vert\begin{matrix}\frac{1}{2}+\nu-z,\frac{1}{2}-\nu-z\\ \frac{1}{2},-\mu-z,0\end{matrix}\right)$ $-1\pm\mathrm{Re}\ \nu < \mathrm{Re}\ z < \frac{1}{2}-\mathrm{Re}\ \mu$
13.77	$\cos(cx^{\frac{1}{2}})$ $\cdot e^{-\frac{1}{2}x}W_{\mu,\nu}(x)$	$[\Gamma(1-\mu+z)]^{-1}\Gamma(\frac{1}{2}+\nu+z)\Gamma(\frac{1}{2}-\nu+z)$ $\cdot {}_2F_2(\frac{1}{2}+\nu+z,\frac{1}{2}-\nu+z;\frac{1}{2},1-\mu+z;-\frac{1}{4}c^2)$ $\mathrm{Re}\ z > -\frac{1}{2}\pm\mathrm{Re}\ \nu$
13.78	$\cos(cx^{\frac{1}{2}})$ $\cdot e^{\frac{1}{2}x}W_{\mu,\nu}(x)$	$\pi^{\frac{1}{2}}[\Gamma(\frac{1}{2}-\mu+\nu)\Gamma(\frac{1}{2}-\mu-\nu)]^{-1}$ $\cdot G^{22}_{23}\left(\frac{1}{4}c^2\middle\vert\begin{matrix}\frac{1}{2}+\nu-z,\frac{1}{2}-\nu-z\\ 0,-\mu-z,\frac{1}{2}\end{matrix}\right)$ $-\frac{1}{2}\pm\mathrm{Re}\ \nu < \mathrm{Re}\ z < \frac{1}{2}-\mathrm{Re}\ \mu$
13.79	$e^{-\frac{1}{2}x}J_{\sigma+\lambda}(ax^{\frac{1}{2}})$ $\cdot J_{\lambda-\sigma}(ax^{\frac{1}{2}})W_{\mu,\nu}(x)$ $\mathrm{Re}\ z > -\frac{1}{2}-\mathrm{Re}(\lambda\pm\nu)$	$(\frac{1}{2}a)^{2\lambda}\dfrac{\Gamma(\frac{1}{2}+\lambda+\nu+z)\Gamma(\frac{1}{2}+\lambda-\nu+z)}{\Gamma(1+\mu+\sigma)\Gamma(1+\lambda-\sigma)\Gamma(1+\lambda-\mu+z)}$ $\cdot {}_4F_4\left(\begin{matrix}1+\lambda,\frac{1}{2}+\lambda,\frac{1}{2}+\lambda+\nu+z,\frac{1}{2}+\lambda-\nu+z;\\ 1+\lambda+\sigma,1+\lambda-\sigma,1+2\lambda,1+\lambda-\mu+z;\end{matrix}-a^2\right)$

	$\phi(x)$	$\Phi(z) = \int_0^\infty \phi(x) x^{z-1} dx$
13.80	$e^{-\frac{1}{2}x} I_{\lambda+\sigma}(ax^{\frac{1}{2}}) \cdot K_{\lambda-\sigma}(ax^{\frac{1}{2}}) W_{\mu,\nu}(x)$	$\frac{1}{2}\pi^{-\frac{1}{2}} G^{24}_{45}\left(a^2 \middle\vert \begin{matrix} 0, \frac{1}{2}, \frac{1}{2}+\nu-z, \frac{1}{2}-\nu-z \\ \lambda, \sigma, -\lambda, -\sigma, \mu-z \end{matrix}\right)$ Re $z > \begin{cases} -\frac{1}{2}-\text{Re}(\lambda\pm\nu) \\ -\frac{1}{2}-\text{Re}(\mu\pm\nu) \end{cases}$
13.81	$W_{K,\mu}(x) W_{\lambda,\nu}(x)$ Re $\rho>-1\pm$Re $\mu\pm$Re ν	$[\Gamma(\frac{1}{2}-\lambda-\nu)\Gamma(\frac{3}{2}-K+\nu+z)]^{-1}$ $\cdot\Gamma(1+\mu+\nu+z)\Gamma(1-\mu+\nu+z)\Gamma(-2\nu)$ $\cdot {}_3F_2(1+\mu+\nu+z, 1-\mu+\nu+z, \frac{1}{2}-\lambda+\nu; 1+2\nu, \frac{3}{2}-K+\nu+z; 1)$ $+[\Gamma(\frac{1}{2}-\lambda+\nu)\Gamma(\frac{3}{2}-K-\nu+z)]^{-1}$ $\cdot\Gamma(1+\mu-\nu+z)\Gamma(1-\mu-\nu+z)\Gamma(2\nu)$ $\cdot {}_3F_2(1+\mu-\nu+z, 1-\mu-\nu+z, \frac{1}{2}-\lambda-\nu; 1-2\nu, \frac{3}{2}-K-\nu+z; 1)$
13.82	$\exp[-\frac{1}{2}x(\alpha+\beta)] \cdot M_{K,\mu}(\alpha x) W_{\lambda,\nu}(\beta x)$ Re $\rho>-1-$Re $\mu\pm$Re ν	$[\Gamma(\frac{3}{2}-\lambda+\mu+z)]^{-1}\Gamma(1+\mu+\nu+z)\Gamma(1+\mu-\nu+z)$ $\alpha^{\frac{1}{2}+\mu}\beta^{-\mu-\frac{1}{2}-z} {}_3F_2(\frac{1}{2}+K+\mu,$ $1+\mu+\nu+z, 1+\mu-\nu+z; 1+2\mu, \frac{3}{2}-\lambda+\mu+z; -\alpha/\beta)$
13.83	$\exp[\frac{1}{2}x(\alpha+\beta)] \cdot W_{K,\mu}(\alpha x) W_{\lambda,\nu}(\beta x)$	$\beta^{-z}[\Gamma(\frac{1}{2}-K+\mu)\Gamma(\frac{1}{2}-K-\mu)\Gamma(\frac{1}{2}-\lambda+\nu)\Gamma(\frac{1}{2}-\lambda-\nu)]^{-1}$ $\cdot G^{33}_{33}\left(\beta/\alpha \middle\vert \begin{matrix} \frac{1}{2}+\mu, \frac{1}{2}-\mu, 1+\lambda+z \\ \frac{1}{2}+\nu+z, \frac{1}{2}-\nu+z, -K \end{matrix}\right)$ $-1\pm$Re $\mu\pm$Re $\nu<$Re $z<-$Re$(K+\lambda)$

	$\phi(x)$	$\Phi(z) = \int_0^\infty \phi(x) x^{z-1} dx$
13.84	$\exp[-\frac{1}{2}x(\alpha-\beta)]$ $\cdot W_{K,\mu}(\alpha x) W_{\lambda,\nu}(\beta x)$	$\beta^{-z}[\Gamma(\frac{1}{2}-\lambda+\nu)\Gamma(\frac{1}{2}-\lambda-\nu)]^{-1}$ $\cdot G^{23}_{33}\left(\beta/\alpha \middle\vert \begin{matrix} \frac{1}{2}+\mu, \frac{1}{2}-\mu, 1+\lambda+z \\ \frac{1}{2}+\nu+z, \frac{1}{2}-\nu+z, K \end{matrix}\right)$ $\mathrm{Re}\ z > -1 \pm \mathrm{Re}\ \mu \pm \mathrm{Re}\ \nu$
13.85	$\exp[-\frac{1}{2}x(\alpha+\beta)]$ $\cdot W_{K,\mu}(\alpha x) W_{\lambda,\nu}(\beta x)$	$\beta^{-z} G^{22}_{33}\left(\beta/\alpha \middle\vert \begin{matrix} \frac{1}{2}+\mu, \frac{1}{2}-\nu, 1-\lambda+z \\ \frac{1}{2}+\nu+z, \frac{1}{2}-\nu+z, K \end{matrix}\right)$ $\mathrm{Re}\ z > -1 \pm \mathrm{Re}\ \nu \pm \mathrm{Re}\ \mu$
13.86	$\exp[-\frac{1}{2}(x/\alpha+\beta/x)]$ $\cdot W_{K,\mu}(x/\alpha) W_{\lambda,\nu}(\beta/x)$	$\beta^{z}\ G^{40}_{24}\left(\beta/\alpha \middle\vert \begin{matrix} 1-K, 1-\lambda-z \\ \frac{1}{2}+\mu, \frac{1}{2}-\mu, \frac{1}{2}+\nu-z, \frac{1}{2}-\nu-z \end{matrix}\right)$ $\mathrm{Re}(\alpha,\beta) > 0$
13.87	$\exp[\frac{1}{2}(x/\alpha-\beta/x)]$ $\cdot W_{K,\mu}(x/\alpha) W_{\lambda,\nu}(\beta/x)$ $\mathrm{Re}\ \beta>0,\ \vert\arg\alpha\vert<3\pi/2$	$\beta^{z}[\Gamma(\frac{1}{2}-K+\mu)\Gamma(\frac{1}{2}-K-\mu)]^{-1}$ $\cdot G^{41}_{24}\left(\beta/\alpha \middle\vert \begin{matrix} 1+K, 1-\lambda-z \\ \frac{1}{2}+\mu, \frac{1}{2}-\mu, \frac{1}{2}+\nu-z, \frac{1}{2}-\nu-z \end{matrix}\right)$ $\mathrm{Re}\ z<-\frac{1}{2}-\vert\mathrm{Re}\ \nu\vert-\mathrm{Re}\ K$
13.88	$\exp[\frac{1}{2}(x/\alpha+\beta/x)]$ $\cdot W_{K,\mu}(x/\alpha) W_{\lambda,\nu}(\beta/x)$ $\vert\arg(\alpha,\beta)\vert<3\pi/2$	$\beta^{z}[\Gamma(\frac{1}{2}-\lambda-\nu)\Gamma(\frac{1}{2}-K+\mu)\Gamma(\frac{1}{2}-K-\mu)\Gamma(\frac{1}{2}-\lambda+\nu)]^{-1}$ $\cdot G^{42}_{24}\left(\beta/\alpha \middle\vert \begin{matrix} 1+K, 1+\lambda-z \\ \frac{1}{2}+\mu, \frac{1}{2}-\mu, \frac{1}{2}+\nu-z, \frac{1}{2}-\nu-z \end{matrix}\right)$ $-\frac{1}{2}+\mathrm{Re}\ \lambda+\vert\mathrm{Re}\ \mu\vert<\mathrm{Re}\ z<\frac{1}{2}-\vert\mathrm{Re}\ \nu\vert-\mathrm{Re}\ K$

1.14 Elliptic Integrals and Elliptic Functions

	$\phi(x)$	$\Phi(z) = \int_0^\infty \phi(x)x^{z-1}dx$
14.1	$K[(\frac{1}{2}-\frac{1}{2}x)^{\frac{1}{2}}]$ $x < 1$ 0 $x > 1$	$\frac{1}{2}\pi^{\frac{1}{2}}2^{-z}\Gamma(z)[\Gamma(\frac{3}{4}+\frac{1}{2}z)\Gamma(\frac{1}{4}+\frac{1}{2}z)]^{-1}$ $\mathrm{Re}\, z > 0$
14.2	$K[(1-x^2/a^2)^{\frac{1}{2}}]$ $x < a$ 0 $x > a$	$\frac{1}{4}\pi a^z\Gamma^2(\frac{1}{2}z)[\Gamma(\frac{1}{2}+\frac{1}{2}z)]^{-2}$ $\mathrm{Re}\, z > 0$
14.3	0 $x < a$ $K[(1-a^2/x^2)^{\frac{1}{2}}]$ $x > a$	$\frac{1}{4}\pi a^2\Gamma^2(-\frac{1}{2}z)[\Gamma(\frac{1}{2}-\frac{1}{2}z)]^{-2}$ $\mathrm{Re}\, z < 0$
14.4	0 $x < a$ $(a+x)^{-\frac{1}{2}}K[(\frac{x-a}{x+a})^{\frac{1}{2}}]$ $x > a$	$\pi^{\frac{1}{2}}a^{z-\frac{1}{2}}2^{-z-5/2}\Gamma^2(\frac{1}{4}-\frac{1}{2}z)$ $\cdot[\Gamma(1-z)]^{-1}$ $\mathrm{Re}\, z < \frac{1}{2}$
14.5	$(a+x)^{-1}K(\lvert\frac{a-x}{a+x}\rvert)$	$\frac{1}{8}\pi^{-1}a^{z-1}[\Gamma(\frac{1}{2}z)\Gamma(\frac{1}{2}-\frac{1}{2}z)]^2$ $0 < \mathrm{Re}\, z < 1$
14.6	$(a^2+x^2)^{-\frac{1}{2}}$ $\cdot K[a(a^2+x^2)^{-\frac{1}{2}}]$	$\frac{1}{4}a^{z-1}\Gamma^2(\frac{1}{2}z)\Gamma(\frac{1}{2}-\frac{1}{2}z)$ $[\Gamma(\frac{1}{2}+\frac{1}{2}z)]^{-1}$ $0 < \mathrm{Re}\, z < 1$
14.7	$(a^2+x^2)^{-\frac{1}{2}}K[x(a^2+x^2)^{-\frac{1}{2}}]$ $0 < \mathrm{Re}\, z < 1$	$\frac{1}{4}a^{z-1}\Gamma(\frac{1}{2}z)\Gamma^2(\frac{1}{2}-\frac{1}{2}z)$ $\cdot[\Gamma(1-\frac{1}{2}z)]^{-1}$
14.8	$(a^2+x^2)^{-\frac{1}{2}}$ $\cdot E[a(a^2+x^2)^{-\frac{1}{2}}]$ $0 < \mathrm{Re}\, z < 1$	$\frac{1}{4}a^{z-1}z\Gamma^2(\frac{1}{2}z)\Gamma(\frac{1}{2}-\frac{1}{2}z)$ $\cdot[\Gamma(\frac{1}{2}+\frac{1}{2}z)]^{-1}$

	$\phi(x)$	$\Phi(z) = \int_0^\infty \phi(\ )x^{z-1}dx$
14.9	$[x^2+(a+b)^2]^{-\frac{1}{2}}$ $\cdot K\{2(ab)^{\frac{1}{2}}[x^2+(a+b)^2]^{-\frac{1}{2}}\}$ $0 < \text{Re } z < 1$	$\pi 2^{-z-1}\Gamma(\frac{1}{2}-\frac{1}{2}z)\Gamma(z)[\Gamma(\frac{1}{2}+\frac{1}{2}z)]^{-1}$ $\cdot(b^2-a^2)^{\frac{1}{2}z-\frac{1}{2}}P^{\nu}_{\frac{1}{2}z-\frac{1}{2}}[\frac{b^2+a^2}{\|b^2-a^2\|})$
14.10	$[b^2+(a+x)^2]^{-\frac{1}{2}}$ $\cdot K\{2(ax)^{\frac{1}{2}}[b^2+(a+x)^2]^{-\frac{1}{2}}\}$	$\frac{1}{4}\pi^{\frac{1}{2}}\Gamma(\frac{1}{2}z)\Gamma(\frac{1}{2}-\frac{1}{2}z)(a^2+b^2)^{\frac{1}{2}z-\frac{1}{2}}$ $\cdot P_{-z}[b(a^2+b^2)^{-\frac{1}{2}}]$ $0 < \text{Re } z < 1$
14.11	$[x^2+(a+b)^2]^{-\frac{1}{2}}$ $\cdot K\{[\frac{x^2+(a-b)^2}{x^2+(a+b)^2}]^{\frac{1}{2}}\}$	$\frac{1}{4}\pi^{-\frac{1}{2}}\Gamma(\frac{1}{2}z)\Gamma(\frac{1}{2}-\frac{1}{2}z)\|a^2-b^2\|^{\frac{1}{2}z-\frac{1}{2}}$ $\cdot q_{-\frac{1}{2}-\frac{1}{2}z}(\frac{a^2+b^2}{\|a^2-b^2\|})$ $0 < \text{Re } z < 1$
14.12	$(a^2+x^2)^{-\frac{1}{2}}$ $\cdot K\{[\frac{1}{2}-\frac{1}{2}x(a^2+x^2)^{-\frac{1}{2}}]^{\frac{1}{2}}\}$	$2^{-3/2-z}\pi^{\frac{1}{2}}a^{z-\frac{1}{2}}\Gamma(\frac{1}{4}-\frac{1}{2}z)$ $\cdot[\Gamma(1-z)\Gamma(3/4+\frac{1}{2}z)]^{-1}$ $0 < \text{Re } z < 1$
14.13	$0 \quad x < a+b$ $[x^2-(a-b)^2]^{-\frac{1}{2}}$ $\cdot K\{[\frac{x^2-(a+b)^2}{x^2-(a-b)^2}]^{\frac{1}{2}}\}$ $x > a+b$	$\pi^{\frac{1}{2}}\Gamma(\frac{1}{2}-\frac{1}{2}z)[\Gamma(1-\frac{1}{2}z)]^{-1}$ $\cdot\|a^2-b^2\|^{\frac{1}{2}z-\frac{1}{2}}q_{-\frac{1}{2}z-\frac{1}{2}}(\frac{a^2+b^2}{\|a^2-b^2\|})$ $\text{Re } z < 1$

	$\phi(x)$	$\Phi(z) = \int_0^\infty \phi(x) x^{z-1} dx$
14.14	$\theta_2(0\|ax)$	$2\pi^{-2z}a^{-z}(2^{2z}-1)\Gamma(z)\zeta(2z)$ Re $z > \frac{1}{2}$
14.15	$\theta_3(0\|ax)-1$	$2\pi^{-2z}a^{-z}\Gamma(z)\zeta(2z)$ Re $z > \frac{1}{2}$
14.16	$\theta_3(\frac{1}{2}\|ax)$	$2\pi^{-\frac{1}{2}}a^{-z}\Gamma(\frac{1}{2}-z)(2^{1-2z}-1)\zeta(1-2z)$ Re $z > 0$
14.17	$\theta_4(0\|ax)$	$2a^{-z}\pi^{-2z}(2^{1-2z}-1)\Gamma(z)\zeta(2z)$ Re $z < 0$
14.18	$\theta_1(y\|x)$ $-\frac{1}{2} < y < \frac{1}{2}$	$\pi^{-\frac{1}{2}}2^{2z-1}\Gamma(\frac{1}{2}-z)$ $\cdot[\zeta(1-2z,\frac{1}{4}-\frac{1}{2}y)-\zeta(1-2z,\frac{1}{4}+\frac{1}{2}y)$ $+\zeta(1-2z,\frac{3}{4}+\frac{1}{2}y)-\zeta(1-2z,\frac{3}{4}-\frac{1}{2}y)]$
14.19	$\theta_2(y\|x)$ $0 < y < 1$	$\pi^{-\frac{1}{2}}2^{2z-1}\Gamma(\frac{1}{2}-z)$ $\cdot[\zeta(1-2z,\frac{1}{2}y)-\zeta(1-2z,\frac{1}{2}+\frac{1}{2}y)$ $-\zeta(1-2z,\frac{1}{2}-\frac{1}{2}y)+\zeta(1-2z,1-\frac{1}{2}y)]$
14.20	$\theta_3(y\|x)$ $0 < y < 1$	$\pi^{-\frac{1}{2}}\Gamma(\frac{1}{2}-z)[\zeta(1-2z,y)+\zeta(1-2z,1-y)]$ Re $z < 0$

	$\phi(x)$	$\Phi(z) = \int_0^\infty \phi(x) x^{z-1} dx$
14.21	$\theta_4(y\|x)$ $-\frac{1}{2} < y < \frac{1}{2}$	$\pi^{-\frac{1}{2}}\Gamma(\frac{1}{2}-z)[\zeta(1-2z,\frac{1}{2}+y)+\zeta(1-2z,\frac{1}{2}-y)]$ $\mathrm{Re}\, z < 0$
14.22	$\hat{\theta}_1(y\|x)$ $-\frac{1}{2} < y < \frac{1}{2}$	$\pi^{-\frac{1}{2}}2^{2z-1}\Gamma(\frac{1}{2}-z)[\zeta(1-2z,\frac{1}{4}+\frac{1}{2}y)-\zeta(1-2z,\frac{3}{4}+\frac{1}{2}y)$ $+\zeta(1-2z,\frac{1}{4}-\frac{1}{2}y)-\zeta(1-2z,\frac{3}{4}-\frac{1}{2}y)]$ $\mathrm{Re}\, z < \frac{1}{2}$
14.23	$\hat{\theta}_2(y\|x)$ $0 < y < 1$	$\pi^{-\frac{1}{2}}2^{2z-1}\Gamma(\frac{1}{2}-z)[\zeta(1-2z,\frac{1}{2}y)-\zeta(1-2z,\frac{1}{2}+\frac{1}{2}y)$ $+\zeta(1-2z,\frac{1}{2}-\frac{1}{2}y)-\zeta(1-2z,1-\frac{1}{2}y)]$ $\mathrm{Re}\, z < \frac{1}{2}$
14.24	$\hat{\theta}_3(y\|x)$ $0 < y < 1$	$\pi^{-\frac{1}{2}}\Gamma(\frac{1}{2}-z)[\zeta(1-2z,y)-\zeta(1-2z,1-y)]$ $\mathrm{Re}\, z < 0$
14.25	$\hat{\theta}_4(y\|x)$ $-\frac{1}{2} < y < \frac{1}{2}$	$\pi^{-\frac{1}{2}}\Gamma(\frac{1}{2}-z)[\zeta(1-2z,\frac{1}{2}+y)-\zeta(1-2z,\frac{1}{2}-y)]$ $\mathrm{Re}\, z < 0$
14.26	$\theta_1(y\|x)$ $-\frac{1}{2} < y < \frac{1}{2}$,	$2^{1+2z}\pi^{-2z}\Gamma(z) \sum_0^\infty (-1)^n(2n+1)^{-2z}$ $\cdot\sin[(2n+1)\pi y]$ $\mathrm{Re}\, z > 0$
14.27	$\theta_2(y\|x)$ $0 < y < 1$	$2^{1+2z}\pi^{-2z}\Gamma(z) \sum_0^\infty (2n+1)^{-2z}\cos[(2n+1)\pi y]$ $\mathrm{Re}\, z > 0$

	$\phi(x)$	$\Phi(z) = \int_0^\infty \phi(x) x^{z-1} dx$
14.28	$\theta_3(y\|x)-1$ $0 < y < 1$	$2\pi^{-2z}\Gamma(z) \sum_1^\infty n^{-2z}\cos(2\pi ny)$ $\mathrm{Re}\ z > 0$
14.29	$\theta_4(y\|x)-1$ $-\frac{1}{2} < y < \frac{1}{2}$	$2\pi^{-2z}\Gamma(z) \sum_1^\infty (-1)^n n^{-2z}\cos(2\pi ny)$ $\mathrm{Re}\ z > 0$

1.15 Hypergeometric Functions

	$\phi(x)$	$\Phi(z) = \int_0^\infty \phi(x) x^{z-1} dx$
15.1	$(1-x)^{\beta-\gamma-n}$ $\cdot {}_2F_1(-n,\beta;\gamma;x) \quad x < 1$ $0 \quad x > 1$ $n = 0,1,2,\cdots$	$[\Gamma(\gamma+n)\Gamma(\gamma-z)\Gamma(1+\beta-\gamma+z)]^{-1}$ $\cdot\Gamma(\gamma)\Gamma(1+\beta-\gamma)\Gamma(z)\Gamma(\gamma+n-z)$ $\mathrm{Re}(\beta-\gamma) > n-1 \quad \mathrm{Re}\, z > 0$
15.2	$(1-x)^{\gamma-1}$ $\cdot {}_2F_1(\alpha,\beta;\gamma;1-x) \quad x < 1$ $0 \quad x > 1$	$[\Gamma(\gamma-\alpha+z)\Gamma(\gamma-\beta+z)]^{-1}$ $\cdot\Gamma(\gamma)\Gamma(z)\Gamma(\gamma-\alpha-\beta+z)$ $\mathrm{Re}\,\gamma > 0, \mathrm{Re}\, z > \begin{cases} 0 \\ \mathrm{Re}(\alpha+\beta-\gamma) \end{cases}$
15.3	$(1-x)^{\rho-1}$ $\cdot {}_2F_1(\alpha,\beta;\gamma;1-x) \quad x < 1$ $0 \quad x > 1$	$\Gamma(\rho)\Gamma(z)[\Gamma(\rho+z)]^{-1}$ $\cdot {}_3F_2(\alpha,\beta,\rho;\gamma,\rho+z;1)$ $\mathrm{Re}\,\rho > 0, \mathrm{Re}\, z > \begin{cases} 0 \\ \mathrm{Re}(\alpha+\beta-\gamma) \end{cases}$
15.4	$0 \quad x < 1$ $(x-1)^{\gamma-1}$ $\cdot {}_2F_1(\alpha,\beta;\gamma;1-x)$ $x > 1$	$[\Gamma(1-z)\Gamma(1+\alpha+\beta-\gamma-z)]^{-1}$ $\cdot\Gamma(\gamma)\Gamma(1+\alpha-\gamma-z)\Gamma(1+\beta-\gamma-z)$ $\mathrm{Re}\,\gamma > 0, \mathrm{Re}\, z > \begin{cases} \mathrm{Re}(1+\alpha-\gamma) \\ \mathrm{Re}(1+\beta-\gamma) \end{cases}$

	$\phi(x)$	$\Phi(z) = \int_0^\infty \phi(x) x^{z-1} dx$
15.5	$(1-x)^{\sigma-1}$ $\cdot {}_2F_1(\alpha,\beta;\gamma;xz) \quad x < 1$ $0 \quad x > 1$	$[\Gamma(\sigma+z)]^{-1}\Gamma(\sigma)\Gamma(z)$ $\cdot {}_3F_2(\alpha,\beta,z;\gamma,\sigma;z)$ $\mathrm{Re}\ \sigma > 0, \mathrm{Re}\ z > 0, \lvert\arg(-z)\rvert < \pi$
15.6	$(1-x)^{\gamma-1}e^{x\delta}$ $\cdot {}_2F_1(\alpha,\beta;\gamma;1-x) \quad x < 1$ $0 \quad x > 1$ $\mathrm{Re}\ \gamma > 0$	$[\Gamma(\gamma-\alpha+z)\Gamma(\gamma-\beta+z)]^{-1}\Gamma(\gamma)\Gamma(z)$ $\cdot\Gamma(\gamma-\alpha-\beta+z)\, {}_2F_2(z,\gamma-\alpha-\beta+z;$ $\gamma-\alpha+z,\gamma-\beta+z;\delta)$ $\mathrm{Re}\ z > \begin{cases} 0 \\ \mathrm{Re}(\alpha+\beta-\gamma) \end{cases}$
15.7	$0 \quad x < a$ $(x-a)^{\gamma-1}$ $\cdot {}_2F_1(\alpha,\beta;\gamma;a-x) \quad x > a$ $\mathrm{Re}\ \gamma > 0$	$[\Gamma(1-z)\Gamma(1+\alpha+\beta-\gamma-z)]^{-1}$ $\cdot\Gamma(\gamma)\Gamma(1+\alpha-\gamma-z)\Gamma(1+\beta-\gamma-z)$ $\cdot {}_2F_1(1+\alpha-\gamma-z,1+\beta-\gamma-z;1+\alpha+\beta-\gamma-z;1-a)$ $\mathrm{Re}\ z > \begin{cases} \mathrm{Re}(1+\alpha-\gamma) \\ \mathrm{Re}(1+\beta-\gamma) \end{cases}$
15.8	${}_2F_1(a,b;c;-x)$ $\mathrm{Re}(z,a-z,b-z) > 0$	$\dfrac{\Gamma(c)\Gamma(a-z)\Gamma(b-z)\Gamma(z)}{\Gamma(a)\Gamma(b)\Gamma(c-z)}$ $c \neq 0,-1,-2,\cdots$
15.9	$e^{-x}\, {}_pF_q(a_1,\cdots,a_p;$ $b_1,\cdots,b_q;ax)$	$\Gamma(z)\, {}_{p+1}F_q(z,a_1,\cdots,a_p;b_1,\cdots,b_q;a)$ $p < a,\ \mathrm{Re}\ z > 0$

	$\phi(x)$	$\Phi(z) = \int_0^\infty \phi(x)x^{z-1}dx$
15.10	$(1-x)^{\nu-1}{}_pF_q(a_1,\cdots,a_p;$ $b_1,\cdots,b_q;ax)$ $x < 1$ 0 $x > 1$	$B(\nu,z)\,{}_{p+1}F_{q+1}(a_1,\cdots,a_p,z;$ $b_1,\cdots,b_q,z+\nu;a)$ $\text{Re}\ \nu > 0,\quad \text{Re}\ z > 0$
15.11	$K_\nu(2x^{\frac{1}{2}})\,{}_pF_q(a_1,\cdots,a_p;$ $b_1,\cdots,b_q;ax)$ $\text{Re}\ z > \frac{1}{2}\lvert\text{Re}\ \nu\rvert$	$\frac{1}{2}\Gamma(z+\frac{1}{2}\nu)\Gamma(z-\frac{1}{2}\nu)\,{}_{p+2}F_q(\frac{1}{2}\nu+z,$ $z-\frac{1}{2}\nu,a_1,\cdots,a_p;b_1,\cdots,b_q;a)$ $p < 9 - 1$
15.12	$K_\lambda(ax)K_\mu(ax)$ $\cdot{}_pF_q(a_1,\cdots,a_p;b_1,\cdots,$ $b_q;bx^2)$	Sinha, S., 1943: Bull. Calcutta Math. Soc. 35, p. 37-42

2.1 General Formulas

	$\Phi(z) = \int_0^\infty x^{z-1}\phi(x)dx$	$\phi(x)$
1.1	$\Phi(z)$	$(2\pi i)^{-1} \int_{c-i\infty}^{c+i\infty} \Phi(z)x^{-z}dz$
1.2	$a^{-z}\Phi(z)$	$\phi(ax)$; $a > 0$
1.3	$\Phi(z+\alpha)$	$x^{\alpha}\phi(x)$
1.4	$\Phi(pz)$	$p^{-1}\Phi(x^{1/p})$; $p > 0$
1.5	$\Phi(-pz)$	$p^{-1}\phi(x^{-1/p})$; $p > a$
1.6	$a^{-z}\Phi(pz+\alpha)$	$p^{-1}(ax)^{\alpha/p}\phi[(ax)^{1/p}]$, $p,a > 0$
1.7	$a^{-z}\Phi(\alpha-pz)$	$p^{-1}(ax)^{-\alpha/p}\phi[(ax)^{-1/p}]$; $p,\alpha > 0$
1.8	$\Phi^{(n)}(z)$	$\phi(x)(\log x)^{n}$

For further formulas see Part I, 1.1.

2.2 Algebraic Functions and Powers of Arbitrary Order

	$\Phi(z) = \int_0^\infty x^{z-1}\phi(x)\,dx$	$\phi(x)$	
2.1	$z^{-1}a^z$	1	$x < a$
	Re $z > 0$	0	$x > a$
2.2	$z^{-1}a^z$	0	$x < a$
	Re $z < 0$	-1	$x > a$
2.3	$(\nu+z)^{-1}a^z$	$(x/a)^\nu$	$x < a$
	Re $z > -$Re ν	0	$x > a$
2.4	$(\nu+z)^{-1}a^z$	0	$x < a$
	Re $z < -$Re ν	$(x/a)^\nu$	$x > a$
2.5	$z^{-2}a^z$	$-\log(x/a)$	$x < a$
	Re $z > 0$	0	$x > a$
2.6	$z^{-2}a^z$	0	$x < a$
	Re $z < 0$	$\log(x/a)$	$x > a$
2.7	$(\nu+z)^{-2}a^z$	$-(x/a)^\nu\log(x/a)$	$x < a$
	Re $z > -$Re ν	0	$x > a$

	$\Phi(z) = \int_0^\infty x^{z-1}\phi(x)\,dx$	$\phi(x)$	
2.8	$(\nu+z)^{-2}a^z$	0	$x < a$
	$\mathrm{Re}\, z < -\mathrm{Re}\,\nu$	$(x/a)^\nu \log(x/a)$	$x > a$
2.9	$[(z+\alpha)(z+\beta)]^{-1}$	$(\beta-\alpha)^{-1}(x^\alpha - x^\beta)$	$x < 1$
	$\mathrm{Re}\, z > -\mathrm{Re}(\alpha,\beta)$	0	$x > 1$
2.10	$[(z+\alpha)(z+\beta)]^{-1}$	$(\beta-\alpha)^{-1}x^\alpha$	$x < 1$
	$-\mathrm{Re}\,\alpha < \mathrm{Re}\, z < -\mathrm{Re}\,\beta$	$(\beta-\alpha)^{-1}x^\beta$	$x > 1$
2.11	$[(z+\alpha)(z+\beta)]^{-1}$	0	$x < 1$
	$\mathrm{Re}\, z < -\mathrm{Re}(\alpha,\beta)$	$(\beta-\alpha)^{-1}(x^\beta - x^\alpha)$	$x > 1$
2.12	$(z^2+\alpha^2)^{-1}$	$\alpha^{-1}\sin[\alpha \log(1/x)]$	$x < 1$
	$\mathrm{Re}\, z > \lvert\mathrm{Im}\,\alpha\rvert$	0	$x > 1$
2.13	$(z^2+\alpha^2)^{-1}$	0	$x < 1$
	$\mathrm{Re}\, z < \lvert\mathrm{Im}\,\alpha\rvert$	$\alpha^{-1}\sin(a \log x)$	$x > 1$
2.14	$(z^2+\alpha^2)^{-1}$	$-\frac{1}{2}i\alpha^{-1}x^{-i\alpha}$	$x < 1$
	$-\mathrm{Im}\,\alpha < \mathrm{Re}\, z < \mathrm{Im}\,\alpha$	$-\frac{1}{2}i\alpha^{-1}x^{i\alpha}$	$x > 1$
	$\mathrm{Im}\,\alpha > 0$		

	$\Phi(z) = \int_0^\infty x^{z-1}\phi(x)dx$	$\phi(x)$	
2.15	$(z^2+\alpha^2)^{-1}$	$\frac{1}{2}i\alpha^{-1}x^{i\alpha}$	$x < 1$
	$\text{Im}\ \alpha < \text{Re}\ z < -\text{Im}\ \alpha$ $\text{Im}\ \alpha < 0$	$\frac{1}{2}i\alpha^{-1}x^{-i\alpha}$	$x > 1$
2.16	$z(z^2+\alpha^2)^{-1}$	$\cos(\alpha \log x)$	$x < 1$
	$\text{Re}\ z > \|\text{Im}\ \alpha\|$	0	$x > 1$
2.17	$z(z^2+\alpha^2)^{-1}$	0	$x < 1$
	$\text{Re}\ z > \|\text{Im}\ \alpha\|$	$-\cos(\alpha \log x)$	$x > 1$
2.18	$z(z^2+\alpha^2)^{-1}$	$\frac{1}{2}x^{-i\alpha}$	$x < 1$
	$\text{Im}\ \alpha > 0$ $-\text{Im}\ \alpha < \text{Re}\ z < \text{Im}\ \alpha$	$-\frac{1}{2}x^{i\alpha}$	$x > 1$
2.19	$z(z^2+\alpha^2)^{-1}$	$\frac{1}{2}x^{i\alpha}$	$x < 1$
	$\text{Im}\ \alpha < 0$ $\text{Im}\ \alpha < \text{Re}\ z < -\text{Im}\ \alpha$	$-\frac{1}{2}x^{-i\alpha}$	$x > 1$
2.20	$z^{-1}(z^2+\alpha^2)^{-1}$	$2\alpha^{-2}\sin^2[\frac{1}{2}\alpha\log(1/x)]$	$x < 1$
	$\text{Re}\ z > \|\text{Im}\ \alpha\|$	0	$x > 1$

	$\Phi(z) = \int_0^\infty x^{z-1}\phi(x)\,dx$	$\phi(x)$	
2.21	$z(z^2+\alpha^2)^{-\frac{1}{2}}$	$\frac{1}{2}\alpha^{-1}\log(1/x)\sin[\alpha\log(1/x)]$	$x < 1$
	$\mathrm{Re}\; z > \lvert \mathrm{Im}\;\alpha \rvert$	0	$x > 1$
2.22	$z^{-1}(z+\alpha)^{-\frac{1}{2}}$	$\alpha^{-\frac{1}{2}}\mathrm{Erf}\{[\alpha\; \log(1/x)]^{\frac{1}{2}}\}$	$x < 1$
	$\mathrm{Re}\; z > \lvert \mathrm{Im}\;\alpha \rvert$	0	$x > 1$
2.23	$(z^2+\alpha^2)^{-\frac{1}{2}}$	$[\frac{1}{2}\pi\log(1/x)]^{-\frac{1}{2}}\cos[(\alpha\log(1/x)]$	
	$\cdot[(z^2+\alpha^2)^{\frac{1}{2}}+z]^{\frac{1}{2}}$		$x < 1$
	$\mathrm{Re}\; z > \lvert \mathrm{Im}\;\alpha \rvert$	0	$x > 1$
2.24	$(z^2+\alpha^2)^{-\frac{1}{2}}$	$[\frac{1}{2}\pi\log(1/x)]^{-\frac{1}{2}}\sin[\alpha\; \log(1/x)]$	
	$\cdot[(z^2+\alpha^2)^{\frac{1}{2}}-z]^{\frac{1}{2}}$		$x < 1$
	$\mathrm{Re}\; z > \lvert \mathrm{Im}\;\alpha \rvert$	0	$x > 1$
2.25	$z^{-\nu}a^z$	$[\Gamma(\nu)]^{-1}[\log(a/x)]^{\nu-1}$	$x < a$
	$\mathrm{Re}\; z > 0,\; \mathrm{Re}\; \nu > 0$	0	$x > a$
2.26	$(-z)^{-\nu}a^z$	0	$x < a$
	$\mathrm{Re}\; z < 0, \mathrm{Re}\; \nu > 0$	$[\Gamma(\nu)]^{-1}[\log(x/a)]^{\nu-1}$	$x > a$

	$\Phi(z) = \int_0^\infty x^{z-1}\phi(x)\,dx$	$\phi(x)$
2.27	$z^{-1}(z+\alpha)^{-\nu}$ $\mathrm{Re}\,\nu > 0$ $\mathrm{Re}\,z > (0,-\mathrm{Re}\,\alpha)$	$\alpha^{-\nu}[\Gamma(\nu)]^{-1}\gamma[\nu,\alpha\log(1/x)]$ $\quad x<1$ 0 $\quad x>1$
2.28	$z^{-1}(z+\alpha)^{-\nu}$ $\mathrm{Re}\,\nu > 0$ $-\mathrm{Re}\,\alpha < \mathrm{Re}\,z < 0$	$-\alpha^{-\nu}[\Gamma(\nu)]^{-1}\Gamma[\nu,\alpha\log(1/x)]$ $\quad x<1$ $-\alpha^{-\nu}$ $\quad x>1$
2.29	$[(z+\alpha)^{\frac{1}{2}}-\alpha^{\frac{1}{2}}]^{\nu}$ $\mathrm{Re}\,z>-\mathrm{Re}\,\alpha, \mathrm{Re}\,\nu>0$	$\frac{1}{2}\nu\alpha^{\frac{1}{2}\nu}[\log(1/x)]^{-1}x^{\frac{1}{2}\alpha}I_{\frac{1}{2}\nu}[\frac{1}{2}\alpha\log(1/x)]$ $x<1$ 0 $\quad x>1$
2.30	$[(z+\alpha)^{\frac{1}{2}}+\beta^{\frac{1}{2}}]^{\nu}$ $\mathrm{Re}\,z>-\mathrm{Re}\,\alpha, \mathrm{Re}\,\nu<0$ $\mathrm{Re}\,\beta > 0$	$-(2/\pi)^{\frac{1}{2}}\nu[2\log(1/x)]^{-1-\frac{1}{2}\nu}x^{\alpha-\frac{1}{2}\beta}$ $\cdot D_{\nu-1}\{[2\beta\log(1/x)]^{\frac{1}{2}}\}$ $\quad x<1$ 0 $\quad x>1$
2.31	$(z+\alpha)^{-\frac{1}{2}}$ $[(z+\alpha)^{\frac{1}{2}}+\beta^{\frac{1}{2}}]^{\nu}$ $\mathrm{Re}\,z>-\mathrm{Re}\,\alpha, \mathrm{Re}\,\nu<1$ $\mathrm{Re}\,\beta > 0$	$(2/\pi)^{\frac{1}{2}}[2\log(1/x)]^{-\frac{1}{2}-\frac{1}{2}\nu}x^{\alpha-\frac{1}{2}\beta}$ $\cdot D_{\nu}\{[2\beta\log(1/x)]^{\frac{1}{2}}\}$ $\quad x<1$ 0 $\quad x>1$

	$\Phi(z) = \int_0^\infty x^{z-1}\phi(x)\,dx$	$\phi(x)$
2.32	$z^{-\nu}[(z+\alpha)^{\frac{1}{2}}-\alpha^{\frac{1}{2}}]^{\nu}$ Re $z>-$Re α, Re $\nu>0$	$(2/\pi)^{\frac{1}{2}}\nu[2\log(1/x)]^{\frac{1}{2}\nu-1}x^{\frac{1}{2}\alpha}$ $\cdot D_{-\nu-1}\{[2\alpha\log(1/x)]^{\frac{1}{2}}\}$ $x<1$ 0 $x>1$
2.33	$z^{-\nu}(z+\alpha)^{-\frac{1}{2}}$ $[(z+\alpha)^{\frac{1}{2}}-\alpha^{\frac{1}{2}}]^{\nu}$ Re $z>-$Re α, Re $\nu>-1$	$(2/\pi)^{\frac{1}{2}}[2\log(1/x)]^{\frac{1}{2}\nu-\frac{1}{2}}x^{\frac{1}{2}\alpha}$ $\cdot D_{-\nu}\{[2\alpha\log(1/x)]^{\frac{1}{2}}\}$ $x<1$ 0 $x>1$
2.34	$(\alpha^{\frac{1}{2}}+z^{\frac{1}{2}})^{\nu}$ Re $z>0$, Re $\nu>0$	$-(2\pi)^{-\frac{1}{2}}\nu 2^{-\frac{1}{2}\nu}[\log(1/x)]^{-1-\frac{1}{2}\nu}x^{-\frac{1}{2}\alpha}$ $\cdot D_{\nu-1}\{[2\alpha\log(1/x)]^{\frac{1}{2}}\}$ $x<1$ 0 $x>1$
2.35	$[(z+\alpha)(z+\beta)]^{-\nu}$ Re $z>-$Re(a,b) Re $\nu>0$	$\pi^{\frac{1}{2}}[\Gamma(\nu)]^{-1}(\alpha-\beta)^{\frac{1}{2}-\nu}[\log(1/x)]^{\nu-\frac{1}{2}}$ $\cdot x^{\frac{1}{2}(\alpha+\beta)}I_{\nu-\frac{1}{2}}[\frac{1}{2}(\alpha-\beta)\log(1/x)]$ $x<1$ 0 $x>1$
2.36	$(z-\alpha)^{\nu}(z+\alpha)^{-\mu}$ Re $z>\|$Re $\alpha\|$ Re$(\mu-\nu)>0$	$(2\alpha)^{\frac{1}{2}\nu-\frac{1}{2}\mu}[\Gamma(\mu-\nu)]^{-1}[\log(1/x)]^{\frac{1}{2}\mu-\frac{1}{2}\nu-1}$ $\cdot M_{\frac{1}{2}\mu+\frac{1}{2}\nu,\frac{1}{2}\mu-\frac{1}{2}\nu-\frac{1}{2}}[2\alpha\log(1/x)]$ $x<1$ 0 $x>1$

	$\Phi(z) = \int_0^\infty x^{z-1}\phi(x)\,dx$	$\phi(x)$
2.37	$(z-\alpha)^{\mu}(z-\beta)^{-\nu}$ $\mathrm{Re}\ z>\mathrm{Re}(\alpha,\beta)$	$(\alpha-\beta)^{\frac{1}{2}\mu-\frac{1}{2}\nu}[\Gamma(\nu-\mu)]^{-1}x^{-\frac{1}{2}\alpha-\frac{1}{2}\beta}$ $\cdot M_{\frac{1}{2}\mu+\frac{1}{2}\nu,\frac{1}{2}\nu-\frac{1}{2}\mu-\frac{1}{2}}[(\alpha-\beta)\log(1/x)]$ $\quad x<1$ $0 \quad x>1$
2.38	$(\alpha-z)^{\mu}(\alpha+z)^{-\nu}$ $\mathrm{Re}\ z>\lvert\mathrm{Re}\ \alpha\rvert$ $\mathrm{Re}(\nu-\mu)>0$	$(2\alpha)^{\frac{1}{2}\mu-\frac{1}{2}\nu}[\Gamma(\nu)]^{-1}[\log(1/x)]^{\frac{1}{2}\nu-\frac{1}{2}\mu-1}$ $\cdot W_{\frac{1}{2}\mu+\frac{1}{2}\nu,\frac{1}{2}\nu-\frac{1}{2}\mu-1}[2\alpha\log(1/x)]$ $\quad x<1$ $0 \quad x>1$
2.39	$(z-\alpha)^{\mu}(\beta-z)^{-\nu}$ $\mathrm{Re}\ \alpha<\mathrm{Re}\ z<\mathrm{Re}\ \beta$ $\mathrm{Re}(\nu-\mu)>0$	$[\Gamma(-\mu)]^{-1}(\beta-\alpha)^{\frac{1}{2}\mu-\frac{1}{2}\nu}x^{-\frac{1}{2}\alpha-\frac{1}{2}\beta}$ $\cdot[\log(1/x)]^{\frac{1}{2}\nu-\frac{1}{2}\mu-1}$ $\cdot W_{-\frac{1}{2}\mu-\frac{1}{2}\nu,\frac{1}{2}+\frac{1}{2}\mu-\frac{1}{2}\nu}[(\beta-\alpha)\log(1/x)]$ $x<1$ $[\Gamma(\nu)]^{-1}(\beta-\alpha)^{\frac{1}{2}\mu-\frac{1}{2}\nu}[\log(1/x)]^{\frac{1}{2}\nu-\frac{1}{2}\mu-1}$ $\cdot x^{-\frac{1}{2}\alpha-\frac{1}{2}\beta}W_{\frac{1}{2}\mu+\frac{1}{2}\nu,\frac{1}{2}+\frac{1}{2}\mu-\frac{1}{2}\nu}[(\beta-\alpha)\log x]$ $x>1$
2.40	$(z^2+\alpha^2)^{-\nu}$ $\mathrm{Re}\ z\ \lvert\mathrm{Im}\ \alpha\rvert$ $\mathrm{Re}\ \nu>0$	$\pi^{\frac{1}{2}}(2\alpha)^{\frac{1}{2}-\nu}[\Gamma(\nu)]^{-1}[\log(1/x)]^{\nu-\frac{1}{2}}$ $\cdot J_{\nu-\frac{1}{2}}[\alpha\log(1/x)]$ $\quad x<1$ $0 \quad x>1$

	$\Phi(z) = \int_0^\infty x^{z-1}\phi(x)dx$	$\phi(x)$
2.41	$(z^2-\alpha^2)^{-\nu}$ $\mathrm{Re}\ z>\lvert\mathrm{Re}\ \alpha\rvert$ $\mathrm{Re}\ \nu > 0$	$\pi^{\frac{1}{2}}(2\alpha)^{\frac{1}{2}-\nu}[\Gamma(\nu)]^{-1}[\log(1/x)]^{\nu-\frac{1}{2}}$ $I_{\nu-\frac{1}{2}}[\alpha \log(1/x)]$ $x < 1$ 0 $x > 1$
2.42	$(\alpha^2-z^2)^{-\nu}$ $-\mathrm{Re}\ \alpha<\mathrm{Re}\ z<\mathrm{Re}\ \alpha$ $\mathrm{Re}\ \nu > 0$	$\pi^{-\frac{1}{2}}(2\alpha)^{\frac{1}{2}-\nu}[\Gamma(\nu)]^{-1}[\log(1/x)]^{\nu-\frac{1}{2}}$ $\cdot K_{\nu-\frac{1}{2}}[\alpha \log(1/x)]$ $x < 1$ $\pi^{-\frac{1}{2}}(2\alpha)^{\frac{1}{2}-\nu}[\Gamma(\nu)]^{-1}(\log x)^{\nu-\frac{1}{2}}$ $\cdot K_{\nu-\frac{1}{2}}(\alpha \log x)$ $x > 1$
2.43	$[z+(z^2+\alpha^2)^{\frac{1}{2}}]^{-\nu}$ $\mathrm{Re}\ z>\lvert\mathrm{Im}\ \alpha\rvert, \mathrm{Re}\ \nu>0$	$\nu\alpha^{-\nu}[\log(\frac{1}{x})]^{-1}J_\nu[\alpha\log(1/x)]$ $x < 1$ 0 $x > 1$
2.44	$(z^2+\alpha^2)^{-\frac{1}{2}}$ $[z+(z^2+\alpha^2)^{\frac{1}{2}}]^{-\nu}$ $\mathrm{Re}\ z>\lvert\mathrm{Im}\ \alpha\rvert, \mathrm{Re}\ \nu>-1$	$\alpha^{-\nu}J_\nu[\alpha\log(1/x)]$ $x < 1$ 0 $x > 1$
2.45	$(z^2-\alpha^2)^{-\frac{1}{2}}$ $\cdot\{[z+(z^2-\alpha^2)^{\frac{1}{2}}]^\nu$ $-[z-(z^2-\alpha^2)^{\frac{1}{2}}]^\nu\}$ $\mathrm{Re}\ z > -\lvert\mathrm{Re}\ \alpha\rvert$	$2\pi^{-1}\alpha^\nu\sin(\pi\nu)K_\nu[\alpha\log(1/x)]$ $x < 1$ 0 $x > 1$ $-1 < \mathrm{Re}\ \nu < 1$

	$\Phi(z) = \int_0^\infty x^{z-1}\phi(x)\,dx$	$\phi(x)$
2.46	$z^{-2\mu}(\alpha^2+z^2)^{-\nu}$ $\mathrm{Re}\, z > \lvert \mathrm{Im}\, \alpha \rvert$ $\mathrm{Re}(\mu+\nu) > 0$	$[\Gamma(2\nu+2\mu)]^{-1}[\log(1/x)]^{2\mu+2\nu-1}$ $\cdot {}_1F_2[\nu;\nu+\mu,\tfrac{1}{2}+\nu+\mu;-\frac{\alpha^2}{4}\log^2(1/x)]$ $\quad x < 1$ $0 \quad x > 1$

2.3 Exponential and Logarithmic Functions

	$\Phi(z) = \int_0^\infty x^{z-1}\phi(x)\,dx$	$\phi(x)$	
3.1	$e^{\alpha z^2}$ Re $\alpha > 0$	$\frac{1}{2}(\pi\alpha)^{-\frac{1}{2}}e^{-\frac{1}{4}y^2/\alpha}$	
3.2	$z^{\nu}e^{\alpha z^2}$ Re $z > 0$, Re $\alpha > 0$	$(2\alpha)^{-\frac{1}{2}\nu}(\pi\alpha)^{-\frac{1}{2}}e^{-\frac{1}{8}y^2/\alpha}$ $\cdot D_{\nu}[(2\alpha)^{-\frac{1}{2}}y]$	
3.3	$z^{-\frac{1}{2}}e^{\alpha/z}$ Re $z > 0$	$(\pi y)^{-\frac{1}{2}}\cosh[2(\alpha y)^{\frac{1}{2}}]$ 0	$x < 1$ $x > 1$
3.4	$z^{-\frac{1}{2}}e^{-\alpha/z}$ Re $z > 0$	$(\pi y)^{-\frac{1}{2}}\cos[2(\alpha y)^{\frac{1}{2}}]$ 0	$x < 1$ $x > 1$
3.5	$z^{-\nu-1}e^{\alpha/z}$ Re $z > 0$, Re $\nu > -1$	$(y/\alpha)^{\frac{1}{2}\nu}I_{\nu}[2(\alpha y)^{\frac{1}{2}}]$ 0	$x < 1$ $x > 1$
3.6	$z^{-\nu-1}e^{-\alpha/z}$ Re $z > 0$, Re $\nu > -1$ $y = \log(1/x)$	$(y/\alpha)^{\frac{1}{2}\nu}J_{\nu}[2(\alpha y)^{\frac{1}{2}}]$ 0	$x < 1$ $x > 1$

	$\Phi(z) = \int_0^\infty x^{z-1}\phi(x)\,dx$	$\phi(x)$
3.7	$e^{-\alpha z^{1/2}}$ Re $z > 0$, Re $\alpha^2 > 0$	$\frac{1}{2}\alpha\pi^{-\frac{1}{2}}y^{-3/2}e^{-\frac{1}{4}\alpha^2/y}$ $\quad x < 1$ 0 $\quad x > 1$
3.8	$z^{-\frac{1}{2}}e^{-\alpha z^{1/2}}$ Re $z > 0$, Re $\alpha^2 > 0$	$(\pi y)^{-\frac{1}{2}}e^{-\frac{1}{4}\alpha^2/y}$ $\quad x < 1$ 0 $\quad x > 1$
3.9	$z^{-1}e^{-\alpha z^{1/2}}$ Re $z > 0$, Re $\alpha^2 > 0$	$\mathrm{Erfc}(\frac{1}{2}\alpha y^{-\frac{1}{2}})$ $\quad x < 1$ 0 $\quad x > 1$
3.10	$z^{-1}(1-e^{\alpha z^{1/2}})$ Re $z > 0$, Re $\alpha^2 > 0$	$\mathrm{Erf}(\frac{1}{2}ay^{-\frac{1}{2}})$ $\quad x < 1$ 0 $\quad x > 1$
3.11	$z^{\nu}e^{-\alpha z^{1/2}}$ Re $z > 0$, Re $\alpha^2 > 0$	$2^{-\nu-\frac{1}{2}}\pi^{-\frac{1}{2}}y^{-\nu-1}e^{-1/8\,\alpha^2/y}$ $\cdot D_{2\nu+1}[a(2y)^{-\frac{1}{2}}]$ $\quad x < 1$ 0 $\quad x > 1$
3.12	$\exp(-3z^{1/3})$ Re $z > 0$ $y = \log(1/x)$	$3^{\frac{1}{2}}\pi^{-1}y^{-3/2}K_{1/3}(2y^{-\frac{1}{2}})$ $\quad x < 1$ 0 $\quad x > 1$

	$\Phi(z) = \int_0^\infty x^{z-1}\phi(x)$	$\phi(x)$	
3.13	$z^{-1/3}\exp(-3z^{1/3})$ Re $z > 0$	$3^{1/2}(\pi y)^{-1}K_{2/3}(2y^{-1/2})$ 0	$x < 1$ $x > 1$
3.14	$z^{-2/3}\exp(-3z^{1/3})$ Re $z > 0$	$\pi^{-1}(\tfrac{1}{3}y)^{-1/2}K_{1/3}(2y^{-1/2})$ 0	$x < 1$ $x > 1$
3.15	$(z^2+\alpha^2)^{-1/2}$ $\cdot\exp[-b(z^2+\alpha^2)^{1/2}]$ Re $z > \|\text{Im } \alpha\|$	$J_o[\alpha(y^2-b^2)^{1/2}]$ 0	$x < e^{-b}$ $x > e^{-b}$
3.16	$(z^2-\alpha^2)$ $\cdot\exp[-b(z^2-\alpha^2)^{1/2}]$ Re $z > \|\text{Re } \alpha\|$	$I_o[\alpha(y^2-b^2)^{1/2}]$ 0	$x < e^{-b}$ $x > e^{-b}$
3.17	$(z^2+\alpha^2)^{-1/2}[(z^2+\alpha^2)^{1/2}-z]^{\nu}$ $\cdot\exp[-b(z^2+\alpha^2)^{1/2}]$ Re $z > \|\text{Im } \alpha\|$, Re $\nu > -1$ $y = \log(1/x)$	$a^{\nu}[(y-b)/(y+b)]^{\frac{1}{2}\nu}$ $\cdot J_{\nu}[a(y^2-b^2)^{1/2}]$ 0	 $x < e^{-b}$ $x > e^{-b}$

	$\Phi(z) = \int_0^\infty x^{z-1}\phi(x)\,dx$	$\phi(x)$
3.18	$(z^2-\alpha^2)^{-\frac{1}{2}}[z-(z^2-\alpha^2)^{\frac{1}{2}}]^{\nu}$ $\cdot\exp[-b(z^2-\alpha^2)^{\frac{1}{2}}]$ $\mathrm{Re}\ z > \lvert\mathrm{Re}\ \alpha\rvert,\ \mathrm{Re}\ \nu > -1$	$a^{\nu}[(y-b)/(y+b)]^{\frac{1}{2}\nu}$ $\cdot I_{\nu}[a(y^2-b^2)^{\frac{1}{2}}]$ $x < e^{-b}$ 0 $x > e^{-b}$
3.19	$z^{-1}e^{\alpha/z}\log z$ $\mathrm{Re}\ z > 0$	$K_0[2(\alpha y)^{\frac{1}{2}}]-\frac{1}{2}I_0[2(\alpha y)^{\frac{1}{2}}]$ $\cdot\log(y/\alpha)$ $x < 1$ 0 $x > 1$
3.20	$z^{-1}e^{-\alpha/z}\log z$ $\mathrm{Re}\ z > 0$	$-\frac{1}{2}\pi Y_0[2(\alpha y)^{\frac{1}{2}}]-\frac{1}{2}J_0[2(\alpha y)^{\frac{1}{2}}]$ $\cdot\log(y/\alpha)$ $x < 1$ 0 $x > 1$
3.21	$z^{-1}\log z$ $\mathrm{Re}\ z > 0$	$-\gamma-\log y$ $x < 1$ 0 $x > 1$
3.22	$z^{-\frac{1}{2}}\log z$ $\mathrm{Re}\ z > 0$ $y = \log(1/x)$	$-(\pi y)^{-\frac{1}{2}}[\log y+\gamma+\log 4]$ $x < 1$ 0 $x > 1$

	$\Phi(z) = \int_0^\infty x^{z-1}\phi(x)\,dx$	$\phi(x)$
3.23	$z^{-n-1}\log z$ $n = 1,2,3,\cdots$ $\mathrm{Re}\ z > 0$	$(n!)^{-1}y^n(1+\frac{1}{2}+\cdots+\frac{1}{n}-\gamma-\log y)$ $\quad x < 1$ $0 \quad x > 1$
3.24	$z^{-n-\frac{1}{2}}\log z$ $n = 1,2,3,\cdots$ $\mathrm{Re}\ z > 0$	$\pi^{-\frac{1}{2}}2^{2n}(n!)[(2n)!]^{-1}y^{n-\frac{1}{2}}$ $\cdot[2(1+\frac{1}{3}+\cdots+\frac{1}{2n-1})-\gamma-\log(4y)]$ $\quad x < 1$ $0 \quad x > 1$
3.25	$z^{-\nu}\log z$ $\mathrm{Re}\ z > 0,\ \mathrm{Re}\ \nu > 0$	$[\Gamma(\nu)]^{-1}y^{\nu-1}[\psi(\nu)-\log y]$ $\quad x < 1$ $0 \quad x > 1$
3.26	$z^{-1}\log(1+z/a)$ $\mathrm{Re}\ z > 0$	$-\mathrm{Ei}(-ay)$ $\quad x < 1$ $0 \quad x > 1$
3.27	$z^{-1}\log(z/a-1)$ $\mathrm{Re}\ z > a$	$-\overline{\mathrm{Ei}}(ay)$ $\quad x < 1$ $0 \quad x > 1$
3.28	$\log(1-a/z)$ $\mathrm{Re}\ z > a$ $y = \log(1/x)$	$y^{-1}(1-x^a)$ $\quad x < 1$ $0 \quad x > 1$

	$\Phi(z) = \int_0^\infty x^{z-1}\phi(x)\,dx$	$\phi(x)$	
3.29	$\log(\frac{z+a}{z-a})$	$y^{-1}(x^{-a}-x^{a})$	$x < 1$
	$\mathrm{Re}\ z > a$	0	$x > 1$
3.30	$\log(\frac{z+b}{z+a})$	$y^{-1}(x^{a}-x^{b})$	$x < 1$
	$\mathrm{Re}\ z > -(a,b)$	0	$x > 1$
3.31	$(z+a)^{-1}\log(z+a)$	$-x^{a}(\gamma+\log y)$	$x < 1$
	$\mathrm{Re}\ z > -a$	0	$x > 1$
3.32	$(z+a)^{-1}\log(z+b)$	$x^{a}\{\log(b-a)-\mathrm{Ei}[-y(b-a)]\}$	$x < 1$
	$b > a,\ \mathrm{Re}\ z > -b$	0	$x > 1$
3.33	$z^{-1}\log(1+a^2/z^2)$	$-2\mathrm{Ci}(ay)$	$x < 1$
	$\mathrm{Re}\ z > 0$	0	$x > 1$
3.34	$z^{-1}\log(z^2+a^2)$	$2\log a-2\mathrm{Ci}(ay)$	$x < 1$
	$\mathrm{Re}\ z > 0$	0	$x > 1$
3.35	$(z^2+a^2)^{-1}\log z$	$a^{-1}\{\cos(ay)\mathrm{Si}(ay)+\sin(ay)$	
	$\mathrm{Re}\ z > 0$	$\cdot[\log a-\mathrm{Ci}(ay)]\}$	$x < 1$
		0	$x > 1$
	$y = \log(1/x)$		

	$\Phi(z) = \int_0^\infty x^{z-1}\phi(x)\,dx$	$\phi(x)$	
3.36	$\log(1+a^2/z^2)$ Re $z > 0$	$2y^{-1}[1-\cos(ay)]$ 0	$x < 1$ $x > 1$
3.37	$\log(1-a^2/z^2)$ Re $z > a$	$2y^{-1}[1-\cosh(ay)]$ 0	$x < 1$ $x > 1$
3.38	$\log(\frac{z^2+a^2}{z^2+b^2})$ Re $z > 0$	$2y^{-1}[\cos(by)-\cos(ay)]$ 0	$x < 1$ $x > 1$
3.39	$\log[\frac{c^2+(z+a)^2}{c^2+(z+b)^2}]$ Re $z > 0$	$2y^{-1}\cos(cy)(x^a-x^b)$ 0	$x < 1$ $x > 1$
3.40	$\log(\frac{z^2-a^2}{z^2-b^2})$ Re $z > (a,b)$	$2y^{-1}[\cosh(by)-\cosh(ay)]$ 0	$x < 1$ $x > 1$
3.41	$(z^2+a^2)^{-1}$ $\cdot\log(z^2+a^2)$ Re $z > 0$ $y = \log(1/x)$	$-a^{-1}\sin(ay)[\gamma+\log(\frac{1}{2}y/a)+\mathrm{Ci}(2ay)]$ $+a^{-1}\cos(ay)\mathrm{Si}(2ay)$ 0	 $x < 1$ $x > 1$

	$\Phi(z) = \int_0^\infty x^{z-1}\phi(x)\,dx$	$\phi(x)$	
3.42	$\log\left(\frac{z^{\frac{1}{2}}+a}{z^{\frac{1}{2}}-a}\right)$ Re $z > 0$	$y^{-1}e^{a^2y}\mathrm{Erf}(ay^{\frac{1}{2}})$ 0	$x < 1$ $x > 1$
3.43	$(z^2+a^2)^{-\frac{1}{2}}$ $\cdot\log[z/a+(1+z^2/a^2)^{\frac{1}{2}}]$ Re $z > 0$	$J_o(ay)\log a-\frac{1}{2}\pi Y_o(ay)$ 0	$x < 1$ $x > 1$
3.44	$(z^2+a^2)^{-\frac{1}{2}}$ $\cdot\log(z^2+a^2)$ Re $z > 0$	$-\frac{1}{2}\pi Y_o(ay)-J_o(ay)[\gamma+\log(2y/a)]$ 0	$x < 1$ $x > 1$
3.45	$(z^2+a^2)^{-\frac{1}{2}}$ $\cdot\log[a/z+(1+a^2/z^2)^{\frac{1}{2}}]$ Re $z > 0$	$-\frac{1}{2}\pi \mathbf{H}_o(ay)$ 0	$x < 1$ $x > 1$
3.46	$(z^2-a^2)^{-\frac{1}{2}}$ $\cdot\log(z^2-a^2)$ Re $z > a$ $y = \log(1/x)$	$K_o(ay)-I_o(ay)[\gamma+\log(2y/a)]$ 0	$x < 1$ $x > 1$

	$\Phi(z) = \int_0^\infty x^{z-1}\phi(x)\,dx$	$\phi(x)$	
3.47	$(z^2-a^2)^{-\frac{1}{2}}$ $\cdot\log(z+(z^2-a^2)^{\frac{1}{2}}]$ Re $z > a$	$I_o(ay)\log a+K_o(ay)$	$x < 1$
		0	$x > 1$
3.48	$(z^2-a^2)^{-\frac{1}{2}}$ $\cdot\log[z/a+(z^2/a^2-1)^{\frac{1}{2}}]$ Re $z > -a$	$K_o(ay)$	$x < 1$
		0	$x > 1$
3.49	$z^{-1}(\log z)^2$ Re $z > 0$	$(\gamma+\log y)-\pi^2/6$	$x < 1$
		0	$x > 1$
3.50	$z^{-\nu}(\log z)^2$ Re $z > 0$	$[\Gamma(\nu)]^{-1}y^{\nu-1}$ $\cdot\{[\psi(\nu)-\log y]^2-\psi'(\nu)\}$	$x < 1$
		0	$x > 1$

2.4 Trigonometric and Hyperbolic Functions

	$\Phi(z) = \int_0^\infty x^{z-1}\phi(x)\,dx$	$\phi(x)$
4.1	$a^z \csc(\pi z)$ $-n < \mathrm{Re}\, z < 1-n$, $n = 0, \pm 1, \pm 2, \cdots$	$(-1)^n \pi^{-1} (x/a)^n (1+x/a)^{-1}$
4.2	$a^z \sec(\pi z)$ $-n-\frac{1}{2} < \mathrm{Re}\, z < -n+\frac{1}{2}$, $n=0, \pm 1, \pm 2, \cdots$	$(-1)^n \pi^{-1} (x/a)^{n+\frac{1}{2}} (1+ x/a)^{-1}$
4.3	$a^z \cot(\pi z)$ $-n < \mathrm{Re}\, z < 1-n$, $n=0, \pm 1, \pm 2, \cdots$ Principal value	$\pi^{-1} (x/a)^n (1-x/a)^{-1}$
4.4	$a^z \tan(\pi z)$ $-n-\frac{1}{2} < \mathrm{Re}\, z < \frac{1}{2}-n$ $n=0, \pm 1, \pm 2, \cdots$; Principal value	$-\pi^{-1} (x/a)^{n+\frac{1}{2}} (1-x/a)^{-1}$
4.5	$z^{-1} a^z \csc(\pi z)$ $0 < \mathrm{Re}\, z < 1$	$\pi^{-1} \log(1+a/x)$

	$\Phi(z) = \int_0^\infty x^{z-1}\phi(x)\,dx$	$\phi(x)$
4.6	$z^{-1}a^{-z}\csc(\pi z)$ $-1 < \mathrm{Re}\, z < 0$	$\pi^{-1}\log(1+ax)$
4.7	$(1-z)^{-1}\csc(\pi z)$ $0 < \mathrm{Re}\, z < 1$	$\pi^{-1}x^{-1}\log(1+x)$
4.8	$z^{-1}c^{-z}(a^z-b^z)$ $\cdot\csc(\pi z)$	$\pi^{-1}\log[(a+cx)/(b+cx)]$ $0 < \mathrm{Re}\, z < 1$
4.9	$z^{-1}a^{-z}\sec(\frac{1}{2}\pi z)$ $-1 < \mathrm{Re}\, z < 0$	$-2\pi^{-1}\arctan(ax)$
4.10	$z^{-1}a^{-z}\sec(\frac{1}{2}\pi z)$ $0 < \mathrm{Re}\, z < 1$	$2\pi^{-1}\mathrm{arccot}(ax)$
4.11	$z^{-1}a^{-z}\cot(\pi z)$	$\log\lvert 1-ax\rvert \quad -1 < \mathrm{Re}\, z < 0$
4.12	$z^{-1}\tanh(\frac{1}{2}\pi z)$	$\pi^{-1}\log\left\lvert\frac{1+x}{1-x}\right\rvert \quad -1 < \mathrm{Re}\, z < 1$

	$\Phi(z) = \int_0^\infty x^{z-1}\phi(x)\,dx$	$\phi(x)$
4.13	$a^z \csc(\pi z)$ $\cdot\sin(z\theta)$ $-\pi < \theta < \pi$	$\pi^{-1} a \sin\theta\, x(x^2+2ax+a^2)^{-1}$ $-1 < \mathrm{Re}\, z < 1$
4.14	$z^{-1}\cos(z\theta)\csc(\pi z)$ $-1 < \mathrm{Re}\, z < 0$	$\tfrac{1}{2}\pi^{-1}\log(1+2x\cos\theta+x^2)$ $-\pi < \theta < \pi$
4.15	$\csc^2(\pi z)$ $-n < \mathrm{Re}\, z < 1-n$, $n=0,\pm1,\pm2,\cdots$; Principal value	$\pi^{-2} x^n (x-1)^{-1} \log x$
4.16	$\sec^2(\pi z)$ $-n-\tfrac{1}{2} < \mathrm{Re}\, z < \tfrac{1}{2}-n$, $n=0,\pm1,\pm2,\cdots$; Principal value	$\pi^{-2} x^{n+\frac{1}{2}} (x-1)^{-1} \log x$
4.17	$\csc^3(\pi z)$ $-n < \mathrm{Re}\, z < 1-n$, $n=0,\pm1,\pm2,\cdots$	$\tfrac{1}{2}\pi^{-3}(-x)^n(\pi^2+\log^2 x)/(1+x)$

	$\Phi(z) = \int_0^\infty x^{z-1}\phi(x)\,dx$	$\phi(x)$
4.18	$a^z \csc(\pi z)$ $\cos(z\ \theta)$ $-\pi < \theta < \pi$	$\pi^{-1} a(a+x\cos\theta)(x^2+2ax\ \cos\theta+a^2)^{-1}$ $0 < \mathrm{Re}\ z < 1$
4.19	$z^{-\frac{1}{2}}\sin(a/z)$ $\mathrm{Re}\ z > 0$	$(\pi y)^{-\frac{1}{2}}\sinh[(2ay)^{\frac{1}{2}}]\sin[(2ay)^{\frac{1}{2}}]$ $x < 1$ 0 $x > 1$
4.20	$z^{-\frac{1}{2}}\cos(a/z)$ $\mathrm{Re}\ z > 0$	$(\pi y)^{-\frac{1}{2}}\cosh[(2ay)^{\frac{1}{2}}]\cos[(2ay)^{\frac{1}{2}}]$ $x < 1$ 0 $x > 1$
4.21	$z^{-\nu}\sin(a/z)$ $\mathrm{Re}\ z > 0,$ $\mathrm{Re}\ \nu > 0$	$(y/a)^{\frac{1}{2}\nu-\frac{1}{2}}\{\sin(3\pi\nu/4+\pi/4)\mathrm{ber}_{\nu-1}[2(ay)^{\frac{1}{2}}]$ $-\cos(3\pi\nu/4+\pi/4)\mathrm{bei}_{\nu-1}[2(ay)^{\frac{1}{2}}]\}$ $x < 1$ 0 $x > 1$
4.22	$z^{-\nu}\cos(a/z)$ $\mathrm{Re}\ z > 0,$ $\mathrm{Re}\ \nu > 0$ $y = \log(1/x)$	$-(y/a)^{\frac{1}{2}\nu-\frac{1}{2}}\{\cos(3\pi\nu/4+\pi/4)\mathrm{ber}_{\nu-1}[2(ay)^{\frac{1}{2}}]$ $+\sin(3\pi\nu/4+\pi/4)\mathrm{bei}_{\nu-1}[2(ay)^{\frac{1}{2}}]\}$ $x < 1$ 0 $x > 1$

	$\Phi(z) = \int_0^\infty x^{z-1}\phi(x)\,dx$	$\phi(x)$	
4.23	$z^{-\frac{1}{2}}e^{-az^{\frac{1}{2}}}$ $\cdot\sin(bz^{\frac{1}{2}})$ $a \geqq b$, Re $z > 0$	$(\pi y)^{-\frac{1}{2}}\exp[-\frac{1}{2}(a^2-b^2)/y]\sin(\frac{1}{2}ab/y)$	$x < 1$
		0	$x > 1$
4.24	$z^{-\frac{1}{2}}e^{-az^{\frac{1}{2}}}$ $\cdot\cos(bz^{\frac{1}{2}})$ $a \geqq b$, Re $z > 0$	$(\pi y)^{-\frac{1}{2}}\exp[-\frac{1}{2}(a^2-b^2)/y]\cos(\frac{1}{2}ab/y)$	$x < 1$
		0	$x > 1$
4.25	$z^{-\nu}\cos(az^{-\frac{1}{2}})$ Re $z > 0$, Re $\nu > 0$	$[\Gamma(\nu)]^{-1}y^{\nu-1}{}_0F_2(\ ;\nu,\frac{1}{2};-\frac{1}{4}a^2y)$	$x < 1$
		0	$x > 1$
4.26	$z^{-\nu}\sin(az^{-\frac{1}{2}})$ Re $z > 0$, Re $\nu > -\frac{1}{2}$	$a[\Gamma(\frac{1}{2}+\nu)]^{-1}{}_0F_2(\ ;\frac{1}{2}+\nu,\frac{3}{2};-\frac{1}{4}a^2y)$	$x < 1$
		0	$x > 1$
4.27	$\arctan(a/z)$ Re $z > 0$	$y^{-1}\sin(ay)$	$x < 1$
		0	$x > 1$
4.28	$z^{-1}\arctan(z/a)$ Re $z > 0$	$-\mathrm{si}(ay)$	$x < 1$
		0	$x > 1$
	$y = \log(1/x)$		

	$\Phi(z) = \int_0^\infty x^{z-1}\phi(x)$	$\phi(x)$	
4.29	$\arctan[2az/(z^2+b^2-a^2)]$	$2y^{-1}\sin(ay)\cos(by)$	$x < 1$
	$\mathrm{Re}\ z > 0$	0	$x > 1$
4.30	$(z^2+a^2)^{-1}$	$\frac{1}{2}a^{-1}\cos(ay)[\mathrm{Ci}(2ay)-\gamma-\log(2ay)]$	
	$\cdot\arctan(a/z)$	$+\frac{1}{2}a^{-1}\sin(ay)\mathrm{Si}(2ay)$	$x < 1$
	$\mathrm{Re}\ z > 0$	0	$x > 1$
4.31	$\log(z^2+a^2)$	$-2y^{-1}\sin(ay)(\gamma+\log y)$	$x < 1$
	$\cdot\arctan(a/z)$	0	$x > 1$
	$\mathrm{Re}\ z > 0$		
4.32	$(z^2-a^2)^{-\frac{1}{2}}$	$K_o(ay)$	$x < 1$
	$\cdot\arccos(a/z)$	0	$x > 1$
	$\mathrm{Re}\ z > -\mathrm{Re}\ a$		
4.33	$(z^2-a^2)^{-\frac{1}{2}}$	$\frac{1}{2}\pi\mathbf{L}_o(ay)$	$x < 1$
	$\cdot\arcsin(a/z)$	0	$x > 1$
	$\mathrm{Re}\ z > \mathrm{Re}\ a$		
4.34	$\sin(z^2/a)$	$\frac{1}{2}\pi^{-\frac{1}{2}}a^{\frac{1}{2}}\sin(\frac{1}{4}ay^2-\frac{1}{4}\pi)$	
	$y = \log(1/x)$		

	$\Phi(z) = \int_0^\infty x^{z-1}\phi(x)\,dx$	$\phi(x)$
4.35	$\cos(z^2/a)$	$\frac{1}{2}\pi^{-\frac{1}{2}}a^{\frac{1}{2}}\cos(\frac{1}{4}ay^2-\frac{1}{4}\pi)$
4.36	$a^z\csc(\pi z)\sinh(bz)$ $-1 < \mathrm{Re}\ z < 1$	$\pi^{-1}ax\ \sinh b(x^2+2ax\ \cosh b+a^2)^{-1}$
4.37	$a^z\csc(\pi z)\cosh(bz)$ $0 < \mathrm{Re}\ z < 1$	$\pi^{-1}a(a+x\cosh\ b)(x^2+2ax\ \cosh\ b+a^2)^{-1}$
4.38	$z^{-\frac{1}{2}}\sinh(a/z)$ $\mathrm{Re}\ z > 0$	$\frac{1}{2}(\pi y)^{-\frac{1}{2}}\{\cosh[2(ay)^{\frac{1}{2}}]-\cos[2(ay)^{\frac{1}{2}}]\}$ $\quad x < 1$ $0 \quad x > 1$
4.39	$z^{-\frac{1}{2}}\cosh(a/z)$ $\mathrm{Re}\ z > 0$	$\frac{1}{2}(\pi y)^{\frac{1}{2}}\{\cosh[2(ay)^{\frac{1}{2}}]+\cos[2(ay)^{\frac{1}{2}}]\}$ $\quad x < 1$ $0 \quad x > 1$
4.40	$z^{-\nu}\sinh(a/z)$ $\mathrm{Re}\ z > 0,$ $\mathrm{Re}\ \nu > -1$	$\frac{1}{2}(y/a)^{\frac{1}{2}\nu-\frac{1}{2}}\{I_{\nu-1}[2(ay)^{\frac{1}{2}}]-J_{\nu-1}[2(ay)^{\frac{1}{2}}]\}$ $x < 1$ $0 \quad x > 1$
4.41	$z^{-\nu}\cosh(a/z)$ $\mathrm{Re}\ z > 0,$ $\mathrm{Re}\ \nu > 0$ $y = \log(1/x)$	$\frac{1}{2}(y/a)^{\frac{1}{2}\nu-\frac{1}{2}}\{I_{\nu-1}[2(ay)^{\frac{1}{2}}]+J_{\nu-1}[2(ay)^{\frac{1}{2}}]\}$ $0 \quad x < 1$ $0 \quad x > 1$

	$\Phi(z) = \int_0^\infty x^{z-1}\phi(x)\,dx$	$\phi(x)$	
4.42	$z^{-\frac{1}{2}}\sinh(\frac{1}{2}ab/z)$	$(\pi y)^{-\frac{1}{2}}\sin(ay^{\frac{1}{2}})\sin(by^{\frac{1}{2}})$	$x < 1$
	$\cdot\exp[-\frac{1}{2}(a^2+b^2)/z]$	0	$x > 1$
	Re $z > 0$		
4.43	$z^{-\frac{1}{2}}\cosh(\frac{1}{2}ab/z)$	$(\pi y)^{-\frac{1}{2}}\cos(ay^{\frac{1}{2}})\cos(by^{\frac{1}{2}})$	$x < 1$
	$\cdot\exp[-\frac{1}{2}(a^2+b^2)/z]$	0	$x > 1$
	Re $z > 0$		
4.44	$(z^2+a^2)^{-1}$	$4a\pi^{-2}\lvert\sin(ay)\rvert$	$x < 1$
	$\coth(2az/\pi)$	0	$x > 1$
	Re $z > 0$		
4.45	$\operatorname{sech}(az^{\frac{1}{2}})$	$-a^{-2}[\frac{\partial}{\partial\nu}\theta_1(\frac{1}{2}\nu\lvert ya^{-2})]_{\nu=0}$	$x < 1$
	Re $z > 0$	0	$x > 1$
4.46	$\operatorname{csch}(az^{\frac{1}{2}})$	$-a^{-2}[\frac{\partial}{\partial\nu}\theta_4(\frac{1}{2}\nu\lvert ya^{-2})]_{\nu=0}$	$x < 1$
	Re $z > 0$	0	$x > 1$
4.47	$z^{-\frac{1}{2}}\tanh(az^{\frac{1}{2}})$	$a^{-1}\theta_2(0\lvert ya^{-2})$	$x < 1$
	Re $z > 0$	0	$x > 1$
	$y = \log(1/z)$		

	$\Phi(z) = \int_0^\infty x^{z-1}\phi(x)\,dx$	$\phi(x)$	
4.48	$z^{-\frac{1}{2}}\coth(az^{\frac{1}{2}})$	$a^{-1}\theta_3(0\|ya^{-2})$	$x < 1$
	$\mathrm{Re}\ z > 0$	0	$x > 1$
4.49	$\sinh(\nu z^{\frac{1}{2}})\,\mathrm{csch}(az^{\frac{1}{2}})$	$a^{-1}[\frac{\partial}{\partial\nu}\theta_4(\frac{1}{2}\nu a^{-1}\|ya^{-2})]$	$x < 1$
	$-a < \nu < a$	0	$x > 1$
	$\mathrm{Re}\ z > 0$		
4.50	$\cosh(\nu z^{\frac{1}{2}})\,\mathrm{sech}(az^{\frac{1}{2}})$	$a^{-1}[\frac{\partial}{\partial\nu}\theta_1(\frac{1}{2}\nu a^{-1}\|ya^{-2})]$	$x < 1$
	$-a < \nu < a$	0	$x > 1$
	$\mathrm{Re}\ z > 0$		
	$y = \log(1/z)$		

2.5 The Gamma Function and Related Functions

	$\Phi(z) = \int_0^\infty x^{z-1}\phi(x)\,dx$	$\phi(x)$
5.1	$\Gamma(z) \qquad \mathrm{Re}\, z > 0$	e^{-x}
5.2	$\Gamma(z) \qquad -1 < \mathrm{Re}\, z < 0$	$e^{-x}-1$
5.3	$\sin(az)\Gamma(z)$ $\mathrm{Re}\, z > -1$	$e^{-x\cos a}\sin(x \sin a)$ $-\tfrac{1}{2}\pi < \mathrm{Re}\, a < \tfrac{1}{2}\pi$
5.4	$\cos(az)\Gamma(z)$ $\mathrm{Re}\, z > 0$	$e^{-x\cos a}\cos(x \sin a)$ $-\tfrac{1}{2}\pi < \mathrm{Re}\, a < \tfrac{1}{2}\pi$
5.5	$\sin(az)\Gamma(z)$ $-m < \mathrm{Re}\, z < 1-m$ $m = 2,3,\cdots$ $-\tfrac{1}{2}\pi < \mathrm{Re}\, a < \tfrac{1}{2}\pi$	$e^{-x\cos a}\sin(x \sin a)$ $+ \sum_{r=1}^{m-1} (-1)^r \sin(ar)x^r/r!$
5.6	$\cos(az)\Gamma(z)$ $-m < \mathrm{Re}\, z < 1-m$ $m = 1,2,\cdots$ $-\tfrac{1}{2}\pi < \mathrm{Re}\, a < \tfrac{1}{2}\pi$	$e^{-x\cos a}\cos(x \sin a)$ $- \sum_{r=0}^{m-1} (-1)^r \cos(ar)x^r/r!$
5.7	$\sec(\pi z)\Gamma(z)$ $0 < \mathrm{Re}\, z < \tfrac{1}{2}$	$e^{x}\mathrm{Erfc}(x^{\frac{1}{2}})$

	$\Phi(z) = \int_0^\infty x^{z-1}\phi(x)\,dx$	$\phi(x)$
5.8	$\sec(\pi z)\Gamma(z)$ $n-\frac{1}{2} < \mathrm{Re}\, z < n+\frac{1}{2}$ $n = 1,2,3,\cdots$	$e^x \mathrm{Erfc}(x^{\frac{1}{2}})$ $-\pi^{-1}\sum_{r=0}^{n-1}(-1)^r\Gamma(\frac{1}{2}+r)x^{-\frac{1}{2}-r}$
5.9	$z^{-1}\Gamma(\frac{1}{2}+\frac{1}{2}z)$ $-1 < \mathrm{Re}\, z < 0$	$-\pi^{\frac{1}{2}}\mathrm{Erf}(ax)$
5.10	$z^{-1}\Gamma(z)$ $\mathrm{Re}\, z > 0$	$-\mathrm{Ei}(-x)$
5.11	$\Gamma(z)\csc(\pi z)$ $0 < \mathrm{Re}\, z < 1$	$-\pi^{-1}e^{ax}\mathrm{Ei}(-x)$
5.12	$\Gamma(z)\cot(\pi z)$ $0 < \mathrm{Re}\, z < 1$	$-\pi^{-1}e^{-ax}\overline{\mathrm{Ei}}(x)$
5.13	$z^{-1}\Gamma(z)\cot(\frac{1}{2}\pi z)$ $0 < \mathrm{Re}\, z < 2$	$\pi^{-1}\mathrm{Ei}(-x)\overline{\mathrm{Ei}}(x)$
5.14	$z^{-1}\sin(\frac{1}{2}\pi z)\Gamma(z)$ $-1 < \mathrm{Re}\, z < 0$	$-\mathrm{Si}(x)$

	$\Phi(z) = \int_0^\infty x^{z-1}\phi(x)\,dx$	$\phi(x)$
5.15	$z^{-1}\cos(\frac{1}{2}\pi z)\Gamma(z)$ $0 < \text{Re } z < 1$	$-\text{Ci}(x)$
5.16	$\sec(\frac{1}{2}\pi z)\Gamma(z)$ $-1 < \text{Re } z < 1$	$2\pi^{-1}[\text{Ci}(x)\sin x - \text{si}(x)\cos x]$
5.17	$\csc(\frac{1}{2}\pi z)\Gamma(z)$ $-1 < \text{Re } z < 1$	$-2\pi^{-1}[\text{Ci}(x)\cos x + \text{si}(x)\sin x]$
5.18	$z^{-1}\dfrac{\Gamma(\frac{1}{2}+z)}{\Gamma(1+z)}$ $\text{Re } z > -\frac{1}{2}$	$2\pi^{-\frac{1}{2}}\arccos(x^{\frac{1}{2}})$ $x < 1$ 0 $x > 1$
5.19	$\Gamma(z)\Gamma(\nu-z)$ $\cdot\sin(\frac{1}{2}\pi z)$	$\Gamma(\nu)(1+x^2)^{-\frac{1}{2}\nu}\sin(\nu\arctan x)$ $-1 < \text{Re } z < \text{Re } \nu$
5.20	$\Gamma(z)\Gamma(\nu-z)$ $\cdot\cos(\frac{1}{2}\pi z)$	$\Gamma(\nu)(1+x^2)^{-\frac{1}{2}\nu}\cos(\nu \arctan x)$ $0 < \text{Re } z < \text{Re } \nu$
5.21	$\Gamma(z)[\Gamma(\frac{1}{2}+\frac{1}{2}z-\frac{1}{2}\nu)$ $\cdot\Gamma(\frac{1}{2}+\frac{1}{2}z+\frac{1}{2}\nu)]^{-1}$ $\text{Re } z > 0$	$2\pi^{-1}(4-x^2)^{-\frac{1}{2}}\cos[\nu\arccos(\frac{1}{2}x)]$ $x < 2$ 0 $x > 2$

	$\Phi(z) = \int_0^\infty x^{z-1}\phi(x)\,dx$	$\phi(x)$
5.22	$z^{-1}\Gamma(\frac{1}{2}+z)$ $\mathrm{Re}\, z > 0$	$\pi^{\frac{1}{2}}\mathrm{Ercf}(x^{\frac{1}{2}})$
5.23	$z^{-1}\sec(\frac{1}{2}\pi z)\Gamma(\frac{1}{2}+z)$ $0 < \mathrm{Re}\, z < 1$	$2\pi^{\frac{1}{2}}\{[\frac{1}{2}-C(ax)]^2+[\frac{1}{2}-S(ax)]^2\}$
5.24	$z^{-1}\csc(\frac{1}{2}\pi z)\Gamma(z)$ $0 < \mathrm{Re}\, z < 2$	$\pi^{-1}[\mathrm{Ci}^2(x)+\mathrm{si}^2(x)]$
5.25	$z^{-1}\Gamma(\nu+z)$ $\mathrm{Re}\, z > 0$	$\Gamma(\nu,x)$
5.26	$\Gamma(z)\csc[\pi(\nu+z)]$ $0 < \mathrm{Re}\, z < 1-\mathrm{Re}\,\nu$	$\pi^{-1}\Gamma(1-\nu)e^{x}\Gamma(\nu,x)$
5.27	$[\Gamma(z)/\Gamma(\frac{1}{2}-z)]^2$ $0 < \mathrm{Re}\, z < 9/8$	$2\pi^{-1}K_o(4x^{\frac{1}{4}})-Y_o(4x^{\frac{1}{4}})$
5.28	$\dfrac{\Gamma(\frac{1}{2}-z)\Gamma(\frac{1}{2}+\frac{1}{2}\nu+\frac{1}{2}z)}{\Gamma(1+\nu-z)\Gamma(1-\frac{1}{2}\nu-\frac{1}{2}z)}$	$2^{1-\nu}\sin x\, J_\nu(x)$ $-1-\mathrm{Re}\,\nu < \mathrm{Re}\, z < \frac{1}{2}$

	$\Phi(z) = \int_0^\infty x^{z-1}\phi(x)\,dx$	$\phi(x)$
5.29	$\frac{\Gamma(\frac{1}{2}-z)\Gamma(\frac{1}{2}\nu+\frac{1}{2}z)}{\Gamma(\frac{1}{2}-\frac{1}{2}\nu-\frac{1}{2}z)\Gamma(1+\nu-z)}$	$2^{1-\nu}\cos x\, J_\nu(x)$ $-\mathrm{Re}\,\nu < \mathrm{Re}\,z < \frac{1}{2}$
5.30	$\frac{\Gamma(\frac{1}{2}-z)\Gamma(z+\nu)}{\Gamma(1+\nu-z)}$	$\pi^{\frac{1}{2}}e^{-\frac{1}{2}x}I_\nu(\frac{1}{2}x)$ $-\mathrm{Re}\,\nu < \mathrm{Re}\,z < \frac{1}{2}$
5.31	$\Gamma(\frac{1}{2}-z)\Gamma(z+\nu)\Gamma(z-\nu)$ $\pm\,\mathrm{Re}\,\nu < \mathrm{Re}\,z < \frac{1}{2}$	$\pi^{\frac{1}{2}}\sec(\pi\nu)e^{\frac{1}{2}x}K_\nu(\frac{1}{2}x)$
5.32	$\frac{\Gamma(z-\nu)\Gamma(z+\nu)}{\Gamma(\frac{1}{2}+z)}$	$\pi^{-\frac{1}{2}}e^{-\frac{1}{2}x}K_\nu(\frac{1}{2}x)$ $\mathrm{Re}\,z > \pm\,\mathrm{Re}\,\nu$
5.33	$\frac{\Gamma(2z)\Gamma(\nu+z)}{\Gamma(1+\nu-z)}$	$2J_{2\nu}(2x^{\frac{1}{4}})K_{2\nu}(2x^{\frac{1}{4}})$ $\mathrm{Re}\,z > 0,\ -\mathrm{Re}\,\nu$
5.34	$\Gamma(z)\Gamma(\frac{1}{2}-z)\Gamma(\nu+z)$ $\pm\mathrm{Re}\,\nu < \mathrm{Re}\,z < \frac{1}{2}$	$\frac{1}{2}\pi^{5/2}\sec(\pi\nu)$ $\cdot[J_\nu^2(x^{\frac{1}{2}})+Y_\nu^2(x^{\frac{1}{2}})]$
5.35	$\Gamma(\alpha+z)/\Gamma(\beta+z)$ $\mathrm{Re}\,z > -\mathrm{Re}\,\alpha, \mathrm{Re}(\beta-\alpha) > 0$	$[\Gamma(\beta-\alpha)]^{-1}x^\alpha(1-x)^{\beta-\alpha-1}$ $\quad x < 1$ $0 \quad x > 1$

	$\Phi(z) = \int_0^\infty x^{z-1}\phi(x)\,dx$	$\phi(x)$
5.36	$\Gamma(\alpha+z)\Gamma(\beta-z)$ $-\mathrm{Re}\ \alpha < \mathrm{Re}\ z < \mathrm{Re}\ \beta$	$\Gamma(\alpha+\beta)x^{\alpha}(1+x)^{-\alpha-\beta}$ $\mathrm{Re}(\alpha+\beta) > 0$
5.37	$\Gamma(\alpha-z)/\Gamma(\beta-z)$ $\mathrm{Re}\ z < \mathrm{Re}\ \alpha,\ \mathrm{Re}(\beta-\alpha) > 0$	$0 \qquad x < 1$ $[\Gamma(\beta-\alpha)]^{-1}x^{1-\beta}(x-1)^{\beta-\alpha-1}$ $x > 1$
5.38	$\Gamma(\alpha+z)/\Gamma(\beta-z)$ $-\mathrm{Re}\ \alpha<\mathrm{Re}\ z<-\frac{1}{4}+\frac{1}{2}\mathrm{Re}(\beta-\alpha)$	$x^{\frac{1}{2}+\frac{1}{2}\alpha-\frac{1}{2}\beta}J_{\alpha+\beta-1}(2x^{\frac{1}{2}})$
5.39	$\Gamma(\alpha+z)\Gamma(\beta+z)$ $\mathrm{Re}\ z > -\mathrm{Re}(\alpha,\beta)$	$2x^{\frac{1}{2}\alpha+\frac{1}{2}\beta}K_{\alpha-\beta}(2x^{\frac{1}{2}})$
5.40	$\Gamma(1-2z)\Gamma(\alpha+z)$ $\cdot[\Gamma(1+\beta-z)\Gamma(1-\beta-z)\Gamma(1+\alpha-z)]^{-1}$	$J_{\alpha+\beta}(2x^{\frac{1}{2}})J_{\alpha-\beta}(2x^{\frac{1}{2}})$ $-\mathrm{Re}\ \alpha < \mathrm{Re}\ z < \frac{1}{2}$
5.41	$\Gamma(1-2z)\Gamma(z+\alpha)\Gamma(z+\beta)$ $\Gamma(z-\beta)/\Gamma(1+\alpha-z)$ $\left.\begin{matrix}-\mathrm{Re}\ \alpha\\ \pm\mathrm{Re}\ \beta\end{matrix}\right\} < \mathrm{Re}\ z < \frac{1}{2}$	$\csc(2\pi\beta)[J_{\alpha+\beta}(2x^{\frac{1}{2}})Y_{\alpha-\beta}(2x^{\frac{1}{2}})$ $-Y_{\alpha+\beta}(2x^{\frac{1}{2}})J_{\alpha-\beta}(2x^{\frac{1}{2}})]$

	$\Phi(z) = \int_0^\infty x^{z-1}\phi(x)\,dx$	$\phi(x)$
5.42	$\cos[\pi(\beta+z)]\Gamma(1-2z)$ $\cdot\Gamma(z+\alpha)\Gamma(z+\beta)$ $\cdot[\Gamma(1+\beta-z)\Gamma(1+\alpha-z)]^{-1}$	$-\pi J_{\frac{1}{2}\alpha+\frac{1}{2}\beta}(2x^{\frac{1}{2}})Y_{\frac{1}{2}\alpha-\frac{1}{2}\beta}(2x^{\frac{1}{2}})$ $-\mathrm{Re}(\alpha,\beta) < \mathrm{Re}\ z < \frac{1}{2}$
5.43	$\Gamma(1-2z)\Gamma(z+\alpha)\Gamma(z+\beta)$ $\cdot[\Gamma(1-z+\alpha)\Gamma(1-z+\beta)]^{-1}$	$2I_{\alpha+\beta}(2x^{\frac{1}{2}})K_{\alpha-\beta}(2x^{\frac{1}{2}})$ $-\mathrm{Re}(\alpha,\beta) < \mathrm{Re}\ z < \frac{1}{2}$
5.44	$[\Gamma(2z)]^{-1}\Gamma(z+\alpha)\Gamma(z-\alpha)$ $\cdot\Gamma(z+\beta)\Gamma(z-\beta)$	$4K_{\alpha-\beta}(2x^{\frac{1}{2}})K_{\alpha+\beta}(2x^{\frac{1}{2}})$ $\mathrm{Re}\ z > \mathrm{Re}(\pm\alpha\pm\beta)$
5.45	$\Gamma(\alpha+z)$ $\cdot\Gamma(\beta-z)/\Gamma(\alpha-z)$	$\Gamma(\alpha+\beta)[\Gamma(2\alpha)]^{-1}e^{-\frac{1}{2}x}M_{\beta,\alpha-\frac{1}{2}}(x)$ $-\mathrm{Re}\ \alpha<\mathrm{Re}\ z < \mathrm{Re}\ \beta,$ $\mathrm{Re}(\alpha+\beta) > 0$
5.46	$\Gamma(1-\alpha+z)$ $\cdot\Gamma(\alpha+z)/\Gamma(\beta+z)$ $\mathrm{Re}\ z > \begin{cases} -1+\mathrm{Re}\ \alpha \\ -\mathrm{Re}\ \alpha \end{cases}$	$e^{-\frac{1}{2}x}W_{1-\beta,\alpha-\frac{1}{2}}(x)$
5.47	$\Gamma(1-\alpha+z)$ $\cdot\Gamma(\alpha+z)\Gamma(\beta-z)$	$\Gamma(\alpha+\beta)\Gamma(1-\alpha+\beta)e^{\frac{1}{2}x}W_{-\beta,\alpha-\frac{1}{2}}(x)$ $\left.\begin{matrix} -\mathrm{Re}\ \alpha \\ -1+\mathrm{Re}\ \alpha \end{matrix}\right\} < \mathrm{Re}\ z<\mathrm{Re}\ \beta$

	$\Phi(z) = \int_0^\infty x^{z-1}\phi(x)\,dx$	$\phi(x)$
5.48	$\frac{\Gamma(\nu+\frac{1}{2}+z)\Gamma(-\mu-z)\Gamma(1+2z)}{\Gamma(\nu+\frac{1}{2}-z)\Gamma(1+z-\mu)}$ $-\frac{1}{2}-\text{Re }\nu < \text{Re } z < -\text{Re }\mu$	$\frac{\Gamma(\frac{1}{2}+\nu-\mu)}{\Gamma(1+2\nu)} M_{-\mu,\nu}(x^{\frac{1}{2}})W_{\mu,\nu}(x^{\frac{1}{2}})$
5.49	$\frac{\Gamma(1+2z)\Gamma(\frac{1}{2}+z+\nu)\Gamma(\frac{1}{2}+z-\nu)}{\Gamma(1+z+\mu)\Gamma(1+z-\mu)}$ $\text{Re } z > \begin{cases} -\frac{1}{2} \\ -\frac{1}{2}\pm\text{Re }\nu \end{cases}$	$W_{\mu,\nu}(x^{\frac{1}{2}})W_{-\mu,\nu}(x^{\frac{1}{2}})$
5.50	$\Gamma(\alpha+z)\Gamma(\beta+z)$ $\cdot[\Gamma(\gamma+z)\Gamma(\delta+z)]^{-1}$ $\text{Re } z > -\text{Re}(\alpha,\beta)$ $\text{Re}(\gamma+\delta-\alpha-\beta) > 0$	$[\Gamma(\gamma+\delta-\alpha-\beta)]^{-1}x^{\alpha}(1-x)^{\gamma+\delta-\alpha-\beta-1}$ $\cdot {}_2F_1(\delta-\beta,\gamma-\beta;\gamma+\delta-\alpha-\beta;1-x)$ $\quad x < 1$ $0 \quad x > 1$
5.51	$\Gamma(\alpha-z)\Gamma(\beta-z)$ $\cdot\Gamma(z)/\Gamma(\gamma-z)$ $0 < \text{Re } z < \text{Re}(\alpha,\beta)$	$\Gamma(\alpha)\Gamma(\beta)[\Gamma(\gamma)]^{-1}$ $\cdot {}_2F_1(\alpha,\beta;\gamma;-x)$
5.52	$\frac{\Gamma(\frac{1}{2}+\alpha-z)\Gamma(\beta-z)}{\Gamma(1-z)}$ $\text{Re } z < \text{Re}(\beta,\frac{1}{2}+\alpha)$	$0 \quad x < 4$ $\pi^{\frac{1}{2}}(x-4)^{-\frac{1}{2}\alpha-\frac{1}{2}\beta}P_{\alpha-\beta}^{\alpha+\beta}(\frac{1}{2}x^{\frac{1}{2}})$ $x > 4$

	$\Phi(z) = \int_0^\infty x^{z-1}\phi(x)\,dx$	$\phi(x)$	
5.53	$z^{-1}\psi(z)$	$-\gamma-\log(x^{-1}-1)$	$x < 1$
	$\mathrm{Re}\ z > 0$	0	$x > 1$
5.54	$z^{-1}\psi(1+z)$	$-\gamma-\log(1-x)$	$x < 1$
	$\mathrm{Re}\ z > 0$	0	$x > 1$
5.55	$z^{-1}[\gamma+\psi(1+z)]$	$-\log(1-x)$	$x < 1$
	$\mathrm{Re}\ z > -1$	0	$x > 1$
5.56	$\psi'(z)$	$(x-1)^{-1}\log x$	$x < 1$
	$\mathrm{Re}\ z > 0$	0	$x > 1$
5.57	$\csc(\pi z)\psi(1+z)$	$\pi^{-1}(1+1/x)^{-1}[\gamma+\log(1+1/x)]$	
	$-1 < \mathrm{Re}\ z < 0$		
5.58	$\csc(\pi z)[\gamma+\psi(1+z)$	$\pi^{-1}(1+1/x)^{-1}\log(1+1/x)$	
	$-1 < \mathrm{Re}\ z < 1$		
5.59	$\psi'(\tfrac{1}{2}+\tfrac{1}{2}z)-\psi'(\tfrac{1}{2}z)$	$2(x+1)^{-1}\log x$	$x < 1$
	$\mathrm{Re}\ z > 0$	0	$x > 1$
5.60	$\psi(z+\alpha)-\psi(z+\beta)$	$(x^{\beta}-x^{\alpha})/(1-x)$	$x < 1$
	$\mathrm{Re}\ z > -\mathrm{Re}(\alpha,\beta)$	0	$x > 1$

	$\Phi(z) = \int_0^\infty x^{z-1}\phi(x)\,dx$	$\phi(x)$
5.61	$\psi(z+\alpha)-\psi(z+\beta)$ $-h<\mathrm{Re}(z+\alpha)<1-h$ $-k<\mathrm{Re}(z+\beta)<1+k$ $h,k = 0,1,2,\cdots$	$(x^{\beta+k}-x^{\alpha+k})/(1-x) \quad x<1$ $(x^{\alpha}-x^{\beta}+x^{\beta+k}-x^{\alpha+h})/(1-x)$ $x>1$
5.62	$\Gamma(z)\psi(z)$ $\mathrm{Re}\, z > 0$	$e^{-x}\log x$
5.63	$B(z,\nu)\psi(z+\nu)$ $\mathrm{Re}(z,\nu) > 0$	$(1-x)^{\nu-1}[\psi(\nu)-\log(1-x)]$ $x<1$ $0 \quad x>1$
5.64	$B(z,\nu)\psi(z)$ $\mathrm{Re}(z,\nu) > 0$	$(1-x)^{\nu-1}[\psi(\nu)-\log(x^{-1}-1)]$ $x<1$ $0 \quad x>1$
5.65	$B(z,\nu-z)\psi(\nu-z)$ $-1 < \mathrm{Re}\, z < \mathrm{Re}\,\nu$	$(1+x^{-\nu})[\psi(\nu)-\log(1+x)]$
5.66	$B(z,\nu-\frac{1}{2}z)$ $\cdot[\psi(\nu-\frac{1}{2}z)-\psi(\nu+\frac{1}{2}z)]$ $0 < \mathrm{Re}\, z < 2\,\mathrm{Re}\,\nu$	$2(1+x^2)^{-\frac{1}{2}}[(1+x^2)^{\frac{1}{2}}-x]^{2\nu-1}$ $\cdot\log[(1+x^2)^{\frac{1}{2}}-x]$

	$\Phi(z) = \int_0^\infty x^{z-1}\phi(x)\,dx$	$\phi(x)$
5.67	$\sin(\frac{1}{2}\pi z)\Gamma(z)\psi(z)$ $0 < \mathrm{Re}\, z < 1$	$\sin x \log x - \frac{1}{2}\pi \cos x$
5.68	$\cos(\frac{1}{2}\pi z)\Gamma(z)\psi(z)$ $0 < \mathrm{Re}\, z < 1$	$\cos x \log x + \frac{1}{2}\pi \sin x$
5.69	$a^{-z}\Gamma(z)\zeta(z)$ $\mathrm{Re}\, z > 1$	$(e^{ax}-1)^{-1}$
5.70	$a^{-z}\Gamma(z)\zeta(z-1)$ $\mathrm{Re}\, z > 2$	$e^{ax}(e^{ax}-1)^{-2}$
5.71	$a^{-z}\Gamma(z)\zeta(1+z)$ $\mathrm{Re}\, z > 0$	$-\log(1-e^{-ax})$
5.72	$(2a)^{-z}\Gamma(z)\zeta(z-1)$ $\mathrm{Re}\, z > 2$	$\frac{1}{4}\mathrm{csch}^2(ax)$
5.73	$\csc(\pi z)\zeta(1-z)$ $-1 < \mathrm{Re}\, z < 0$	$-\pi^{-1}[\gamma+\psi(x+1)]$
5.74	$\csc(\pi z)\zeta(1-z)$ $0 < \mathrm{Re}\, z < 1$	$-\pi^{-1}[\psi(1+x)-\log x]$

	$\Phi(z) = \int_0^\infty x^{z-1}\phi(x)\,dx$	$\phi(x)$
5.75	$\csc(\pi z)$ $[\zeta(1+z)-z^{-1}]$ $-1 < \mathrm{Re}\, z < 0$	$\pi^{-1}[\psi(1+1/x)-\log(1+1/x)]$
5.76	$\Gamma(z)[\zeta(z)-(z-1)^{-1}]$ $\mathrm{Re}\, z > 0$	$e^{-x}[1-x^{-1}+(e^{x}-1)^{-1}]$
5.77	$\Gamma(z)\Gamma(n+1-z)$ $\zeta(n+1-z)$	$(-1)^{n-1}\psi^{n}(1+x)$ $0 < \mathrm{Re}\, z < n, n=1,2,3,\cdots$
5.78	$a^{-z}\Gamma(z)\Gamma(\nu-z)$ $\cdot\zeta(\nu-z)$	$\Gamma(\nu)\zeta(\nu,1+ax)$ $0 < \mathrm{Re}\, z < \mathrm{Re}\,\nu-1$
5.79	$a^{-z}\Gamma(z)\zeta(2z)$ $\mathrm{Re}\, z > \frac{1}{2}$	$\frac{1}{2}[\Theta_3(0\|ax\pi^{-2})-1]$
5.80	$z^{-1}\zeta(z)$ $\mathrm{Re}\, z > 1-\mathrm{Re}\,\alpha$	$n,\ (n+1)^{-1}<x<n^{-1}$ $x < 1$, $n = 1,2,3,\cdots$; 0 $x > 1$; $= \sum_{n=1}^{\infty} H(\frac{1}{x} - n)$

	$\Phi(z) = \int_0^\infty x^{z-1}\phi(x)\,dx$	$\phi(x)$
5.81	$z^{-1}\zeta(z+\alpha)$ $\operatorname{Re} z > 1-\operatorname{Re}\alpha$	$\sum_{1\leqq n\leqq \frac{1}{x}} n^{-\alpha}$
5.82	$z^{-\nu}\zeta(z)$ $\operatorname{Re} z > 1, \operatorname{Re}\nu > 0$	$[\Gamma(\nu)]^{-1} \sum_{1\leqq n\leqq \frac{1}{x}} [\log(\frac{1}{nx})]^{\nu-1}$
5.83	$c^{-z}\Gamma(z)\zeta(z,\alpha)$ $\operatorname{Re} z > 1, \operatorname{Re}\alpha > 0$	$(e^{cx}-1)^{-1}e^{cx(1-\alpha)}$
5.84	$a^{-z}\Gamma(z)\Gamma(\nu-z)$ $\cdot\zeta(\nu-z,\alpha)$	$\Gamma(\nu)\zeta(\nu,\alpha+ax)$ $0 < \operatorname{Re} z < \operatorname{Re}\nu-1, \ \operatorname{Re}\alpha > 0$
5.85	$\zeta(\nu,z/\alpha)$ $\operatorname{Re} z > 0, \operatorname{Re}\alpha > 0$ $\operatorname{Re}\nu > 1$	$[\Gamma(\nu)]^{-1}\alpha^{\nu}[\log(1/x)]^{\nu-1}(1-x^{\alpha})^{-1}$ $\quad x < 1$ $0 \quad x > 1$
5.86	$\zeta(\nu,\frac{1}{2}+\frac{1}{2}z/\alpha)$ $\operatorname{Re} z > -\operatorname{Re}\alpha, \operatorname{Re}\nu>1$	$(2\alpha)^{\nu}[\Gamma(\nu)]^{-1}(\log\frac{1}{x})^{\nu-1}(x^{-\alpha}-x^{\alpha})^{-1}$ $\quad x < 1$ $0 \quad x > 1$

	$\Phi(z) = \int_0^\infty x^{z-1}\phi(x)\,dx$	$\phi(x)$
5.87	$\zeta(\nu,1+z/\alpha)$ $\mathrm{Re}\, z > -\mathrm{Re}\,\alpha, \mathrm{Re}\,\nu > 1$	$\alpha^{\nu}[\Gamma(\nu)]^{-1}(\log\frac{1}{x})^{\nu-1}(x^{-\alpha}-1)^{-1}$ $x < 1$ 0 $x > 1$
5.88	$\csc(\pi z)$ $\cdot[\zeta(1-z,b)-\zeta(1-z,a)]$	$\pi^{-1}[\psi(a+x)-\psi(b+x)]$ $0 < \mathrm{Re}\, z < 1$

2.6 Orthogonal Polynomials and Legendre Functions

	$\Phi(z) = \int_0^\infty x^{z-1}\phi(x)dx$	$\phi(x)$
6.1	$z^{-n}T_n(1-z^{-1})$ Re $z > 0$	$(-1)^n[(2n-1)!]^{-1}(2y)^{n-1}He_{2n}(y^{\frac{1}{2}})$
6.2	$z^{-n-1}U_n(z^{-1}-1)$ Re $z > 0$	$2^{n-1}[(2n-1)!]^{-1}y^{n-\frac{1}{2}}He_{2n-1}(y^{\frac{1}{2}})$ $\quad x < 1$ $0 \quad x > 1$
6.3	$z^{-\frac{1}{2}-\frac{1}{2}n}He_n[(2z)^{\frac{1}{2}}]$ Re $z > 0$	$2^{\frac{1}{2}n-1}(\pi y)^{-\frac{1}{2}}[(1+iy^{\frac{1}{2}})^n+(1-iy^{\frac{1}{2}})^n]$ $\quad x < 1$ $0 \quad x > 1$
6.4	$z^{-\frac{1}{2}-n}He_{2n}[(2z)^{\frac{1}{2}}]$ Re $z > 0$	$2^n(\pi y)^{-\frac{1}{2}}(1+y)^n\ T_n[(1-y)/(1+y)]$ $\quad x < 1$ $0 \quad x > 1$
6.5	$z^{-n-2}He_{2n+1}[(2z)^{\frac{1}{2}}]$ Re $z > 0$ $y = \log(1/x)$	$2^{n+3/2}(n+1)^{-1}(y/\pi)^{\frac{1}{2}}(y+1)^n$ $\cdot U_{n+1}[(1-y)/(1+y)]$ $\quad x < 1$ $0 \quad x > 1$

	$\Phi(z) = \int_0^\infty x^{z-1}\phi(x)\,dx$	$\phi(x)$
6.6	$z^{-n-\frac{1}{2}}e^{-a/z}He_{2n}[(2a/z)]^{\frac{1}{2}}$ Re $z > 0$	$(-2)^n\pi^{-\frac{1}{2}}y^{n-\frac{1}{2}}\cos[2(ay)^{\frac{1}{2}}]$ $x < 1$ 0 $x > 1$
6.7	$z^{-n-1}e^{-a/z}He_{2n+1}[(2a/z)^{\frac{1}{2}}]$ Re $z > 0$	$(-2)^n(2/\pi)^{\frac{1}{2}}y^n\sin[2(ay)^{\frac{1}{2}}]$ $x < 1$ 0 $x > 1$
6.8	$z^{-\frac{1}{2}}(1-\frac{1}{2}/z)He_n[a(1-\frac{1}{2}/z)^{-\frac{1}{2}}]$ Re $z > 0$	$(2\pi y)^{-\frac{1}{2}}[He_n(a+y^{\frac{1}{2}})+He_n(a-y^{\frac{1}{2}})]$ $x < 1$ 0 $x > 1$
6.9	$z^{-n-1}P_n(1-a/z)$ Re $z > 0$	$(\frac{1}{2}a)^{n+1}y^nL_n(\frac{1}{2}ay)/n!$ $x < 1$ 0 $x > 1$
6.10	$z^{-\nu}P_n(1-a/z)$ Re $z > 0$, Re $\nu > 0$	$y^{n-1}{}_2F_2(-n,n+1;1,\nu;\frac{1}{2}ay)$ $x < 1$ 0 $x > 1$
6.11	$z^{-\frac{1}{2}}P_n(a/z)$ Re $z > 0$ $y = \log(1/x)$	$(n!)^{-1}i^{-n}(\pi y)^{-\frac{1}{2}}He_n[(2ay)^{\frac{1}{2}}]He_n[(-2ay)^{\frac{1}{2}}]$ $x < 1$ 0 $x > 1$

	$\Phi(z) = \int_0^\infty x^{z-1}\phi(x)\,dx$	$\phi(x)$
6.12	$z^{-\frac{1}{2}-\frac{1}{2}n} P_n[(2z)^{-\frac{1}{2}}]$ $\mathrm{Re}\ z > 0$	$(n!)^{-1} 2^{\frac{1}{2}n} \pi^{-\frac{1}{2}} y^{\frac{1}{2}n-\frac{1}{2}} He_n(y^{\frac{1}{2}})$ $x < 1$ 0 $x > 1$
6.13	$(z+b)^{-n-1} P_n(\frac{z+a}{z+b})$ $\mathrm{Re}\ z > -b$	$(n!)^{-1} y^n e^{-by} L_n(\frac{1}{2}by-\frac{1}{2}ay)$ $x < 1$ 0 $x > 1$
6.14	$(z+b)^{-\nu} P_n(\frac{z+a}{z+b})$ $\mathrm{Re}\ z > -b,\ \mathrm{Re}\ \nu > 0$	$[\Gamma(\nu)]^{-1} y^{\nu-1} e^{-by}\, {}_2F_2(-n,n+1;1,\nu;\frac{1}{2}by-\frac{1}{2}ay)$ $x < 1$ 0 $x > 1$
6.15	$(z^2+a^2)^{-\frac{1}{2}n-\frac{1}{2}}$ $\cdot P_n[z(z^2+a^2)^{-\frac{1}{2}}]$ $\mathrm{Re}\ z > 0$	$(n!)^{-1} y^n J_o(ay)$ $x < 1$ 0 $x > 1$
6.16	$(z^2-a^2)^{-\frac{1}{2}n-\frac{1}{2}}$ $\cdot P_n[z(z^2-a^2)^{-\frac{1}{2}}]$ $\mathrm{Re}\ z > a$ $y = \log(1/x)$	$(n!)^{-1} y^n I_o(ay)$ $x < 1$ 0 $x > 1$

	$\Phi(z) = \int_0^\infty x^{z-1}\phi(x)\,dx$	$\phi(x)$
6.17	$z^{-n-2\nu}C_n^\nu(1-a/z)$ $\mathrm{Re}\, z > 0$	$\pi^{\frac{1}{2}}2^{1-n-4\nu}[\Gamma(\nu)\Gamma(\frac{1}{2}+\nu+n)]^{-1}a^{n+2\nu}$ $\cdot y^{n+2\nu-1}L_n^{\nu-\frac{1}{2}}(\frac{1}{2}ay) \quad x < 1$ $0 \quad x > 1$
6.18	$z^{-\mu}C_n^\nu(1-a/z)$ $\mathrm{Re}\, z > 0,\ \mathrm{Re}\,\mu > 0$	$[nB(n,2\nu)\Gamma(\mu)]^{-1}y^{\mu-1}$ $\cdot {}_2F_2(-n,n+2\nu;\frac{1}{2}+\nu,\mu;\frac{1}{2}ay) \quad x < 1$ $0 \quad x > 1$
6.19	$(z+b)^{-\mu}C_n^\nu(\frac{z+a}{z+b})$ $\mathrm{Re}\, z > 0,\ \mathrm{Re}(\mu,\nu) > 0$	$[nB(n,2\nu)\Gamma(\mu)]^{-1}y^{\mu-1}e^{-by}$ $\cdot {}_2F_2(-n,n+2\nu;\frac{1}{2}+\nu,\mu;\frac{1}{2}by-\frac{1}{2}ay)$ $x < 1$ $0 \quad x > 1$
6.20	$z^{-\nu-\frac{1}{2}n}C_n^\nu(z^{-\frac{1}{2}})$ $\mathrm{Re}\, z > 0$	$2^{\frac{1}{2}n}[n!\Gamma(\nu)]^{-1}y^{\nu+\frac{1}{2}n-1}$ $\cdot He_n[(2y)^{\frac{1}{2}}] \quad x < 1$ $0 \quad x > 1$
6.21	$z^{-n-1}L_n(z)$ $\mathrm{Re}\, z > 0$ $y = \log(1/x)$	$(n!)^{-1}(1+y)^nP_n(\frac{z-1}{z+1}) \quad x < 1$ $0 \quad x > 1$

	$\Phi(z) = \int_0^\infty x^{z-1}\phi(x)\,dx$	$\phi(x)$
6.22	$z^{-n-\nu}L_n^{\alpha}(z)$ Re $z > 0$	$[\Gamma(n+\nu)]^{-1}(1+y)^n y^{\nu-1}$ $\cdot P_n^{\alpha,\nu-1}(\frac{z-1}{z+1})$ $\quad x < 1$ 0 $\quad x > 1$
6.23	$z^{-\nu}L_n^{\alpha}(a/z)$ Re $z > 0$, Re $\nu > 0$	$(\alpha+1)_n[n!\Gamma(\nu)]^{-1}y^{\nu-1}$ $\cdot {}_1F_2(-n;\alpha+1,\nu;ay)$ $\quad x < 1$ 0 $\quad x > 1$
6.24	$z^{-n-\nu-1}e^{-a/z}L_n^{\nu}(a/z)$ Re $z > 0$, Re$(\nu+n) > -1$	$(n!)^{-1}a^{-\frac{1}{2}\nu}y^{n+\frac{1}{2}\nu}J_\nu[2(ay)^{\frac{1}{2}}]$ $\quad x < 1$ 0 $\quad x > 1$
6.25	$z^{-n-\nu-1}e^{-a/z}(z-1)^n$ $\cdot L_n^{\nu}[az^{-1}(1-z)^{-1}]$ Re $z > 0$, Re $\nu > -1$	$(y/a)^{\frac{1}{2}\nu}L_n^{\nu}(y)J_\nu[2(ay)^{\frac{1}{2}}]$ $\quad x < 1$ 0 $\quad x > 1$
6.26	$O_n(z/a)$ Re $z > 0$ $y = \log(1/x)$	$\frac{1}{2}a\{[ay+(1+a^2y^2)^{\frac{1}{2}}]^n+[ay-(1+a^2y^2)^{\frac{1}{2}}]^n\}$ $x < 1$ 0 $\quad x > 1$

	$\Phi(z) = \int_0^\infty x^{z-1}\phi(x)\,dx$	$\phi(x)$
6.27	$S_n(z/a)$ $\mathrm{Re}\ z > 0$	$(1+a^2y^2)^{-\frac{1}{2}}\{[ay+(1+a^2y^2)^{\frac{1}{2}}]^n$ $-[ay-(1+a^2y^2)^{\frac{1}{2}}]^n\}$ $x < 1$ 0 $x > 1$
6.28	$P_\nu(z)$ $\mathrm{Re}\ z > -1, -1 < \mathrm{Re}\ \nu < 0$	$-2^{\frac{1}{2}}\pi^{-\frac{3}{2}}\sin(\pi\nu)y^{-\frac{1}{2}}K_{\nu+\frac{1}{2}}(y)$ $x < 1$ 0 $x > 1$
6.29	$q_\nu(z)$ $\mathrm{Re}\ z > 1, \mathrm{Re}\ \nu > -1$	$(\frac{1}{2}\pi/y)^{\frac{1}{2}}I_{\nu+\frac{1}{2}}(y)$ $x < 1$ 0 $x > 1$
6.30	$(z^2-a^2)^{-\frac{1}{2}\mu}$ $\cdot p_\nu^\mu(z/a)$ $\mathrm{Re}\ z > -a, \mathrm{Re}(\mu+\nu) > -1$ $\mathrm{Re}(\mu-\nu) > 0$	$(2a/\pi)^{\frac{1}{2}}[\Gamma(\mu-\nu)\Gamma(\mu+\nu+1)]^{-1}$ $\cdot y^{\mu-\frac{1}{2}}K_{\nu+\frac{1}{2}}(ay)$ $x < 1$ 0 $x > 1$
6.31	$(z^2-a^2)^{-\frac{1}{2}\mu}$ $\cdot e^{-i\pi\mu}q_\nu^\mu(z/a)$ $\mathrm{Re}\ z > a$ $y = \log(1/x)$	$(\frac{1}{2}\pi a)^{\frac{1}{2}}y^{\mu-\frac{1}{2}}I_{\nu+\frac{1}{2}}(ay)$ $x < 1$ 0 $x > 1$ $\mathrm{Re}(\mu+\nu) > -1$

	$\Phi(z) = \int_0^\infty x^{z-1}\phi(x)dx$	$\phi(x)$
6.32	$[(z+a)(z-a)]^{-\frac{1}{2}\mu}$ $\cdot p_\nu^\mu(z/a)$ Re $z > -a$	$-\pi^{-1}\sin(\pi\nu)y^{-1}W_{\mu,\frac{1}{2}+\nu}(2ay)$ $x < 1$ 0 $x > 1$ $-\frac{1}{2} <$ Re $\nu < \frac{1}{2}$
6.33	$[(2+a)/(z-a)]^{-\frac{1}{2}\mu}$ $e^{-i\pi\mu}q_\nu^\mu(z/a)$ Re $z > a$	$\frac{1}{2}[\Gamma(2+2\nu)]^{-1}\Gamma(1+\nu+\mu)$ $\cdot y^{-1}M_{\mu,\frac{1}{2}+\nu}(2ay)$ $x < 1$ 0 $x > 1$ Re $\nu > -\frac{1}{2}$
6.34	$p_\nu[(2ab)^{-1}(z^2+a^2+b^2)]$ Re $z > 0, -1 <$ Re $\nu < 0$	$(ab)^{\frac{1}{2}}\sin(\pi\nu)[J_{\nu+\frac{1}{2}}(by)Y_{-\nu-\frac{1}{2}}(ay)$ $+J_{-\nu-\frac{1}{2}}(ay)Y_{\nu+\frac{1}{2}}(by)]$ $x < 1$ 0 x $x > 1$
6.35	$p_\nu[(2ab)^{-1}(a^2+b^2-z^2)]$ $-(a+b) <$ Re $z < a+b$ $-1 <$ Re $\nu < 0$	$-2\pi^{-2}(ab)^{\frac{1}{2}}\sin(\pi\nu)$ $\cdot K_{\nu+\frac{1}{2}}(a\|y\|)K_{\nu+\frac{1}{2}}(b\|y\|)$
6.36	$q_\nu[(2ab)^{-1}(a^2+b^2-z^2)]$ $-(a-b) <$ Re $z < a-b$ $y = \log(1/x)$	$(ab)^{\frac{1}{2}}I_{\nu+\frac{1}{2}}(b\|y\|)K_{\nu+\frac{1}{2}}(a\|y\|)$ $a > b$, Re $\nu > -1$

	$\Phi(z) = \int_0^\infty x^{z-1}\phi(x)\,dx$	$\phi(x)$
6.37	$q_\nu[(2ab)^{-1}(z^2+a^2+b^2)]$ $\mathrm{Re}\, z > 0,\ \mathrm{Re}\, \nu > -1$	$\pi(ab)^{\frac{1}{2}} J_{\nu+\frac{1}{2}}(ay) J_{\nu+\frac{1}{2}}(by) \quad x < 1$ $0 \quad x > 1$
6.38	$p_\nu[(2ab)^{-1}(z^2-a^2-b^2)]$ $\mathrm{Re}\, z > a+b$ $-1 < \mathrm{Re}\, \nu < 0$	$(ab)^{\frac{1}{2}} \tan(\pi\nu)[I_{\nu+\frac{1}{2}}(ay) I_{\nu+\frac{1}{2}}(by)$ $-I_{-\nu-\frac{1}{2}}(ay) I_{-\nu-\frac{1}{2}}(by)] \quad x < 1$ $0 \quad x > 1$
6.39	$q_\nu[(2ab)^{-1}(z^2-a^2-b^2)]$ $\mathrm{Re}\, z > a+b,\ \mathrm{Re}\, \nu > -\frac{1}{2}$	$\pi(ab)^{\frac{1}{2}} I_{\nu+\frac{1}{2}}(ay) I_{\nu+\frac{1}{2}}(by) \quad x < 1$ $0 \quad x > 1$
6.40	$q_n[(z/a)^{\frac{1}{2}}]$ $\mathrm{Re}\, z > a$ $n = 1,2,3,\cdots$	$\frac{1}{2}a^{-\frac{1}{2}}\Gamma(1+\frac{1}{2}n)[\Gamma(\frac{3}{2}+n)^{-1}]y^{-\frac{5}{4}}$ $\cdot e^{\frac{1}{2}ay} M_{-\frac{1}{4},\frac{1}{2}n+\frac{1}{4}}(ay) \quad x < 1$ $0 \quad x > 1$
6.41	$(2z+a)^{-\frac{1}{2}}(2z-a)^{\frac{1}{2}\mu}$ $\cdot p_\nu^\mu[(\frac{1}{2}+z/a)^{\frac{1}{2}}]$ $\mathrm{Re}\, z > \frac{1}{2}a$ $\mathrm{Re}(\nu+\mu) < 1, \mathrm{Re}(\nu-\mu) > -1$ $y = \log(1/x)$	$a^{\frac{1}{4}} 2^{\frac{3}{2}\mu-\frac{1}{2}}[\Gamma(1+\frac{1}{2}\nu-\frac{1}{2}\mu)\Gamma(\frac{1}{2}-\frac{1}{2}\nu-\frac{1}{2}\mu)]^{-1}$ $\cdot y^{-\frac{1}{2}\mu-\frac{3}{4}} W_{\frac{1}{2}\mu+\frac{1}{4},\frac{1}{2}\nu+\frac{1}{4}}(ay) \quad x < 1$ $0 \quad x > 1$

	$\Phi(z) = \int_0^\infty x^{z-1}\phi(x)\,dx$	$\phi(x)$
6.42	$(2z-a)^{\frac{1}{2}\mu} p_\nu^\mu[(\frac{1}{2}+z/a)^{\frac{1}{2}}]$ $\mathrm{Re}\ z > \frac{1}{2}a$ $\mathrm{Re}(\nu-\mu) > -1$ $\mathrm{Re}(\nu+\mu) < 0$	$2^{3/2-3/2\,\mu} a^{-\frac{1}{4}}[\Gamma(\frac{1}{2}\mu-\frac{1}{2}\nu-\frac{1}{2})\,\Gamma(-\frac{1}{2}\nu-\frac{1}{2}\mu)]^{-1}$ $\cdot y^{-\frac{1}{2}\mu-5/4} W_{\frac{1}{2}\mu-\frac{1}{2},\,\frac{1}{2}\nu+\frac{1}{2}}(ay) \quad x < 1$ $0 \quad x > 1$
6.43	$(2z-a)^{\frac{1}{2}\nu} p_\nu^\mu[(\frac{2z+a}{2z-a})^{\frac{1}{2}}]$ $\mathrm{Re}\ z > a$ $\mathrm{Re}(\nu+\mu) < 0$	$2^{\frac{1}{2}\nu+\mu} a^{-\frac{1}{2}}[\Gamma(-\frac{1}{2}\mu-\frac{1}{2}\nu)\,\Gamma(1-\mu)]^{-1}$ $\cdot y^{-\frac{1}{2}\nu-3/2} M_{\frac{1}{2}\nu,\,-\frac{1}{2}\mu}(ay) \quad x < 1$ $0 \quad x > 1$
6.44	$(z^2+a^2)^{-\frac{1}{2}\nu-\frac{1}{2}}$ $\cdot P_\nu^{-\mu}[z(z^2+a^2)^{-\frac{1}{2}}]$ $\mathrm{Re}\ z > 0$	$[\Gamma(\nu+\mu+1)]^{-1} y^\nu J_\mu(ay) \quad x < 1$ $0 \quad x > 1$ $\mathrm{Re}(\mu+\nu) > -1$
6.45	$(z^2-a^2)^{-\frac{1}{2}\nu-\frac{1}{2}}$ $\cdot p_\nu^{-\mu}[z(z^2-a^2)^{-\frac{1}{2}}]$ $\mathrm{Re}\ z > a$	$[\Gamma(\nu+\mu+1)]^{-1} y^\nu I_\mu(ay) \quad x < 1$ $0 \quad x > 1$ $\mathrm{Re}(\nu+\mu) > -1$
6.46	$(z^2-a^2)^{-\frac{1}{2}\nu-\frac{1}{2}} e^{-i\pi\mu}$ $\cdot q_\nu^\mu[z(z^2-a^2)^{-\frac{1}{2}}]$ $\mathrm{Re}\ z > a$ $y = \log(1/x)$	$[\Gamma(\nu-\mu+1)]^{-1} y^\nu K_\mu(ay) \quad x < 1$ $0 \quad x > 1$ $\mathrm{Re}(\nu\pm\mu) > -1$

	$\Phi(z) = \int_0^\infty x^{z-1}\phi(x)dx$	$\phi(x)$
6.47	$a^z \csc(\pi z) P_z(\cos\theta)$ $-1 < \mathrm{Re}\, z < 0$	$-\pi^{-1}x(x^2+2ax\cos\theta+a^2)^{-\frac{1}{2}}$ $-\pi < \theta < \pi$
6.48	$a^z \csc(\pi z) p_z(\cosh\alpha)$ $-1 < \mathrm{Re}\, z < 0$	$-\pi^{-1}x(x^2+2ax\cosh\alpha+a^2)^{-\frac{1}{2}}$
6.49	$a^z \Gamma(1+z-\nu)\Gamma(-\nu-z)$ $\cdot P_z^{\nu}(\cos\theta)$ $-1+\mathrm{Re}\,\nu < \mathrm{Re}\, z < -\mathrm{Re}\,\nu$	$\pi^{-\frac{1}{2}}\Gamma(\frac{1}{2}-\nu)(2a\sin\theta)^{-\nu}x^{1-\nu}$ $\cdot(x^2+2ax\cos\theta+a^2)^{\nu-\frac{1}{2}}$ $-\pi < \theta < \pi$
6.50	$a^z \Gamma(1+z-\nu)\Gamma(-\nu-z)$ $\cdot p_z^{\nu}(\cosh\alpha)$ $-1+\mathrm{Re}\,\nu < \mathrm{Re}\, z < -\mathrm{Re}\,\nu$	$\pi^{-\frac{1}{2}}\Gamma(\frac{1}{2}-\nu)(2a\sinh\alpha)^{-\nu}$ $\cdot(x^2+2ax\cosh\alpha+a^2)^{\nu-\frac{1}{2}}$
6.51	$\Gamma(1+\nu+z)a^z$ $\cdot P_z^{-\nu}(a), \mathrm{Re}\, z > -1-\mathrm{Re}\,\nu$	$a^{-1}xe^{-x}J_\nu[x(a^{-2}-1)^{\frac{1}{2}}]$ $-1 < a < 1$
6.52	$\Gamma(1+\nu+z)(\zeta^2-1)^{\frac{1}{2}z+\frac{1}{2}}$ $\cdot p_z^{-\nu}(\zeta)$, $\mathrm{Re}\, z > -1-\mathrm{Re}\,\nu$	$x\exp[-(1-\zeta^{-2})^{-\frac{1}{2}}x]I_\nu(x)$ $\zeta > 1$

	$\Phi(z) = \int_0^\infty x^{z-1}\phi(x)\,dx$	$\phi(x)$
6.53	$2^z[\Gamma(1-\frac{1}{2}z)]^{-1}\eta^{-\frac{1}{2}z}$ $\cdot\Gamma(\nu+\frac{1}{2}z)p^{-\nu}_{-\frac{1}{2}z}(\zeta)$	$2J_\nu[\eta^{\frac{1}{2}}(\frac{1}{2}\zeta+\frac{1}{2})^{\frac{1}{2}}x]J_\nu[\eta^{\frac{1}{2}}(\frac{1}{2}\zeta-\frac{1}{2})^{\frac{1}{2}}x]$ $-2\ \mathrm{Re}\ \nu < \mathrm{Re}\ z < 2$
6.54	$2^z\Gamma(\frac{1}{2}z)\Gamma(\nu+\frac{1}{2}z)\eta^{-\frac{1}{2}z}$ $\cdot p^{-\nu}_{-\frac{1}{2}z}(\zeta)$	$4I_\nu[\eta^{\frac{1}{2}}(\frac{1}{2}\zeta-\frac{1}{2})^{\frac{1}{2}}x]K_\nu[\eta^{\frac{1}{2}}(\frac{1}{2}\zeta+\frac{1}{2})^{\frac{1}{2}}x]$ $\mathrm{Re}\ z > (0,-2\ \mathrm{Re}\ \nu)$
6.55	$2^z\Gamma(\frac{1}{2}z)\Gamma(\nu+\frac{1}{2}z)$ $\cdot\eta^{-\frac{1}{2}z}e^{i\pi\nu}q^{-\nu}_{\frac{1}{2}z-1}(\zeta)$	$4K_\nu[\eta^{\frac{1}{2}}(\frac{1}{2}\zeta-\frac{1}{2})^{\frac{1}{2}}x]K_\nu[\eta^{\frac{1}{2}}(\frac{1}{2}\zeta+\frac{1}{2})^{\frac{1}{2}}x]$ $\mathrm{Re}\ z > (0,\ \pm\ 2\ \mathrm{Re}\ \nu)$
6.56	$e^{-i\pi\mu}q^{\mu}_{z-\frac{1}{2}}(\cosh\ \alpha)$ $\mathrm{Re}\ z < -\frac{1}{2}-\mathrm{Re}\ \mu$ $\mathrm{Re}\ \mu < \frac{1}{2}$	$(\frac{1}{2}\pi)^{\frac{1}{2}}[\Gamma(\frac{1}{2}-\mu)]^{-1}(\sinh\ a)^{\mu}$ $\cdot(\cosh y - \cosh a)^{-\mu-\frac{1}{2}} \quad x < e^{-a}$ $0 \quad x > e^{-a}$
6.57	$\Gamma(\frac{1}{2}+z)\Gamma(-\mu-z)$ $\cdot\eta^z p^{\mu}_z(\zeta)$ $-\frac{1}{2} < \mathrm{Re}\ z < -\mathrm{Re}\ \mu$	$(\frac{1}{2}\pi\eta)^{-\frac{1}{2}}(\zeta^2-1)^{-\frac{1}{4}}x^{\frac{1}{2}}$ $\cdot q_{-\mu-\frac{1}{2}}\{(1-\zeta^{-2})^{-\frac{1}{2}}[1+x(\eta\zeta)^{-1}]\}$
6.58	$\Gamma(z-\frac{1}{2})\Gamma(1-\mu-z)$ $\cdot\eta^z e^{-i\pi\mu}q^{\mu}_{-z}(\zeta)$ $\frac{1}{2} < \mathrm{Re}\ z < 1\pm\mathrm{Re}\ \mu$	$\pi(\frac{1}{2}\pi\eta)^{\frac{1}{2}}\sec(\pi\mu)(\zeta^2-1)^{-\frac{1}{4}}x^{-\frac{1}{2}}$ $\cdot\ p_{-\mu-\frac{1}{2}}\{(1-\zeta^{-2})^{-\frac{1}{2}}[1+x(\eta\zeta)^{-1}]\}$

2.7 Bessel Functions and Related Functions

	$\Phi(z) = \int_0^\infty x^{z-1}\phi(x)\,dx$	$\phi(x)$
7.1	$z^{-\nu}[\sin(az+b)J_\nu(az)$ $-\cos(az+b)Y_\nu(az)$ Re $z > 0$, Re $\nu > -\frac{1}{2}$	$2\pi^{-\frac{1}{2}}(2a)^{-\nu}[\Gamma(\frac{1}{2}+\nu)]^{-1}y^{\nu-\frac{1}{2}}$ $\cdot(4a^2+y^2)^{\frac{1}{2}\nu-\frac{1}{4}}\cos[b+(\nu-\frac{1}{2})\operatorname{arccot}(\frac{1}{2}y/a)]$ $x < 1$; 0 $x > 1$
7.2	$z^{-\nu}[\cos(az+b)J_\nu(az)$ $+\sin(az+b)Y_\nu(az)]$ Re $z > 0$, Re $\nu > -\frac{1}{2}$	$-2\pi^{-\frac{1}{2}}(2a)^{-\nu}[\Gamma(\frac{1}{2}+z)]^{-1}y^{\nu-\frac{1}{2}}$ $\cdot(4a^2+y^2)^{\frac{1}{2}\nu-\frac{1}{4}}\sin[b+(\nu-\frac{1}{2})\operatorname{arccot}(\frac{1}{2}y/a)]$ $x < 1$; 0 $x > 1$
7.3	$J_\nu^2(az)+Y_\nu^2(az)$ Re $z > 0$	$2(\pi a)^{-1}P_{\nu-\frac{1}{2}}(1+\frac{1}{2}y^2/a^2)$ $x < 1$; 0 $x > 1$
7.4	$J_\nu^2(az\ )+Y_\nu^2(az^{\frac{1}{2}})$ Re $z > 0$	$2\pi^{-2}y^{-1}\exp(\frac{1}{2}a^2/y)K_\nu(\frac{1}{2}a^2/y)$ $x < 1$; 0 $x > 1$
7.5	$J_0[a(b^2-z^2)^{\frac{1}{2}}]$ $y = \log(1/x)$	0 $x < e^{-a}$; $\pi^{-1}(a^2-y^2)^{-\frac{1}{2}}\cos[b(a^2-y^2)^{\frac{1}{2}}]$ $e^{-a} < x < e^a$; 0 $x > e^a$

	$\Phi(z) = \int_0^\infty \phi(x)x^{z-1}(x)dx$	$\phi(x)$
7.6	$\sin(az^2)J_o(az^2)$ $-\cos(az^2)Y_o(az^2)$ $\mathrm{Re}\ z > 0$	$(\pi a)^{-\frac{1}{2}}[\cos\alpha J_o(\alpha)+\sin\alpha Y_o(\alpha)]\quad x < 1$ $0\quad x > 1$ $\alpha = y^2/(16a)$
7.7	$\cos(az^2)J_o(az^2)$ $+\sin(az^2)Y_o(az^2)$ $\mathrm{Re}\ z > 0$	$(\pi a)^{-\frac{1}{2}}[\cos\alpha J_o(\alpha)-\sin\alpha Y_o(\alpha)]\quad x < 1$ $0\quad x > 1$ $\alpha = y^2/(16a)$
7.8	$(b^2-z^2)^{\frac{1}{2}\nu}$ $\cdot J_\nu[a(b^2-z^2)^{\frac{1}{2}}]$	$-\pi^{-1}(2b/\pi)^{\frac{1}{2}}(ab)^\nu\sin(\pi\nu)(y^2-a^2)^{-\frac{1}{2}\nu-\frac{1}{4}}$ $\cdot K_{\nu+\frac{1}{2}}[b(y^2-a^2)^{\frac{1}{2}}]$, $x < e^{-a}$ and $x > e^a$
7.9	$-\mathrm{Re}\ b < \mathrm{Re}\ z < \mathrm{Re}\ b$ $\mathrm{Re}\ \nu < \frac{1}{2}$ $y = \log y(1/x)$	$-(2\pi/b)^{-\frac{1}{2}}(ab)^\nu(a^2-y^2)^{-\frac{1}{2}\nu-\frac{1}{4}}$ $\cdot Y_{\nu+\frac{1}{2}}[b(a^2-y^2)^{\frac{1}{2}}]$, $e^{-a} < x < e^a$

	$\Phi(z) = \int_0^\infty x^{z-1}\phi(x)\,dx$	$\phi(x)$
7.10	$(b^2-z^2)^{-\frac{1}{2}\nu}$ $\cdot J_\nu[a(b^2-z^2)^{\frac{1}{2}}]$ $\operatorname{Re}\nu > -\frac{1}{2}$	$0 \qquad x < e^{-a}$ $(\frac{1}{2}b/\pi)^{\frac{1}{2}}(ab)^{-\nu}(a^2-y^2)^{\frac{1}{2}\nu-\frac{1}{4}}$ $\cdot J_{\nu-\frac{1}{2}}[b(a^2-y^2)^{\frac{1}{2}}] \qquad e^{-a} < x < e^a$ $0 \qquad x > e^a$
7.11	$e^{-az}J_0[a(b^2-z^2)^{\frac{1}{2}}]$	$0 \qquad x < e^{-2a}$ $\pi^{-1}(2ay-y^2)^{-\frac{1}{2}}\cos[b(2ay-y^2)^{\frac{1}{2}}]$ $x > e^{-2a}$
7.12	$z^{-\mu}J_\nu(az^{-\frac{1}{2}})$ $\operatorname{Re} z > 0, \operatorname{Re}(\nu+2\mu) > 0$	$(\frac{1}{2}a)^\nu[\Gamma(1+\nu)\Gamma(\frac{1}{2}\nu+\mu)]^{-1}y^{\mu+\frac{1}{2}\nu-1}$ ${}_0F_2(\ ;\mu+\frac{1}{2}\nu,1+\nu;-\frac{1}{4}a^2y) \qquad x < 1$ $0 \qquad x > 1$
7.13	$z^\mu J_\nu(a/z)$ $\operatorname{Re} z > 0, \operatorname{Re}(\nu-\mu) > 0$ $y = \log$	$(\frac{1}{2}a)^\nu[\Gamma(1+\nu)\Gamma(\nu-\mu)]^{-1}y^{\nu-\mu-1}$ $\cdot\, {}_0F_3(\ ;1+\nu,\frac{1}{2}\nu-\frac{1}{2}\mu,\frac{1}{2}+\frac{1}{2}\nu-\frac{1}{2}\mu;-\frac{1}{16}a^2y^2)$ $x < 1$ $0 \qquad x > 1$

	$\Phi(z) = \int_0^\infty x^{z-1}\phi(x)\,dx$	$\phi(x)$
7.14	$z^{-\frac{1}{2}}[\sin(a/z)J_\nu(a/z)$ $-\cos(a/z)Y_\nu(a/z)]$ $\mathrm{Re}\ z > 0, -\frac{1}{2} < \mathrm{Re}\ \nu < \frac{1}{2}$	$4\pi^{-3/2}y^{-\frac{1}{2}}\cos(\pi\nu)[\cos(\frac{1}{2}\pi\nu)\mathrm{ker}_{2\nu}(\alpha)$ $-\sin(\frac{1}{2}\pi\nu)\mathrm{kei}_{2\nu}(\alpha)]$ $x < 1$ 0 $x > 1$ $\alpha = 2(2ay)^{\frac{1}{2}}$
7.15	$z^{-\frac{1}{2}}[\cos(a/z)J_\nu(a/z)$ $+\sin(a/z)Y_\nu(a/z)]$ $\mathrm{Re}\ z > 0, -\frac{1}{2} < \mathrm{Re}\ \nu < \frac{1}{2}$	$-4\pi^{-3/2}y^{-\frac{1}{2}}\cos(\pi\nu)[\sin(\frac{1}{2}\pi\nu)\mathrm{ker}_{2\nu}(\alpha)$ $+\cos(\frac{1}{2}\pi\nu)\mathrm{kei}_{2\nu}(\alpha)]$ $x < 1$ 0 $x > 1$ $\alpha = 2(2ay)^{\frac{1}{2}}$
7.16	$z^{-1}e^{-a/z}J_\nu(b/z)$ $\mathrm{Re}\ z > 0, \mathrm{Re}\ \nu > -1$	$J_\nu(Ay^{\frac{1}{2}})I_\nu(By^{\frac{1}{2}})$ $x < 1$ 0 $x > 1$ $\begin{matrix}A\\B\end{matrix} = 2^{\frac{1}{2}}[(a^2+b^2)^{\frac{1}{2}}\pm a)$
7.17	$z^{-1}e^{a/z}J_\nu(b/z)$ $\mathrm{Re}\ z > 0, \mathrm{Re}\ \nu > -1$ $y = \log(1/x)$	$I_\nu(Ay^{\frac{1}{2}})J_\nu(By^{\frac{1}{2}})$ $x < 1$ 0 $x > 1$ $\begin{matrix}A\\B\end{matrix} = 2^{\frac{1}{2}}[(a^2+b^2)^{\frac{1}{2}}\pm a]$

	$\Phi(z) = \int_0^\infty x^{z-1}\phi(x)\,dx$	$\phi(x)$
7.18	$(a/b)^{\frac{1}{2}z}\Gamma(\frac{1}{2}-\frac{1}{2}z)$ $\cdot J_{\frac{1}{2}z}(2ab)$ $\pm 2\,\mathrm{Re}\,\nu < \mathrm{Re}\,z < 1$	$\pi^{3/2}\sec(\pi\nu)Y_\nu(u)J_\nu(v)-J_\nu(u)Y_\nu(v)]$ $\begin{matrix}u\\v\end{matrix} = b[(a^2+x^2)^{\frac{1}{2}}\pm a]$
7.19	$(a/b)^{\frac{1}{2}z}\Gamma(\frac{1}{2}-\frac{1}{2}z)$ $Y_{\frac{1}{2}z}(2ab)$ $\pm 2\,\mathrm{Re}\,\nu < \mathrm{Re}\,z < 1$	$-\pi^{3/2}\sec(\pi\nu)\,[J_\nu(u)J_\nu(v)+Y_\nu(u)Y_\nu(v)]$ $\begin{matrix}u\\v\end{matrix} = b[(a^2+x^2)^{\frac{1}{2}}\pm a]$
7.20	$[\Gamma(1+\nu-\frac{1}{2}z)]^{-1}\Gamma(\frac{1}{2}-\frac{1}{2}z)$ $\cdot\Gamma(\nu+\frac{1}{2}z)(a/b)^{\frac{1}{2}z}J_{-\frac{1}{2}z}(2ab)$ $-2\,\mathrm{Re}\,\nu < \mathrm{Re}\,z < 1$	$2\pi^{\frac{1}{2}}J_\nu(u)J_\nu(v)$ $\begin{matrix}u\\v\end{matrix} = b[(a^2+x^2)^{\frac{1}{2}}\pm a]$
7.21	$[\Gamma(1+\nu-\frac{1}{2}z)]^{-1}\Gamma(\frac{1}{2}-\frac{1}{2}z)$ $\cdot\Gamma(\nu+\frac{1}{2}z)(a/b)^{\frac{1}{2}z}Y_{-\frac{1}{2}z}(2ab)$ $-2\,\mathrm{Re}\,\nu < \mathrm{Re}\,z < 1$	$2\pi^{\frac{1}{2}}J_\nu(v)Y_\nu(u)$ $\begin{matrix}u\\v\end{matrix} = b[a^2+x^2)^{\frac{1}{2}}\pm a]$
7.22	$b^z\csc(\frac{1}{2}\pi z)$ $[J_{-z}(2ab)-J_z(2ab)]$	$2\pi^{-1}\cos[a(x+b^2/x)]$ $-1 < \mathrm{Re}\,z < 1$

	$\Phi(z) = \int_0^\infty x^{z-1}\phi(x)\,dx$	$\phi(x)$
7.23	$b^z \sec(\frac{1}{2}\pi z)$ $\cdot[J_z(2ab)+J_{-z}(2ab)]$	$2\pi^{-1}\sin[a(x+b^2/x)]$ $-1 < \mathrm{Re}\, z < 1$
7.24	$(2a/b)^{\frac{1}{2}z}\Gamma(\frac{1}{2}z)$ $\cdot J_{\nu-\frac{1}{2}z}(ab)$	$2a^{\nu}(a^2+x^2)^{-\frac{1}{2}\nu}J_{\nu}[b(a^2+x^2)^{\frac{1}{2}}]$ $0 < \mathrm{Re}\, z < \frac{3}{2}+\mathrm{Re}\,\nu$
7.25	$(2a/b)^{\frac{1}{2}z}\Gamma(\frac{1}{2}z)$ $\cdot Y_{\nu-\frac{1}{2}z}$	$2a^{\nu}(a^2+x^2)^{-\frac{1}{2}\nu}Y_{\nu}[b(a^2+x^2)^{\frac{1}{2}}]$ $0 < \mathrm{Re}\, z < \frac{3}{2}+\mathrm{Re}\,\nu$
7.26	$(2a/b)^{\frac{1}{2}z}\Gamma(\frac{1}{2}z)$ $\cdot J_{\nu+\frac{1}{2}z}(ab)$ $\mathrm{Re}\, z > 0,\ \mathrm{Re}\,\nu > -1$	$2a^{-\nu}(a^2-x^2)^{\frac{1}{2}\nu}J_{\nu}[b(a^2-x^2)^{\frac{1}{2}}]$ $x < a$; 0 $x > a$
7.27	$(\frac{1}{2}a)^{-z}\Gamma(\frac{1}{2}+z)$ $\cdot J_z(a)$ $\mathrm{Re}\, z > -\frac{1}{2}$	$\pi^{-\frac{1}{2}}(1/x-1)^{-\frac{1}{2}}\cos[a(1-x)^{\frac{1}{2}}]$ $x < 1$; 0 $x > 1$
7.28	$(\frac{1}{2}a)^{z}\Gamma(z)$ $\cdot J_{\nu+z}(a)$ $\mathrm{Re}\, z > 0$	$(1-x)^{\frac{1}{2}\nu}J_{\nu}[a(1-x)^{\frac{1}{2}}]$ $x < 1$; 0 $x > 1$

	$\Phi(z) = \int_0^\infty x^{z-1}\phi(x)\,dx$	$\phi(x)$
7.29	$(2a/b)^{\frac{1}{2}z}\Gamma(\tfrac{1}{2}z)$ $\cdot[J_{\nu+\frac{1}{2}z}(ab)\cos(\tfrac{1}{2}\pi z)$ $-Y_{\nu+\frac{1}{2}z}(ab)\sin(\tfrac{1}{2}\pi z)]$	$2a^{-\nu}(a^2+x^2)^{\frac{1}{2}\nu}J_\nu[b(a^2+x^2)^{\frac{1}{2}}]$ $0 < \mathrm{Re}\, z < \tfrac{3}{2}-\mathrm{Re}\,\nu$
7.30	$(2a/b)^{\frac{1}{2}z}\Gamma(\tfrac{1}{2}z)$ $\cdot[Y_{\nu+\frac{1}{2}z}(ab)\cos(\tfrac{1}{2}\pi z)$ $+J_{\nu+\frac{1}{2}z}(ab)\sin(\tfrac{1}{2}\pi z)]$	$2a^{-\nu}(a^2+x^2)^{\frac{1}{2}\nu}Y_\nu[b(a^2+x^2)^{\frac{1}{2}}]$ $0 < \mathrm{Re}\, z < \tfrac{3}{2}-\nu$
7.31	$J^2_{\frac{1}{2}z}(a)+Y^2_{\frac{1}{2}z}(a)$	$4\pi^{-2}K_o(a\|x-x^{-1}\|)$ Principal value
7.32	$J_{\nu-\frac{1}{2}z}(a)Y_{-\nu-\frac{1}{2}z}(a)$ $-J_{-\nu-\frac{1}{2}z}(a)Y_{\nu-\frac{1}{2}z}(a)$	$0 \qquad x < 1$ $4\pi^{-2}\sin(2\pi\nu)K_{2\nu}[a(x-x^{-1})]$ $x > 1$ $-\tfrac{1}{2} < \mathrm{Re}\, z < \tfrac{1}{2}$
7.33	$J_{\nu-\frac{1}{2}}(a)Y_{\nu+\frac{1}{2}z}(a)$ $+J_{\nu+\frac{1}{2}z}(a)Y_{\nu-\frac{1}{2}z}(a)$ $y = \log y(1/x)$	$-2\pi^{-1}J_{2\nu}[a(x+x^{-1})]$ $-\tfrac{3}{2} < \mathrm{Re}\, z < \tfrac{3}{2}$

	$\Phi(z) = \int_0^\infty x^{z-1}\phi(x)\,dx$	$\phi(x)$
7.34	$J_{\nu-\frac{1}{2}z}(a)J_{\nu+\frac{1}{2}z}(a)$ $-Y_{\nu-\frac{1}{2}z}(a)Y_{\nu+\frac{1}{2}z}(a)$	$2\pi^{-1}Y_{2\nu}[a(x+x^{-1})]$ $-\frac{3}{2} < \mathrm{Re}\ z < \frac{3}{2}$
7.35	$J_{\nu-z}(a)Y_{\nu+z}(b)$ $+J_{\nu+z}(b)Y_{\nu-z}(a)$ $-\frac{3}{2} < \mathrm{Re}\ z < \frac{3}{2}$	$-\pi^{-1}(a+bx)^{\nu}(b+ax)^{-\nu}$ $\cdot J_{2\nu}\{[a^2+b^2+ab(x+x^{-1})]^{\frac{1}{2}}\}$
7.36	$J_{\nu-z}(a)J_{\nu+z}(b)$ $-Y_{\nu-z}(a)Y_{\nu+z}(b)$ $-\frac{3}{2} < \mathrm{Re}\ z < \frac{3}{2}$	$\pi^{-1}(a+bx)^{\nu}(b+ax)^{-\nu}$ $\cdot Y_{2\nu}\{[a^2+b^2+ab(x+x^{-1})]^{\frac{1}{2}}\}$
7.37	$\cosh(\frac{1}{2}\pi z)J_{\nu-\frac{1}{2}iz}(a)$ $\cdot J_{\nu+\frac{1}{2}iz}(a)$ $\mathrm{Re}\ \nu > -\frac{1}{2}$	$0 \qquad x < e^{-\pi}$ $\frac{1}{2}\pi^{-1}J_{2\nu}(2a\sin y)\quad e^{-\pi} < x < e^{\pi}$ $0 \qquad x > e^{\pi}$
7.38	$J_{\nu-iz}(a)J_{\nu+iz}(a)$ $\mathrm{Re}\ \nu > -\frac{1}{2}$ $y = \log(1/x)$	$0 \qquad x < e^{-\pi}$ $\frac{1}{2}\pi^{-1}J_{2\nu}[2a\cos(\frac{1}{2}x)]$ $e^{-\pi} < x < e^{\pi}$ $0 \qquad x > e^{\pi}$

	$\Phi(z) = \int_0^\infty x^{z-1}\phi(x)\,dx$	$\phi(x)$	
7.39	$e^{-az}I_o(az)$	0	$x < e^{-2a}$
		$\pi^{-1}(2ay-y^2)^{-\frac{1}{2}}$	$x > e^{-2a}$
7.40	$z^{-1}I_o(z)$	1	$x < e^{-1}$
	Re $z > 0$	$\pi^{-1}\arccos(-y)$	$e^{-1} < x < e$
		0	$x > e$
7.41	$I_{2n}(az)$	0	$x < e^{-a}$
	$n = 0,1,2,\cdots$	$\pi^{-1}(a^2-y^2)^{-\frac{1}{2}}T_{2n}(y/a)$	$e^{-a} < x < e^a$
		0	$x > e^a$
7.42	$z^{-\nu}I_{\nu+2n}(az)$	0	$x < e^{-a}$
	Re $\nu > -\frac{1}{2}$, $n=0,1,2,\cdots$	$\frac{1}{2}\pi^{-1}(\frac{1}{2}a)^{-\nu}(2n)!\Gamma(\nu)[\Gamma(2n+2\nu)]^{-1}$ $\cdot(a^2-y^2)^{\nu-\frac{1}{2}}C_{2n}^{\nu}(y/a)$	$e^{-a}<x<e^a$
		0	$x > e^a$
7.43	$I_\nu(z)$	$-\pi^{-1}2^\nu\sin(\pi\nu)(y^2-1)^{-\frac{1}{2}}$ $\cdot[(y-1)^{\frac{1}{2}}+(y+1)^{\frac{1}{2}}]^{-2\nu}$	$x < e^{-1}$
	Re $z > 0$	$\pi^{-1}(1-y^2)^{-\frac{1}{2}}\cos[\nu\arccos(-y)]$	$e^{-1} < x < e$
	$y = \log$	0	$x > e$

	$\Phi(z) = \int_0^\infty x^{z-1}\phi(x)dx$	$\phi(x)$
7.44	$z^{-1}I_\nu(z)$ Re $z > 0$	$2^\nu(\pi\nu)^{-1}\sin(\pi\nu)$ $\cdot[(y-1)^{\frac{1}{2}}+(y+1)^{\frac{1}{2}}]^{-2\nu}$ $\quad x < e^{-1}$ $(\pi\nu)^{-1}\sin[\nu \arccos(-y)]$ $e^{-1} < x < e$ $0 \quad x > e$
7.45	$z^{-\nu}I_\nu(az)$ Re $\nu > -\frac{1}{2}$	$0 \quad x < e^{-a}$ $\pi^{-\frac{1}{2}}(2a)^{-\nu}[\Gamma(\frac{1}{2}+\nu)]^{-1}(a^2-y^2)^{\nu-\frac{1}{2}}$ $e^{-a} < x < e^a$ $0 \quad x > e^a$
7.46	$z^{-\nu}\cosh(az)I_\nu(az)$ Re $\nu > -\frac{1}{2}$	$0 \quad x < e^a$ $\frac{1}{2}\pi^{-\frac{1}{2}}(2a)^{-\nu}[\Gamma(\frac{1}{2}+\nu)]^{-1}(2ay-y^2)^{\nu-\frac{1}{2}}$ $e^{-a} < x < e^a$ $0 \quad x > e^a$
7.47	$I_\nu\{\frac{1}{2}a[z+(z^2-b^2)^{\frac{1}{2}}]\}$ $\cdot I_\nu\{\frac{1}{2}a[z-(z^2-b^2)^{\frac{1}{2}}]\}$ Re $\nu > -\frac{1}{2}$ $y = \log$	$0 \quad x < e^{-a}$ $\pi^{-1}(a^2-y^2)^{-\frac{1}{2}}J_{2\nu}[b(a^2-y^2)^{\frac{1}{2}}]$ $e^{-a} < x < e^a$ $0 \quad x > e^a$

	$\Phi(z) = \int_0^\infty x^{z-1}\phi(x)\,dx$	$\phi(x)$
7.48	$I_\nu\{\frac{1}{2}a[(z^2+b^2)^{\frac{1}{2}}-z]\}$ $\cdot I_\nu\{\frac{1}{2}a[(z^2+b^2)^{\frac{1}{2}}+z]\}$ $\mathrm{Re}\ \nu > -\frac{1}{2}$	$0 \qquad x < e^{-a}$ $\pi^{-1}(a^2-y^2)^{-\frac{1}{2}}I_{2\nu}[b(a^2-y^2)^{\frac{1}{2}}]\ e^{-a}<x<e^a$ $0 \qquad x > e^a$
7.49	$I_\nu(a/z)$ $\mathrm{Re}\ z > 0,\ \mathrm{Re}\ \nu > 0$	$(\frac{1}{2}y/a)^{-\frac{1}{2}}\{\mathrm{ber}_\nu[(2ay)^{\frac{1}{2}}]\mathrm{bei}'_\nu[(2ay)^{\frac{1}{2}}]$ $+\mathrm{bei}_\nu[(2ay)^{\frac{1}{2}}]\mathrm{ber}'_\nu[(2ay)^{\frac{1}{2}}]\}\ x<1$ $0 \qquad x > 1$
7.50	$z^{-\mu}I_\nu(a/z)$ $\mathrm{Re}\ z > 0, \mathrm{Re}(\mu+\nu > 0$	$(\frac{1}{2}a)^\nu[\Gamma(1+\nu)\Gamma(\mu+\nu)]^{-1}y^{\mu+\nu-1}$ $\cdot {}_0F_3(\ ;\nu+1,\frac{1}{2}\mu+\frac{1}{2}\nu,\frac{1}{2}+\frac{1}{2}\nu+\frac{1}{2}\mu;a^2y^2/16)$ $x < 1$ $0 \qquad x > 1$
7.51	$z^{-\frac{1}{2}}e^{-a/z}I_\nu(a/z)$ $\mathrm{Re}\ z > 0, \mathrm{Re}\ \nu > -\frac{1}{2}$	$(\pi y)^{-\frac{1}{2}}J_{2\nu}[(8ay)^{\frac{1}{2}}] \qquad x < 1$ $0 \qquad x > 1$
7.52	$z^{-1}e^{a/z}I_\nu(b/z)$ $\mathrm{Re}\ z > 0, \mathrm{Re}\ \nu > -1$ $a > b$ $y = \log(1/x)$	$I_\nu(Ay^{\frac{1}{2}})I_\nu(By^{\frac{1}{2}})$ ${A \atop B} = (a+b)^{\frac{1}{2}} \pm (a-b)^{\frac{1}{2}} \qquad x < 1$ $0 \qquad x > 1$

	$\Phi(z) = \int_0^\infty x^{z-1}\phi(x)\,dx$	$\phi(x)$	
7.53	$z^{-1}e^{-a/z}I_\nu(b/z)$	$J_\nu(Ay^{\frac{1}{2}})J_\nu(By^{\frac{1}{2}})$	$x < 1$
	$\mathrm{Re}\, z > 0, \mathrm{Re}\, \nu > -1$	0	$x > 1$
	$a > b$	$\begin{matrix}A\\B\end{matrix} = (a+b)^{\frac{1}{2}} \pm (a-b)^{\frac{1}{2}}$	
7.54	$z^{-\frac{1}{2}}e^{a/z}I_\nu(a/z)$	$(\pi y)^{-\frac{1}{2}}I_{2\nu}[(8ay)^{\frac{1}{2}}]$	$x < 1$
	$\mathrm{Re}\, z > 0,\ \mathrm{Re}\, \nu > -\frac{1}{2}$	0	$x > 1$
7.55	$z^{-2}\exp(-z^{-2})I_\nu(z^{-2})$	$2^{\frac{1}{2}-\nu}[\Gamma(1+2\nu)\Gamma(1+\nu)]^{-1}y^{2\nu}$	
	$\mathrm{Re}\, z > 0$	$\cdot\, {}_0F_2(\ ;1+\nu, 1+2\nu; -\frac{1}{2}y^2)$	$x < 1$
		0	$x > 1$
7.56	$z^{-\frac{3}{2}}e^{a/z}$	$(\frac{1}{2}\pi a)^{-\frac{1}{2}}I_{2\nu}[(8ay)^{\frac{1}{2}}]$	$x < 1$
	$\cdot[I_{\nu-\frac{1}{2}}(a/z) - I_{\nu+\frac{1}{2}}(a/z)]$	0	$x > 1$
		$\mathrm{Re}\, z > 0$	
	$y = \log(1/x)$		

	$\Phi(z) = \int_0^\infty x^{z-1}\phi(x)\,dx$	$\phi(x)$
7.57	$K_o(az)$ $\mathrm{Re}\ z > 0$	$(y^2-a^2)^{-\frac{1}{2}}$ $\quad x < e^{-a}$ 0 $\quad x > e^{-a}$
7.58	$z^{-1}K_o(az)$ $\mathrm{Re}\ z > 0$	$\log[y/a+(y^2/a^2-1)^{\frac{1}{2}}]$ $\quad x < e^{-a}$ 0 $\quad x > e^{-a}$
7.59	$K_\nu(az)$ $\mathrm{Re}\ z > 0$	$\frac{1}{2}(y^2-a^2)^{-\frac{1}{2}}\ [y/a+(y^2/a^2-1)^{\frac{1}{2}}]^\nu$ $+[y/a-(y^2/a^2)^{\frac{1}{2}}]^\nu$ $\quad x < e^{-a}$ 0 $\quad x > e^{-a}$
7.60	$z^{-1}K_\nu(az)$ $\mathrm{Re}\ z > 0$	$\frac{1}{2}\ \nu^{-1}\{[y/a+(y^2/a^2-1)^{\frac{1}{2}}]^\nu$ $-[y/a-(y^2/a^2-1)^{\frac{1}{2}}]^\nu\}$ $x < e^{-a}$ 0 $\quad x > e^{-a}$
7.61	$z^{-\nu}K_\nu(az)$ $\mathrm{Re}\ z > 0, \mathrm{Re}\ \nu > -\frac{1}{2}$	$\pi^{\frac{1}{2}}(2a)^{-\nu}[\Gamma(\frac{1}{2}+\nu)]^{-1}(y^2-a^2)^{\nu-\frac{1}{2}}$ $x < e^{-a}$ 0 $\quad x > e^{-a}$
7.62	$z^{-\mu}K_\nu(az)$ $\mathrm{Re}\ z > 0$ $y = \log$	$(\frac{1}{2}\pi/a)^{\frac{1}{2}}(y^2-a^2)^{\frac{1}{2}\mu-\frac{1}{4}}P_{\nu-\frac{1}{2}}^{\frac{1}{2}-\mu}(y/a)$ $x < e^{-a}$ 0 $\quad x > e^{-a}$

	$\Phi(z) = \int_0^\infty x^{z-1}\phi(x)\,dx$	$\phi(x)$	
7.63	$e^{az}K_o(az)$	$(y^2+2ay)^{-\frac{1}{2}}$	$x < 1$
	Re $z > 0$	0	$x > 1$
7.64	$z^{-\nu}e^{az}K_\nu(az)$	$\pi^{\frac{1}{2}}(2a)^{-\nu}[\Gamma(\frac{1}{2}+\nu)]^{-1}(y^2+2ay)^{\nu-\frac{1}{2}}$	
	Re $z > 0$, Re $\nu > -\frac{1}{2}$		$x < 1$
		0	$x > 1$
7.65	$z^{-\mu}e^{az}K_\nu(az)$	$(\frac{1}{2}\pi/a)^{\frac{1}{2}}(y^2+2ay)^{\frac{1}{2}\mu-\frac{1}{2}}$	
	Re $z > 0$	$\cdot P_{\nu-\frac{1}{2}}^{\frac{1}{2}-\mu}(1+y/a)$	$x < 1$
		0	$x > 1$
7.66	$z^{-1}K_o(az^{\frac{1}{2}})$	$-\frac{1}{2}\mathrm{Ei}(-\frac{1}{2}a^2/y)$	$x < 1$
	Re $z > 0$	0	$x > 1$
7.67	$z^{\frac{1}{2}\nu}K_\nu(az^{\frac{1}{2}})$	$a^\nu(2y)^{-\nu-1}e^{-\frac{1}{4}a^2/y}$	$x < 1$
	Re $z > 0$	0	$x > 1$
7.68	$z^\mu K_\nu(az^{\frac{1}{2}})$	$a^{-1}y^{-\mu-\frac{1}{2}}\exp(-\frac{1}{8}a^2/y)$	
	Re $z > 0$	$\cdot W_{\frac{1}{2}+\mu,\frac{1}{2}\nu}(\frac{1}{4}a^2/y)$	$x < 1$
		0	$x > 1$
	$y = \log(1/x)$		

	$\Phi(z) = \int_0^\infty x^{z-1}\phi(x)\,dx$	$\phi(x)$
7.69	$z^{-\frac{1}{2}}K_\nu(az^{\frac{1}{2}})$ Re $z > 0$	$\frac{1}{2}(\pi y)^{-\frac{1}{2}}\exp(-\frac{1}{8}a^2/y)K_{\frac{1}{2}\nu}(\frac{1}{8}a^2/y)$ $x < 1$ 0 $x > 1$
7.70	$K_o[a(b^2+z^2)^{\frac{1}{2}}]$ Re $z > \lvert\text{Im } b\rvert$	$(y^2-a^2)^{-\frac{1}{2}}\cos[b/y^2-a^2)^{\frac{1}{2}}]$ $x<e^{-a}$ 0 $x > e^{-a}$
7.71	$e^{az}K_o[a(b^2+z^2)^{\frac{1}{2}}]$ Re $z > \lvert\text{Im } b\rvert$	$(y^2+2ay)^{-\frac{1}{2}}\cos[b(y^2+2ay)^{\frac{1}{2}}]$ $x < 1$ 0 $x > 1$
7.72	$(b^2+z^2)^{-\frac{1}{2}\nu}$ $\cdot K_\nu[a(b^2+z^2)^{\frac{1}{2}}]$ Re $z > \lvert\text{Im } b\rvert$, Re $\nu > -\frac{1}{2}$	$(\frac{1}{2}\pi b)^{\frac{1}{2}}(ab)^{-\nu}(y^2-a^2)^{\frac{1}{2}\nu-\frac{1}{4}}$ $\cdot J_{\nu-\frac{1}{2}}[b(y^2-a^2)^{\frac{1}{2}}]$ $x < e^{-a}$ 0 $x > e^{-a}$
7.73	$e^{az}(b^2+z^2)^{-\frac{1}{2}\nu}$ $\cdot K_\nu[a(b^2+z^2)^{\frac{1}{2}}]$ Re $z > \lvert\text{Im } b\rvert$, Re $\nu > -\frac{1}{2}$	$(\frac{1}{2}\pi b)^{\frac{1}{2}}(ab)^{-\nu}(y^2+2ay)^{\frac{1}{2}\nu-\frac{1}{4}}$ $\cdot J_{\nu-\frac{1}{2}}[b(y^2+2ay)^{\frac{1}{2}}]$ $x < 1$ 0 $x > 1$
7.74	$K_o[a(b^2-z^2)^{\frac{1}{2}}]$ $-$Re $b<$Re $z<$Re b, Re $b>0$	$\frac{1}{2}(a^2+y^2)^{-\frac{1}{2}}\exp[-b(a^2+y^2)^{\frac{1}{2}}]$

	$\Phi(z) = \int_0^\infty x^{z-1}\phi(x)\,dx$	$\phi(x)$
7.75	$(b^2-z^2)^{-\frac{1}{2}}$ $\cdot K_1[a(b^2-z^2)^{\frac{1}{2}}]$	$2(ab)^{-1}\exp[-b(a^2+y^2)^{\frac{1}{2}}]$ $-\text{Re } b < \text{Re } z < \text{Re } b, \text{Re } b>0$
7.76	$(b^2-z^2)^{-\frac{1}{2}}$ $\cdot K_\nu[a(b^2-z^2)^{\frac{1}{2}}]$ $-\text{Re } b<\text{Re } z<\text{Re } b$ $\text{Re } b > 0$	$K_{\frac{1}{2}\nu}\{\frac{1}{2}b[(a^2+y^2)^{\frac{1}{2}}-y]\}$ $\cdot K_{\frac{1}{2}\nu}\{\frac{1}{2}b[(a^2+y^2)^{\frac{1}{2}}+y]\}$
7.77	$(b^2-z^2)^{-\frac{1}{2}\nu}$ $\cdot K_\nu[a(b^2-z^2)^{\frac{1}{2}}]$ $-\text{Re } b<\text{Re } z<\text{Re } b$	$(2\pi/b)^{-\frac{1}{2}}(ab)^{-\nu}(a^2+y^2)^{\frac{1}{2}\nu-\frac{1}{4}}$ $\cdot K_{\nu-\frac{1}{2}}[b(a^2+y^2)^{\frac{1}{2}}]$ $\text{Re } b > 0$
7.78	$K_0[a(z^2-b^2)^{\frac{1}{2}}]$ $\text{Re } z > \lvert\text{Re } b\rvert$	$(y^2-a^2)^{-\frac{1}{2}}\cosh[b(y^2-a^2)^{\frac{1}{2}}]$ $\quad x < e^{-a}$ $0 \quad x > e^{-a}$
7.79	$(z^2-b^2)^{-\frac{1}{2}\nu}$ $\cdot K_\nu[a(z^2-b^2)^{\frac{1}{2}}]$ $\text{Re } z > \lvert\text{Re } b\rvert$ $\text{Re } \nu > -\frac{1}{2}$ $y = \log(1/x)$	$(\frac{1}{2}\pi b)^{\frac{1}{2}}(ab)^{-\nu}(y^2-a^2)^{\frac{1}{2}\nu-\frac{1}{4}}$ $\cdot I_{\nu-\frac{1}{2}}[b(y^2-a^2)^{\frac{1}{2}}] \quad x < e^{-a}$ $0 \quad x > e^{-a}$

	$\Phi(z) = \int_0^\infty x^{z-1}\phi(x)\,dx$	$\phi(x)$
7.80	$[(z+b)/(z-b)]^{\frac{1}{2}\nu}$ $\cdot K_\nu[a(z^2-b^2)^{\frac{1}{2}}]$ $\operatorname{Re} z > \lvert\operatorname{Re} b\rvert$ $u = (y^2-a^2)^{\frac{1}{2}}$	$\frac{1}{2}a^{-\nu}(y^2-a^2)^{-\frac{1}{2}}[(y+u)^\nu e^{ub}+(y-u)^\nu e^{-ub}]$ $\quad x < e^{-a}$ $0 \quad x > e^{-a}$
7.81	$I_\nu(bz)K_\nu(az)$ $\operatorname{Re} z > 0,\ a \geqq b$	$\pi^{-1}(ab)^{-\frac{1}{2}}\cos(\pi\nu)$ $\cdot q_{\nu-\frac{1}{2}}[(2ab)^{-1}(y^2-a^2-b^2)]$ $\quad x < e^{-(a+b)}$ $\frac{1}{2}(ab)^{-\frac{1}{2}}P_{\nu-\frac{1}{2}}[(2ab)^{-1}(a^2+b^2-y^2)]$ $\quad e^{-(a+b)} < x < e^{-(a-b)}$ $0 \quad x > e^{-a-b)}$
7.82	$K_\nu(az)K_\nu(bz)$ $\operatorname{Re} z > 0$	$\frac{1}{2}\pi(ab)^{\frac{1}{2}}P_{\nu-\frac{1}{2}}[(2ab)^{-1}(y^2-a^2-b^2)]$ $\quad x < e^{-(a+b)}$ $0 \quad x > e^{-(a+b)}$
7.83	$J_\nu(bz^{\frac{1}{2}})K_\nu(az^{\frac{1}{2}})$ $\operatorname{Re} z > 0,\ a \geqq b$ $y = \log(1/x)$	$\frac{1}{2}y^{-1}\exp[-\frac{1}{4}(a^2-b^2)/y]J_\nu(\frac{1}{2}ab/y)$ $\quad x < 1$ $0 \quad x > 1$

	$\Phi(z) = \int_0^\infty x^{z-1}\phi(x)\,dx$	$\phi(x)$
7.84	$Y_\nu(bz^{\frac{1}{2}})K_\nu(az^{\frac{1}{2}})$	$\frac{1}{2}y^{-1}\exp[-\frac{1}{4}(a^2-b^2)/y]Y_\nu(\frac{1}{2}ab/y)$ $\quad x < 1$
	$\operatorname{Re} z > 0,\ a \geqq b$	$0 \quad x > 1$
7.85	$I_\nu(bz^{\frac{1}{2}})K_\nu(az^{\frac{1}{2}})$	$\frac{1}{2}y^{-1}\exp[-\frac{1}{4}(a^2+b^2)/y]I_\nu(\frac{1}{2}ab/y)$ $\quad x < 1$
	$\operatorname{Re} z > 0$ $a \geqq b$	$0 \quad x > 1$
7.86	$K_\nu(bz^{\frac{1}{2}})K_\nu(az^{\frac{1}{2}})$	$\frac{1}{2}y^{-1}\exp[-\frac{1}{4}(a^2+b^2)/y]K_\nu(\frac{1}{2}ab/y)$ $\quad x < 1$
	$\operatorname{Re} z > 0$	$0 \quad x > 1$
7.87	$z^{-\nu}[K_\nu(az^{\frac{1}{2}})]^2$	$\frac{1}{2}\pi^{\frac{1}{2}}a^{\nu-1}y^{-\frac{1}{2}-\frac{3}{2}\nu}\exp(-\frac{1}{2}a^2/y)W_{\frac{1}{2}\nu,\frac{1}{2}\nu}(a^2/y)$ $\quad x < 1$
	$\operatorname{Re} z > 0$	$0 \quad x > 1$
7.88	$K_o\{a[(z^2+b^2)^{\frac{1}{2}}-z]\}$ $\cdot K_o\{a[(z^2+b^2)^{\frac{1}{2}}+z]\}$	$-\pi(y^2-4a^2)^{-\frac{1}{2}}Y_o[b(y^2-4a^2)^{\frac{1}{2}}]$ $\quad x < e^{-2a}$
	$\operatorname{Re} z > \lvert\operatorname{Im} b\rvert$	$0 \quad x > e^{-2a}$
	$y = \log(1/x)$	

	$\Phi(z) = \int_0^\infty x^{z-1}\phi(x)\,dx$	$\phi(x)$
7.89	$e^{2az}K_o\{a[(z^2+b^2)^{\frac{1}{2}}-z]\}$ $\cdot K_o\{a[(z^2+b^2)^{\frac{1}{2}}+z]\}$ Re $z > \lvert\text{Im } b\rvert$	$-\pi(y^2+4ay)^{-\frac{1}{2}}Y_o[b(y^2+4ay)^{\frac{1}{2}}]$ $\quad x < 1$ $0 \quad x > 1$
7.90	$I_\nu\{a[(z^2+b^2)^{\frac{1}{2}}-z]\}$ $\cdot K_\nu\{a[(z^2+b^2)^{\frac{1}{2}}+z]\}$ Re $z > \lvert\text{Im } b\rvert$	$(y^2-4a^2)^{-\frac{1}{2}}J_{2\nu}[b(y^2-4a^2)^{\frac{1}{2}}]$ $\quad x < e^{-2a}$ $0 \quad x > e^{-2a}$ Re $\nu > -\frac{1}{2}$
7.91	$e^{2az}I_\nu\{a[(z^2+b^2)^{\frac{1}{2}}-z]\}$ $\cdot K_\nu\{a[(z^2+b^2)^{\frac{1}{2}}+z]\}$ Re $z > \lvert\text{Im } b\rvert$	$(y^2+4ay)^{-\frac{1}{2}}J_{2\nu}[b(y^2+4ay)]$ $\quad x < 1$ $0 \quad x > 1$ Re $\nu > -\frac{1}{2}$
7.92	$I_\nu\{a[z-(z^2-b^2)^{\frac{1}{2}}]\}$ $\cdot K_\nu\{a[z+(z^2-b^2)^{\frac{1}{2}}]\}$ Re $z > \lvert\text{Re } b\rvert$ $y = \log(1/x)$	$(y^2-4a^2)^{-\frac{1}{2}}I_{2\nu}[b(y^2-4a^2)^{\frac{1}{2}}]$ $\quad x < e^{-2a}$ $0 \quad x > e^{-2a}$ Re $\nu > -\frac{1}{2}$

	$\Phi(z) = \int_0^\infty x^{z-1}\phi(x)\,dx$	$\phi(x)$
7.93	$e^{2az}I_\nu\{a[z-(z^2-b^2)^{\frac{1}{2}}]\}$ $\cdot K_\nu\{a[z+(z^2-b^2)^{\frac{1}{2}}]\}$ $\mathrm{Re}\ z > \lvert\mathrm{Re}\ b\rvert$	$(y^2+4ay)^{-\frac{1}{2}}I_{2\nu}[b(y^2+4ay)^{\frac{1}{2}}]\ x < 1$ $0 \qquad x < 1$ $\mathrm{Re}\ \nu > -\frac{1}{2}$
7.94	$K_\nu\{a[z-(z^2-a^2)^{\frac{1}{2}}]\}$ $\cdot K_\nu\{a[z+(z^2-a^2)^{\frac{1}{2}}]\}$ $\mathrm{Re}\ z>\lvert\mathrm{Im}\ b\rvert, -\frac{1}{2}<\mathrm{Re}\ \nu<\frac{1}{2}$	$2\cos(\pi\nu)(y^2-4a^2)^{-\frac{1}{2}}$ $\cdot K_{2\nu}[b(y^2-4ay)^{\frac{1}{2}}] \qquad x < e^{-2a}$ $0 \qquad x > e^{-2a}$
7.95	$e^{2az}K_\nu\{a[z-(z^2-b^2)^{\frac{1}{2}}]\}$ $\cdot K_\nu\{a[z+(z^2-b^2)^{\frac{1}{2}}]\}$ $\mathrm{Re}\ z>\lvert\mathrm{Im}\ b\rvert, -\frac{1}{2}<\mathrm{Re}\ \nu<\frac{1}{2}$	$2\cos(\pi\nu)(y^2+4ay)^{-\frac{1}{2}}$ $\cdot K_{2\nu}[b(y^2+4ay)^{\frac{1}{2}}] \qquad x < 1$ $0 \qquad x > 1$
7.96	$z^{-\frac{1}{2}}e^{a/z}K_\nu(a/z)$ $-\frac{1}{2} < \mathrm{Re}\ \nu < \frac{1}{2}$	$2(\pi y)^{-\frac{1}{2}}\cos(\pi\nu)K_{2\nu}[(8ay)^{\frac{1}{2}}]\}$ $x < 1$ $0 \qquad x > 1$
7.97	$z^{-\frac{1}{2}}e^{-a/z}K_\nu(a/z)$ $\mathrm{Re}\ z > 0, -\frac{1}{2} < \mathrm{Re}\ \nu < \frac{1}{2}$ $\mathrm{Re}\ z > 0, -\frac{1}{2} < \mathrm{Re}\ \nu < \frac{1}{2}$ $y = \log(1/x)$	$-(\pi y)^{\frac{1}{2}}\{\sin(\pi\nu)J_{2\nu}[(8ay)^{\frac{1}{2}}]$ $+\cos(\pi\nu)Y_{2\nu}[(8ay)^{\frac{1}{2}}]\}\ x < 1$ $0 \qquad x > 1$

	$\Phi(z) = \int_0^\infty x^{z-1}\phi(x)\,dx$	$\phi(x)$
7.98	$z^{-1}e^{a/z}K_\nu(b/z)$ $\operatorname{Re} z > 0, -1<\operatorname{Re}\nu<1$ $\begin{matrix}A\\B\end{matrix} = (a+b)^{\frac12} \pm (a-b)^{\frac12}$	$2\pi^{-1}\sin(\pi\nu)\,[K_\nu(Ay^{\frac12})K_\nu(By^{\frac12})$ $+K_\nu(Ay^{\frac12})I_\nu(By^{\frac12})]+I_\nu(Ay^{\frac12})K_\nu(By^{\frac12})$ $\quad x < 1$ $0 \quad x > 1$
7.99	$z^{-1}e^{-a/z}K_\nu(b/z)$ $\operatorname{Re} z > 0, -1 < \operatorname{Re}\nu < 1$ $\begin{matrix}A\\B\end{matrix} = (a+b)^{\frac12} \pm (a-b)^{\frac12}$	$-\frac12\pi\{\sin(\pi\nu)\,[J_\nu(Ay^{\frac12})J_\nu(By^{\frac12})$ $-Y_\nu(Ay^{\frac12})Y_\nu(By^{\frac12})]$ $+\cos(\pi\nu)\,[J_\nu(Ay^{\frac12})Y_\nu(By^{\frac12})$ $+J_\nu(By^{\frac12})Y_\nu(Ay^{\frac12})]\}$ $\quad x < 1$ $0 \quad x > 1$
7.100	$z^{-2\nu}e^{az^2}K_\nu(az^2)$ $\operatorname{Re} z > 0, \operatorname{Re}\nu > -\frac12$ $y = \log(1/x)$	$2\pi^{\frac12}[\Gamma(1+2\nu)]^{-1}(8a)^{\frac12\nu}y^{\nu-1}$ $\cdot\exp(-\frac{1}{16}y^2/a)\,M_{-\frac32\nu,\frac12\nu}(\frac18 y^2/a)$ $\quad x < 1$ $0 \quad x > 1$

	$\Phi(z) = \int_0^\infty x^{z-1}\phi(x)\,dx$	$\phi(x)$
7.101	$(2a/b)^{\frac{1}{2}z}[\Gamma(1-\frac{1}{2}z)]^{-1}$ $\cdot K_{\nu+\frac{1}{2}z}(ab)$ Re $z < \frac{3}{2}-$Re ν	$0 \quad x < a$ $a^{-\nu}(x^2-a^2)^{\frac{1}{2}\nu}J_\nu[b(x^2-a^2)^{\frac{1}{2}}] \quad x > a$ Re $\nu > -1$
7.102	$(2a/b)^{\frac{1}{2}z}\Gamma(\frac{1}{2}z)$ $K_{\nu-\frac{1}{2}z}(ab)$	$2a^{\nu}(a^2+x^2)^{-\frac{1}{2}\nu}K_\nu[b(a^2+x^2)^{\frac{1}{2}}]$ Re $z > 0$
7.103	$(2a/b)^{\frac{1}{2}z}\Gamma(\frac{1}{2}z)$ $\cdot K_{\nu+\frac{1}{2}z}(ab)$	$2a^{-\nu}(a^2+z^2)^{\frac{1}{2}\nu}K_\nu[b(a^2+x^2)^{\frac{1}{2}}]$ Re $z > 0$
7.104	$(2a/b)^{\frac{1}{2}z}\Gamma(\frac{1}{2}z)$ $\cdot I_{\nu+\frac{1}{2}z}(ab)$ Re $z > 0$	$2a^{-\nu}(a^2-x^2)^{\frac{1}{2}\nu}I_\nu[b(a^2-x^2)^{\frac{1}{2}}] \quad x < a$ $0 \quad x > a$
7.105	$b^z\cos(\frac{1}{2}\pi z)K_z(2ab)$ $-1 <$ Re $z < 1$	$\frac{1}{2}\cos[a(x-b^2/x)]$
7.106	$b^z\sin(\frac{1}{2}\pi z)K_z(2ab)$ $-1 <$ Re $z < 1$	$\frac{1}{2}\sin[a(x-b^2/x)]$

	$\Phi(z) = \int_0^\infty x^{z-1}\phi(x)\,dx$	$\phi(x)$
7.107	$a^z K_z(b)$	$\frac{1}{2}\exp[-\frac{1}{2}b(x/a+a/x)]$
7.108	$\Gamma(\nu+\frac{1}{2}z)[\Gamma(1+\nu-\frac{1}{2}z)]^{-1}$ $\cdot(4a/b)^{\frac{1}{2}z}K_z[2(ab)^{\frac{1}{2}}]$ $\mathrm{Re}\ z > -2\ \mathrm{Re}\ \nu$	$2J_{2\nu}(\alpha)K_{2\nu}(\beta)$ ${\beta \atop \alpha} = (2b)^{\frac{1}{2}}[(a^2+x^2)^{\frac{1}{2}}\pm a]^{\frac{1}{2}}$
7.109	$\Gamma(\nu+\frac{1}{2}z)[\Gamma(1+\nu-\frac{1}{2}z)]^{-1}$ $\cdot\Gamma(\frac{1}{2}-\frac{1}{2}z)(a/b)^{\frac{1}{2}z}K_{\frac{1}{2}z}(2ab)$ $-\mathrm{Re}\ \nu < \mathrm{Re}\ z < 1$	$2\pi^{\frac{1}{2}}I_\nu(\alpha)K_\nu(\beta)$ ${\beta \atop \alpha} = b[(a^2+x^2)^{\frac{1}{2}}\pm a]$
7.110	$\Gamma(\frac{1}{2}z+\nu)\Gamma(\frac{1}{2}z-\nu)$ $[\Gamma(\frac{1}{2}+\frac{1}{2}z)]^{-1}(a/b)^{\frac{1}{2}z}K_{\frac{1}{2}z}(2ab)$ $\mathrm{Re}\ z > \pm 2\mathrm{Re}\ \nu$	$2\pi^{-\frac{1}{2}}K_\nu(\alpha)K_\nu(\beta)$ ${\beta \atop \alpha} = b[(a^2+x^2)^{\frac{1}{2}}\pm a]$
7.111	$\cos(\frac{1}{2}\pi z)K^2_{\frac{1}{2}z}(a)$ $-\frac{3}{2} < \mathrm{Re}\ z < \frac{3}{2}$	$\frac{1}{2}\pi Y_o[a\lvert x-x^{-1}\rvert]$ Principal value
7.112	$J_z(a)K_z(a)$ $\mathrm{Re}\ z > -\frac{5}{4}$	$\frac{1}{2}J_o[a(1/x-x)^{\frac{1}{2}}]$ $x < 1$ 0 $x > 1$

	$\Phi(z) = \int_0^\infty x^{z-1}\phi(x)\,dx$	$\phi(x)$
7.113	$I_{\frac{1}{2}z}(a)K_{\frac{1}{2}z-\frac{1}{2}}(a)$ $\operatorname{Re} z > -1$	$-(\tfrac{1}{2}\pi a)^{-\frac{1}{2}}(1-x^2)^{-\frac{1}{2}}\sin[a(x-x^{-1})]$ $\quad x < 1$ $0 \quad x > 1$
7.114	$K_{\frac{1}{2}z}(a)I_{\frac{1}{2}z-\frac{1}{2}}(a)$ $\operatorname{Re} z > -1$	$(\tfrac{1}{2}\pi a)^{-\frac{1}{2}}(1-x^2)^{-\frac{1}{2}}\cos[a(x-x^{-1})]$ $\quad x < 1$ $0 \quad x > 1$
7.115	$K_{\nu+\frac{1}{2}z}(a)K_{\nu-\frac{1}{2}z}(a)$	$K_{2\nu}[a(x+x^{-1})]$
7.116	$I_{\nu+z}(a)K_{\nu-z}(a)$ $\operatorname{Re} z > -\tfrac{3}{4}, \operatorname{Re} \nu > -\tfrac{1}{2}$	$\tfrac{1}{2}J_{2\nu}[a(x^{-\frac{1}{2}}-x^{\frac{1}{2}})] \quad x < 1$ $0 \quad x > 1$
7.117	$K_{\frac{1}{2}\nu+\frac{1}{2}z}(a)I_{\frac{1}{2}\nu-\frac{1}{2}z}(a)$ $\operatorname{Re} z < \tfrac{3}{2}, \operatorname{Re} \nu > -1$	$0 \quad x < 1$ $J_\nu[a(x-x^{-1})] \quad x > 1$
7.118	$I_{\frac{1}{2}\nu-\frac{1}{2}z}(a)K_{\frac{1}{2}\nu+\frac{1}{2}z}(a)$ $+I_{\frac{1}{2}\nu+\frac{1}{2}z}(a)K_{\frac{1}{2}\nu-\frac{1}{2}z}(a)$	$J_\nu[a\lvert x-x^{-1}\rvert]$ $-\tfrac{3}{2} < \operatorname{Re} z < \tfrac{3}{2}, \operatorname{Re} \nu > -1$
7.119	$K_{\nu-z}(a)K_{\nu-z}(b)$	$\tfrac{1}{2}(a+bx)^{\nu}(b+ax)^{-\nu}$ $\cdot K_{2\nu}\{[a^2+b^2+ab(x+x^{-1})]^{\frac{1}{2}}\}$

	$\Phi(z) = \int_0^\infty x^{z-1}\phi(x)\,dx$	$\phi(x)$
7.120	$\cosh(\frac{1}{2}\pi z)$ $\cdot I_{\nu-\frac{1}{2}iz}(a)\,I_{\nu+\frac{1}{2}iz}(a)$ $\operatorname{Re}\nu > -\frac{1}{2}$	$0 \quad x < e^{-\pi}$ $\frac{1}{2}\pi^{-1}I_{2\nu}(2a\sin y) \quad e^{-\pi} < x < e^{\pi}$ $0 \quad x > e^{\pi}$
7.121	$I_{\nu-iz}(a)\,I_{\nu+iz}(a)$ $\operatorname{Re}\nu > -\frac{1}{2}$	$0 \quad x < e^{-\pi}$ $\frac{1}{2}\pi^{-1}I_{2\nu}[2a\cos(\frac{1}{2}x)] \quad e^{-\pi} < x < e^{\pi}$ $0 \quad x > e^{\pi}$
7.122	$z^{-1}[\mathbf{H}_0(az) - Y_0(az)]$ $\operatorname{Re} z > 0$	$2\pi^{-1}\log[y(a+(1+y^2/a^2)^{\frac{1}{2}}] \quad x < 1$ $0 \quad x > 1$
7.123	$z^{-1}[I_0(az) - \mathbf{L}_0(az)]$ $\operatorname{Re} z > 0$	$1 \quad x < e^{-a}$ $2\pi^{-1}\arcsin(y/a) \quad e^{-a} < x < 1$ $0 \quad x > 1$
7.124	$z^{-\nu}[\mathbf{H}_\nu(az) - Y_\nu(az)]$ $\operatorname{Re} z > 0$ $y = \log(1/x)$	$2\pi^{-\frac{1}{2}}(2a)^{-\nu}[\Gamma(\frac{1}{2}+\nu)]^{-1}$ $\cdot(y^2+a^2)^{\nu-\frac{1}{2}} \quad x < 1$ $0 \quad x > 1$

	$\Phi(z) = \int_0^\infty x^{z-1}\phi(x)\,dx$	$\phi(x)$
7.125	$z^{-\nu}[I_\nu(az) - \mathbf{L}_\nu(az)]$ $\mathrm{Re}\ z > 0,\ \mathrm{Re}\ \nu > -\frac{1}{2}$	$0 \qquad x < e^{-a}$ $2\pi^{-\frac{1}{2}}(2a)^{-\nu}[\Gamma(\frac{1}{2}+\nu)]^{-1}$ $\cdot(a^2-y^2)^{\nu-\frac{1}{2}} \qquad e^{-a} < x < 1$ $0 \qquad x > 1$
7.126	$z^{\frac{1}{2}\nu}[\mathbf{H}_\nu(az^{\frac{1}{2}}) - Y_\nu(az^{\frac{1}{2}})]$ $\mathrm{Re}\ z > 0,\ \mathrm{Re}\ \nu < \frac{1}{2}$	$\pi^{-1}\cos(\pi\nu)(\frac{1}{2}a)^\nu y^{-\nu-1}$ $\cdot\exp(\frac{1}{4}a^2/y)\,\mathrm{Erfc}(\frac{1}{2}ay^{-\frac{1}{2}}) \quad x < 1$ $0 \qquad x > 1$
7.127	$(\frac{1}{2}a)^{-z}\Gamma(\frac{1}{2}+z)$ $\cdot[\mathbf{H}_z(a) - Y_z(a)]$	$0 \qquad x < 1$ $\pi^{-\frac{1}{2}}x^{\frac{1}{2}}(x-1)^{-\frac{1}{2}}\exp[-a(x-1)^{\frac{1}{2}}] \quad x > 1$
7.128	$(\frac{1}{2}a)^{-z}\Gamma(\frac{1}{2}+z)$ $\cdot[I_z(a) - \mathbf{L}_z(a)]$ $\mathrm{Re}\ z > -\frac{1}{2}$	$\pi^{-\frac{1}{2}}x^{\frac{1}{2}}(1-x)^{-\frac{1}{2}}\exp[-a(1-x)^{\frac{1}{2}}] \quad x < 1$ $0 \qquad x > 1$
7.129	$(\frac{1}{2}a)^{-z}\Gamma(z)\mathbf{H}_{\nu+z}(a)$ $\mathrm{Re}\ z > 0,\ \mathrm{Re}\ \nu > -\frac{3}{2}$	$(1-x)^{\frac{1}{2}\nu}\mathbf{H}_\nu[a(1-x)^{\frac{1}{2}}] \qquad x < 1$ $0 \qquad x > 1$
7.130	$(\frac{1}{2}a)^{-z}\Gamma(z)\mathbf{L}_{\nu+z}(a)$ $\mathrm{Re}\ z > 0,\ \mathrm{Re}\ \nu > -\frac{3}{2}$ $y = \log(1/x)$	$(1-x)^{\frac{1}{2}\nu}\mathbf{L}_\nu[a(1-x)^{\frac{1}{2}}] \qquad x < 1$ $0 \qquad x > 1$

	$\Phi(z) = \int_0^\infty x^{z-1}\phi(x)\,dx$	$\phi(x)$
7.131	$(2/a)^{z-1}\Gamma(\frac{1}{2}+z)$ $\cdot\mathbf{H}_z(ab)$ Re $z > -\frac{1}{2}$	$\pi^{-\frac{1}{2}}x^{\frac{1}{2}}(1-x)^{-\frac{1}{2}}\sin[a(1-x)^{\frac{1}{2}}]$ $\quad x < 1$ 0 $\quad x > 1$
7.132	$(2a)^{-z}\Gamma(\frac{1}{2}+z)$ $[I_{-z}(a)-\mathbf{L}_z(a)]$ Re $z > -\frac{1}{2}$	0 $\quad x < 1$ $\pi^{-\frac{1}{2}}x^{\frac{1}{2}}(x-1)^{-\frac{1}{2}}\sin[a(x-1)^{\frac{1}{2}}]$ $\quad x > 1$
7.133	$(2b/a)^z\Gamma(z)\Gamma(\frac{1}{2}-\nu-z)$ $\cdot\Gamma(\frac{1}{2}+\nu+z)$ $\cdot[\mathbf{H}_{\nu+z}(ab)-Y_{\nu+z}(ab)]$ $0 <$ Re $z < \frac{1}{2}-$Re ν	$\pi b^{-\nu}\sec(\pi\nu)(b^2+x^2)^{\frac{1}{2}\nu}$ $\cdot\{\mathbf{H}_\nu[a(b^2+x^2)^{\frac{1}{2}}]-Y_\nu[a(b^2+x^2)^{\frac{1}{2}}]\}$
7.134	$\mathbf{J}_\nu(az)-J_\nu(az)$ Re $z > 0$	$\pi^{-1}\sin(\pi\nu)a^\nu(y^2+a^2)^{-\frac{1}{2}}$ $\cdot[(y^2+a^2)^{\frac{1}{2}}+y]^{-\nu}$ $\quad x < 1$ 0 $\quad x > 1$
7.135	$\csc(\pi z)[\mathbf{J}_z(a)-J_z(a)]$ $y = \log(1/x)$	$\pi^{-1}\exp[-\frac{1}{2}a(x^{-1}-x)]$ $\quad x < 1$ 0 $\quad x > 1$

	$\Phi(z) = \int_0^\infty x^{z-1}\phi(x)$	$\phi(x)$
7.136	$z^{-\mu-\frac{1}{2}}S_{2\mu,2\nu}[2(az)^{\frac{1}{2}}]$ $\mathrm{Re}\ z > 0,\ \mathrm{Re}(\mu\pm\nu) > -\frac{1}{2}$	$2^{2\mu-1}a^{-\frac{1}{2}}y^{\mu}\exp(\frac{1}{2}a^2/y)$ $\cdot W_{\mu,\nu}(a/y) \qquad x < 1$ $0 \qquad x > 1$
7.137	$(2b)^{z-2}\Gamma(\frac{1}{2}z-\frac{1}{2}\nu)$ $\cdot\Gamma(\frac{1}{2}z+\frac{1}{2}\nu)S_{1-z,\nu}(ab)$	$(b^2+x^2)^{-1}K_\nu(ax)$ $\mathrm{Re}\ z > \pm\ \mathrm{Re}\ \mu$
7.138	$(b/a)^z\Gamma(z)\Gamma(\frac{1}{2}-z)$ $S_{z+\nu,z-\nu}(ab)$ $0 < \mathrm{Re}\ z < \frac{1}{2}$	$\frac{1}{2}\pi(2b)^{\nu}\Gamma(\frac{1}{2}+z)(b^2+x)^{-\frac{1}{2}\nu}$ $\cdot\{\mathbf{H}_\nu[a(b^2+x)^{\frac{1}{2}}]-Y_\nu[a(b^2+x)]^{\frac{1}{2}}\}$
7.139	$\Gamma(z)[\Gamma(\frac{1}{2}+z)]^{-1}$ $\cdot a^{-z}S_{\nu+z,\nu-z}(a)$ $\mathrm{Re}\ z > 0$	$[\Gamma(\frac{1}{2}+\nu)]^{-1}2^{\nu-1}$ $\cdot(1-z)^{-\frac{1}{2}\nu}H_\nu[a(1-x)^{\frac{1}{2}}] \qquad x < 1$ $0 \qquad x > 1$
7.140	$a^{-z}\Gamma(z)\Gamma(\frac{1}{2}-\frac{1}{2}\mu-\frac{1}{2}\nu-z)$ $\cdot S_{z+\mu,z+\nu}(a)$ $0 < \mathrm{Re}\ z < \frac{1}{2}-\frac{1}{2}\mathrm{Re}(\nu+\mu)$ $y = \log(1/x)$	$\Gamma(\frac{1}{2}-\frac{1}{2}\nu-\frac{1}{2}\mu)(1+x)^{\frac{1}{2}\nu}$ $\cdot\ S_{\mu,\nu}[a(1+x)^{\frac{1}{2}}]$

2.8 Whittaker Functions and Special Cases

	$\Phi(z) = \int_0^\infty x^{z-1}\phi(x)\,dx$	$\phi(x)$	
8.1	$Ei[-b(a+z)]$ $\mathrm{Re}\ z > -a$	$-y^{-1}x^a$ 0	$x < e^{-b}$ $x > e^{-b}$
8.2	$z^{-1}Ei(-az)$ $\mathrm{Re}\ z > 0$	$\log(a/y)$ 0	$x < e^{-a}$ $x > e^{-a}$
8.3	$e^{az}Ei(-az)$	$-(a+y)^{-1}$ 0	$x < 1$ $x > 1$
8.4	$z^{-1}e^{az}Ei(-az)$ $\mathrm{Re}\ z > 0$	$-\log(1+y/a)$ 0	$x < 1$ $x > 1$
8.5	$e^{-az}Ei(-az-bz)$	$-(a+y)^{-1}$ 0	$x < e^{-b}$ $x > e^{-b}$
8.6	$e^{az}[Ei(-az)]^2$ $\mathrm{Re}\ z > 0$ $y = \log(1/x)$	$2(a+y)^{-1}\log(y/a)$ 0	$x < e^{-a}$ $x > e^{-a}$

	$\Phi(z) = \int_0^\infty x^{z-1}\phi(x)\,dx$	$\phi(x)$
8.7	$e^{(a+b)z}$ $\cdot Ei(-az)Ei(-bz)$ $Re\ z > 0$	$(y+a+b)^{-1}\log[(ab)^{-1}(y+a)(y+b)]$ $x < 1$ 0 $x > 1$
8.8	$Ei(-az)Ei(-bz)$ $Re\ z > 0$	$y^{-1}\log[(ab)^{-1}(y-a)(y-b)]$ $x < e^{-(a+b)}$ 0 $x > e^{-(a+b)}$
8.9	$\overline{Ei}(az)Ei(-bz)$ $Re\ z > 0$	$y^{-1}\log\|1-y^2/a^2\|$ $x < 1$ 0 $x > 1$
8.10	$z^{-1}e^{a/z}Ei(-a/z)$ $Re\ z > 0$	$-2K_o[2(ay)^{\frac{1}{2}}]$ $x < 1$ 0 $x > 1$
8.11	$z^{-1}e^{-a/z}\overline{Ei}(a/z)$ $Re\ z > 0$	$\pi Y_o[2(ay)^{\frac{1}{2}}]$ $x < 1$ 0 $x > 1$
8.12	$e^{az^2}Ei(-az^2)$ $Re\ z > 0$ $y = \log(1/x)$	$i(\pi a)^{-\frac{1}{2}}e^{-\frac{1}{2}y^2/a}Erf(\frac{1}{2}ia^{\frac{1}{2}}y)$ $x < 1$ 0 $x > 1$

	$\Phi(z) = \int_0^\infty x^{z-1}\phi(x)\,dx$	$\phi(x)$
8.13	$z^{-\frac{1}{2}}\mathrm{Ei}(-a/z)$ $\mathrm{Re}\ z > 0$	$2(\pi y)^{-\frac{1}{2}}\mathrm{Ci}[2(ay)^{\frac{1}{2}}]$ $x < 1$ 0 $x > 1$
8.14	$z^{-\frac{1}{2}}\overline{\mathrm{Ei}}(a/z)$ $\mathrm{Re}\ z > 0$	$(\pi y)^{-\frac{1}{2}}\{\mathrm{Ei}[-2(ay)^{\frac{1}{2}}]+\overline{\mathrm{Ei}}[2(ay)^{\frac{1}{2}}]\}$ $x < 1$ 0 $x > 1$
8.15	$z^{-\frac{1}{2}}e^{a/z}\mathrm{Ei}(-a/z)$ $\mathrm{Re}\ z > 0$	$(\pi y)^{-\frac{1}{2}}\{\exp[2(ay)^{\frac{1}{2}}]\mathrm{Ei}[-2(ay)^{\frac{1}{2}}]$ $+\exp[-2(ay)^{\frac{1}{2}}]\overline{\mathrm{Ei}}[2(ay)^{\frac{1}{2}}]\}$ $x < 1$ 0 $x > 1$
8.16	$z^{-\frac{1}{2}}e^{-a/z}\overline{\mathrm{Ei}}(a/z)$ $\mathrm{Re}\ z > 0$	$2(\pi y)^{-\frac{1}{2}}\{\cos[2(ay)^{\frac{1}{2}}]\mathrm{Ci}[2(ay)^{\frac{1}{2}}]$ $+\sin[2(ay)^{\frac{1}{2}}]\mathrm{Si}[2(ay)^{\frac{1}{2}}]\}$ $x < 1$ 0 $x > 1$
8.17	$z^{-\frac{1}{2}}\mathrm{Ei}(-az^{\frac{1}{2}})$ $\mathrm{Re}\ z > 0$	$\frac{1}{2}(\pi y)^{-\frac{1}{2}}\mathrm{Ei}(-\frac{1}{4}a^2/y)$ $x < 1$ 0 $x > 1$
8.18	$\exp(a^2z^2)$ $\cdot\mathrm{Erfc}(b+az)$ $y = \log(1/x)$	$\pi^{-\frac{1}{2}}a^{-1}\exp(-\frac{1}{4}y^2/a^2)$ $x < e^{-2ab}$ 0 $x > e^{-2ab}$

	$\Phi(z) = \int_0^\infty x^{z-1} (x)dx$	$\phi(x)$	
8.19	$z^{-1}\exp(a^2z^2)$ $\cdot\mathrm{Erfc}(az)$ $\mathrm{Re}\ z > 0$	$\mathrm{Erf}(\tfrac{1}{2}y/a)$ 0	$x < 1$ $x > 1$
8.20	$z^{-1}e^{z^2}\mathrm{Erfc}(a+z)$ $\mathrm{Re}\ z > 0$	$\mathrm{Erf}(\tfrac{1}{2}y)-\mathrm{Erf}a$ 0	$x < e^{-2a}$ $x > e^{-2a}$
8.21	$\mathrm{Erfc}[(az)^{\frac{1}{2}}]$ $\mathrm{Re}\ z > 0$	$\pi^{-1}a^{\frac{1}{2}}y^{-1}(y-a)^{-1}$ 0	$x < e^{-a}$ $x > e^{-a}$
8.22	$z^{-\frac{1}{2}}\mathrm{Erf}[(az)^{\frac{1}{2}}]$ $\mathrm{Re}\ z > 0$	0 $(\pi y)^{-\frac{1}{2}}$ 0	$x < e^{-a}$ $e^{-a} < x < 1$ $x > 1$
8.23	$z^{-\frac{1}{2}}\mathrm{Erfc}[(az)^{\frac{1}{2}}]$ $\mathrm{Re}\ z > 0$	$(\pi y)^{-\frac{1}{2}}$ 0	$x < e^{-a}$ $x > e^{-a}$
8.24	$e^{az}\mathrm{Erfc}[(az)^{\frac{1}{2}}]$ $\mathrm{Re}\ z > 0$ $y = \log(1/x)$	$\pi^{-1}(y/a)^{-\frac{1}{2}}(y+a)^{-1}$ 0	$x < 1$ $x > 1$

	$\Phi(z) = \int_0^\infty x^{z-1}\phi(x)\,dx$	$\phi(x)$	
8.25	$z^{-\frac{1}{2}}e^{a^2/z}\mathrm{Erfc}(az^{-\frac{1}{2}})$	$(\pi y)^{-\frac{1}{2}}\exp(-2ay^{\frac{1}{2}})$	$x < 1$
	$\mathrm{Re}\ z > 0$	0	$x > 1$
8.26	$\mathrm{Erf}(az^{-\frac{1}{2}})$	$(\pi y)^{-1}\sin(2ay^{\frac{1}{2}})$	$x < 1$
	$\mathrm{Re}\ z > 0$	0	$x > 1$
8.27	$z^{-\nu}e^{a^2/z}$	$a^{1-\nu}y^{\frac{1}{2}\nu-\frac{1}{2}}\mathbf{L}_{\nu-1}(2ay^{\frac{1}{2}})$	$x < 1$
	$\cdot\mathrm{Erf}(az^{-\frac{1}{2}})$	0	$x > 1$
	$\mathrm{Re}\ z > 0$		
8.28	$z^{-\nu}e^{a^2/z}$	$a^{1-\nu}y^{\frac{1}{2}\nu-\frac{1}{2}}[I_{\nu-1}(2ay^{\frac{1}{2}})-\mathbf{L}_{\nu-1}(2ay^{\frac{1}{2}})]$	
	$\cdot\mathrm{Erfc}(az^{-\frac{1}{2}})$		$x < 1$
	$\mathrm{Re}\ z > 0$	0	$x > 1$
8.29	$b^{-z}\gamma(z,ab)$	e^{-bx}	$x < a$
	$\mathrm{Re}\ z > 0$	0	$x > a$
8.30	$b^{-z}\Gamma(z,ab)$	0	$x < a$
		e^{-bx}	$x > a$
	$y = \log(1/x)$		

	$\Phi(z) = \int_0^\infty x^{z-1}\phi(x)$	$\phi(x)$	
8.31	$\Gamma(z)\Gamma(1-z,a)$ Re $z > 0$	$(1+x)^{-1}e^{-a(x+1)}$	
8.32	$\Gamma(\nu,az)$ Re $z > 0$, Re $\nu < 1$	$a^{\nu}[\Gamma(1-\nu)]^{-1}y^{-1}(y-a)^{-\nu}$ 0	$x < e^{-a}$ $x > e^{-a}$
8.33	$e^{az}\Gamma(\nu,az)$ Re $z > 0$, Re $\nu < 1$	$a^{\nu}[\Gamma(1-\nu)]^{-1}y^{-\nu}(y+a)^{-1}$ 0	$x < 1$ $x > 1$
8.34	$a^{z}\gamma(z,a)$ Re $z > 0$	e^{-ax} 0	$x < 1$ $x > 1$
8.35	$a^{-z}\Gamma(-z,a)$	$e^{-a/x}$ 0	$x < 1$ $x > 1$
8.36	$z^{-\nu}\gamma(\nu,az)$ Re $\nu > -1$	0 $y^{\nu-1}$ 0	$x < e^{-a}$ $e^{-a} < x < 1$ $x > 1$
8.37	$z^{-\nu}\Gamma(\nu,az)$ Re $\nu > -1$ $y = \log(1/x)$	$y^{\nu-1}$ 0	$x < e^{-a}$ $x > e^{-a}$

	$\Phi(z) = \int_0^\infty x^{z-1}\phi(x)dx$	$\phi(x)$	
8.38	$z^{-\nu}e^{az}\Gamma(\nu,az)$	$(y+a)^{\nu-1}$	$x < 1$
	$\mathrm{Re}\ z > 0$	0	$x > 1$
8.39	$z^{-\nu}e^{-az}\gamma(\nu,-az)$	0	$x < e^{-a}$
		$(a-y)^{\nu-1}$	$e^{-a} < x < 1$
		0	$x > 1$
8.40	$\gamma(\nu,a/z)$	$a^{\frac{1}{2}\nu}y^{\frac{1}{2}\nu-1}J_\nu[2(ay)^{\frac{1}{2}}]$	$x < 1$
	$\mathrm{Re}\ z > 0,\ \mathrm{Re}\ \nu > 0$	0	$x > 1$
8.41	$z^{\nu-1}\gamma(\nu,a/z)$	$a^{\frac{1}{2}\nu}\Gamma(\nu)y^{-\frac{1}{2}\nu}I_\nu[2(ay)^{\frac{1}{2}}]$	$x < 1$
	$\mathrm{Re}\ z > 0,\ \mathrm{Re}\ \nu > 0$	0	$x > 1$
8.42	$z^{\nu-1}e^{a/z}\Gamma(\nu,a/z)$	$2[\Gamma(1-\nu)]^{-1}(y/a)^{-\frac{1}{2}\nu}K_\nu[2(ay)^{\frac{1}{2}}]$	
	$\mathrm{Re}\ z > 0,\ \mathrm{Re}\ \nu < 1$		$x < 1$
		0	$x > 1$
8.43	$z^\mu e^{a/z}\gamma(\nu,a/z)$	$[\nu\Gamma(\nu-\mu)]^{-1}a^\nu y^{\nu-\mu-1}$	
	$\mathrm{Re}\ z > 0,\ \mathrm{Re}(\nu,\mu) > 0$	$\cdot\ {}_1F_2(1;\nu+1,\nu-\mu;ay)$	$x < 1$
		0	$x > 1$
	$y = \log(1/x)$		

	$\Phi(z) = \int_0^\infty x^{z-1}\phi(x)\,dx$	$\phi(x)$
8.44	$\exp(\frac{1}{4}a^2z^2)D_\nu(az)$ $\operatorname{Re}\nu < 0$	$a^\nu[\Gamma(-\nu)]^{-1}y^{-\nu-1}\exp(-\frac{1}{2}y^2/a^2)$ $x < 1$ 0 $x > 1$
8.45	$D_\nu[2(az^{\frac{1}{2}})]$ $\operatorname{Re} z > 0, \operatorname{Re}\nu < 0$	$2^{\frac{1}{2}\nu+\frac{1}{2}}a^{\frac{1}{2}}[\Gamma(-\frac{1}{2}\nu)]^{-1}(y-a)^{-1-\frac{1}{2}\nu}$ $\cdot(y+a)^{\frac{1}{2}\nu-\frac{1}{2}}$ $x < e^{-a}$ 0 $x > e^{-a}$
8.46	$z^{-\frac{1}{2}}D_\nu[2(az)^{\frac{1}{2}}]$ $\operatorname{Re} z > 0, \operatorname{Re}\nu < 1$	$2^{\frac{1}{2}\nu}[\Gamma(\frac{1}{2}-\frac{1}{2}\nu)]^{-1}(y-a)^{-\frac{1}{2}\nu-\frac{1}{2}}(y+a)^{\frac{1}{2}\nu}$ $x < e^{-a}$ 0 $x > e^{-a}$
8.47	$z^{\frac{1}{2}\nu}e^{\frac{1}{2}az}$ $\cdot D_\nu[(2az)^{\frac{1}{2}}]$ $\operatorname{Re} z > 0, \operatorname{Re}\nu < 0$	$2^{-1-\frac{1}{2}\nu}[\Gamma(-\nu)]^{-1}(a+y)^{-\frac{1}{2}}$ $\cdot[(a+y)^{\frac{1}{2}}-a^{\frac{1}{2}}]^{-\nu-1}$ $x < 1$ 0 $x > 1$
8.48	$z^{\frac{1}{2}\nu-1}D_\nu[(2az)^{\frac{1}{2}}]$ $\operatorname{Re} z > 0, \operatorname{Re}\nu < 1$	$2^{-\frac{1}{2}\nu}[\Gamma(1-\nu)]^{-1}[(a+y)^{\frac{1}{2}}-a^{\frac{1}{2}}]^{-\nu}$ $x < 1$ 0 $x > 1$
8.49	$z^\nu\exp(a/z)$ $\cdot D_{2\nu}[2(a/z)^{\frac{1}{2}}]$ $\operatorname{Re} z > 0, \operatorname{Re}\nu < 0$ $y = \log(1/x)$	$[\Gamma(-2\nu)]^{-1}(2y)^{-\nu-1}\exp[-2(2ay)^{\frac{1}{2}}]$ $x < 1$ 0 $x > 1$

	$\Phi(z) = \int_0^\infty x^{z-1}\phi(x)\,dx$	$\phi(x)$
8.50	$D_\nu(z_1^{\frac{1}{2}})D_\nu(z_2^{\frac{1}{2}})$ $\text{Re } z > -b$ $z_{\frac{1}{2}} = 2a[b \pm (b^2-z^2)^{\frac{1}{2}}]$	$(\frac{1}{2}a)^{\frac{1}{2}}[\Gamma(-\nu)]^{-1}y^{-\nu-1}(a^2+y^2)^{-\frac{1}{2}}$ $\cdot[(a^2+y^2)^{\frac{1}{2}}+a]^{\frac{1}{2}+\nu}\exp[-b(a^2+y^2)^{\frac{1}{2}}]$ $\quad x < 1$ $0 \quad x > 1$ $\text{Re } \nu < 0$
8.51	$D_\nu(z_1^{\frac{1}{2}})D_\nu(z_2^{\frac{1}{2}})$ $\text{Re } z > -b$ $z_{\frac{1}{2}} = 2a[z \pm (z^2-b^2)^{\frac{1}{2}}]$	$(\frac{1}{2}a)^{\frac{1}{2}}[\Gamma(-\nu)]^{-1}(y^2-a^2)^{-1-\frac{1}{2}\nu}$ $\cdot(y+a)^{\frac{1}{2}+\nu}\exp[-b(y^2-a^2)]$ $\quad x < e^{-a}$ $0 \quad x > e^{-a}$ $\text{Re } \nu < 0$
8.52	$\Gamma(z)D_{-z}(a)$ $\text{Re } z > 0$	$e^{-\frac{1}{4}a^2}e^{-\frac{1}{2}x^2-ax}$
8.53	$\Gamma(z)D_{-z}^2[(2a)^{\frac{1}{2}}]$ $\text{Re } z > 0$ $y = \log(1/x)$	$(1-x^2)^{-\frac{1}{2}}\exp[-a(\frac{1+x}{1-x})]$ $\quad x < 1$ $0 \quad x > 1$

	$\Phi(z) = \int_0^\infty x^{z-1}\phi(x)\,dx$	$\phi(x)$
8.54	$z^{-\mu-\frac{1}{2}}e^{\frac{1}{2}az}$ $\cdot W_{\nu,\mu}(az)$ Re $z > 0$, Re$(\mu-\nu) > -\frac{1}{2}$	$a^{\frac{1}{2}-\mu}[\Gamma(\frac{1}{2}+\mu-\nu)]^{-1}y^{\mu-\nu-\frac{1}{2}}$ $(y+a)^{\nu+\mu-\frac{1}{2}}$ $\quad x < 1$ 0 $\quad x > 1$
8.55	$z^{-\mu-\frac{1}{2}}W_{\nu,\mu}(2az)$ Re $z > 0$ Re$(\mu-\nu) > -\frac{1}{2}$	$(2a)^{\frac{1}{2}-\mu}[\Gamma(\frac{1}{2}+\mu-\nu)]^{-1}(y-a)^{\mu-\nu-\frac{1}{2}}$ $\cdot(y+a)^{\nu+\mu-\frac{1}{2}}$ $\quad x < e^{-a}$ 0 $\quad x > e^{-a}$
8.56	$z^{-1}W_{\nu,\mu}(2az)$ Re $z > 0$, Re $\nu > 1$	$[(y+a)/(y-a)]^{\frac{1}{2}\nu}P^{\nu}_{\mu-\frac{1}{2}}(y/a)$ $\quad x<e^{-a}$ 0 $\quad x > e^{-a}$
8.57	$z^{-\alpha}e^{az}W_{\nu,\mu}(az)$ Re $z > 0$, Re$(\alpha-\nu) > 0$	$a^{\nu}[\Gamma(\alpha-\nu)]^{-1}y^{\alpha-\nu-1}$ $\cdot{}_2F_1(\frac{1}{2}-\nu+\mu,\frac{1}{2}-\nu-\mu;\alpha-\nu;y/a)$ $x < e^{-a}$ 0 $\quad x > e^{-a}$
8.58	$M_{\nu,\mu}(z_2)W_{\nu,\mu}(z_1)$ Re $z > b$ Re$(\mu-\nu) > -\frac{1}{2}$ $z_{\frac{1}{2}}=a[z\pm(z^2-b^2)^{\frac{1}{2}}]$ $y = \log(1/x)$	$ab\Gamma(1+2\mu)[\Gamma(\frac{1}{2}+\mu-\nu)]^{-1}$ $\cdot(y-a)^{-\nu-\frac{1}{2}}(y+a)^{\nu-\frac{1}{2}}$ $\cdot I_{2\mu}[b(y^2-a^2)^{\frac{1}{2}}]$ $\quad x < e^{-a}$ 0 $\quad x > e^{-a}$

	$\Phi(z) = \int_0^\infty x^{z-1}\phi(x)\,dx$	$\phi(x)$
8.59	$W_{\nu,\mu}(z_2)W_{\nu,\mu}(z_1)$ Re $z > b$ $\mathrm{Re}(\mu\pm\nu) > -\frac{1}{2}$ $z_{\frac{1}{2}} = a[z+(z^2-b^2)^{\frac{1}{2}}]$	$2ab[\Gamma(\frac{1}{2}+\mu-\nu)\Gamma(\frac{1}{2}-\mu-\nu)]^{-1}$ $\cdot(y-a)^{-\nu-\frac{1}{2}}(y+a)^{\nu-\frac{1}{2}}$ $\cdot K_{2\mu}[b(y^2-a^2)^{\frac{1}{2}}]$ $\quad x < e^{-a}$ 0 $\quad x > e^{-a}$
8.60	$z^{-2\mu}\exp(\frac{1}{2}az^2)$ $\cdot W_{\nu,\mu}(az^2)$ Re $z > 0$ $\mathrm{Re}(\mu-\nu) > 0$	$2^{\mu-\nu}a^{\frac{1}{2}\nu+\frac{1}{2}\mu}[\Gamma(2\mu-2\nu)]^{-1}y^{\mu-\nu-1}$ $\cdot\exp(-\frac{1}{8}y^2/a)M_{\alpha,\beta}(\frac{1}{4}y^2/a)$ $x < 1$ 0 $\quad x > 1$ $\alpha = \frac{1}{2}(\mu-\nu-1)$, $\beta = \frac{1}{2}(1-\nu-3\mu)$
8.61	$z^{-2\mu-1}\exp(\frac{1}{2}az^2)$ $\cdot W_{\nu,\mu}(az^2)$ Re $z > 0$ $\mathrm{Re}(\mu-\nu) > -\frac{1}{2}$	$2^{1+\mu-\nu}a^{\frac{1}{2}+\frac{1}{2}\nu+\frac{1}{2}\mu}[\Gamma(1+2\mu-2\nu)]^{-1}$ $\cdot y^{\mu-\nu-1}\exp(-\frac{1}{8}y^2/a)M_{\alpha,\beta}(\frac{1}{4}y^2/a)$ $x < 1$ 0 $\quad x > 1$ $\alpha = \frac{1}{2}(\mu-\nu)$, $\beta = -\frac{1}{2}(\nu+3\mu)$
8.62	$z^{-\nu}e^{-\frac{1}{2}a/z}$ $\cdot M_{\nu,\mu}(a/z)$ Re $z > 0$, $\mathrm{Re}(\nu+\mu) > -\frac{1}{2}$ $y = \log(1/x)$	$a^{\frac{1}{2}}\Gamma(1+2\mu)[\Gamma(\frac{1}{2}+\nu+\mu)]^{-1}y^{\nu-\frac{1}{2}}$ $\cdot J_{2\mu}[2(ay^{\frac{1}{2}})]$ $\quad x < 1$ 0 $\quad x > 1$

	$\Phi(z) = \int_0^\infty x^{z-1}\phi(x)dx$	$\phi(x)$
8.63	$z^{-\nu}e^{-\frac{1}{2}a/z}$ $\cdot W_{\nu,\mu}(a/z)$ $\mathrm{Re}\ z > 0$ $\mathrm{Re}(\nu\pm\mu) > -\frac{1}{2}$	$-a^{\frac{1}{2}}\csc(2\pi\mu)y^{\nu-\frac{1}{2}}\{\cos(\pi\nu+\pi\mu)J_{2\mu}[2(ay)^{\frac{1}{2}}]$ $-\cos(\pi\nu-\pi\mu)J_{-2\mu}[2(ay)^{\frac{1}{2}}]\}$ $x < 1$ 0 $x > 1$
8.64	$z^{\nu}e^{\frac{1}{2}a/z}$ $\cdot M_{\nu,\mu}(a/z)$ $\mathrm{Re}\ z > 0, \mathrm{Re}(\mu-\nu) > -\frac{1}{2}$	$a^{\frac{1}{2}}\Gamma(1+2\mu)[\Gamma(\frac{1}{2}+\mu-\nu)]^{-1}y^{-\nu-\frac{1}{2}}$ $\cdot I_{2\mu}[2(ay)^{\frac{1}{2}}]$ $x < 1$ 0 $x > 1$
8.65	$z^{\nu}e^{\frac{1}{2}a/z}$ $\cdot W_{\nu,\mu}(a/z)$ $\mathrm{Re}\ z > 0,\ \mathrm{Re}(\nu\pm\mu) < \frac{1}{2}$	$2a^{\frac{1}{2}}[\Gamma(\frac{1}{2}+\mu-\nu)\Gamma(\frac{1}{2}-\mu-\nu)]^{-1}$ $\cdot y^{-\nu-\frac{1}{2}}K_{2\mu}[2(ay)^{\frac{1}{2}}]$ $x < 1$ 0 $x > 1$
8.66	$\Gamma(z+\mu)W_{-z,\nu}(a)$ $\mathrm{Re}\ z > -\mathrm{Re}\ \mu$ $y = \log(1/x)$	$ax^{\mu}(1-x)^{-\mu-1}\exp[-\frac{1}{2}a(x^{-1}-1)^{-1}]$ $\cdot W_{\mu,\nu}[a(x^{-1}-1)^{-1}]$ $x < 1$ 0 $x > 1$

	$\Phi(z) = \int_0^\infty x^{z-1}\phi(x)\,dx$	$\phi(x)$
8.67	$\frac{\Gamma(\frac{1}{2}+\nu+z)\Gamma(\frac{1}{2}-\nu+z)}{\Gamma(1-\mu+z)}$ $\cdot W_{-z,\nu}(a)$ $\mathrm{Re}\, z > -\frac{1}{2}\pm\mathrm{Re}\,\nu$	$(1-x)^{-\mu}\exp[-\frac{1}{2}a(1-x)^{-1}]$ $\cdot W_{\mu,\nu}[a(x^{-1}-1)^{-1}]$ $x < 1$ 0 $x > 1$
8.68	$a^z W_{\nu-z,z}(a)$ $\mathrm{Re}\,\nu < \frac{1}{2}$	$\frac{1}{2}a^{\frac{1}{2}}[\Gamma(\frac{1}{2}-\mu)]^{-1}(1-x^{\frac{1}{2}})^{\nu-1}$ $e^{-ax^{-\frac{1}{2}}}$ $x < 1$ 0 $x > 1$
8.69	$a^{-z}[\Gamma(2\nu+1-2z)]^{-1}$ $\cdot\Gamma(2z)M_{\mu-z,\nu-z}(a)$ $0 < \mathrm{Re}\, z < \frac{1}{4}+\frac{1}{2}\mathrm{Re}(\nu+\mu)$	$\frac{1}{2}\Gamma(\frac{1}{2}+\mu+\nu)[\Gamma(1+2\nu)]^{-1}$ $\cdot(1+x^{\frac{1}{2}})^{-\nu-\frac{1}{2}}e^{-\frac{1}{2}ax^{\frac{1}{2}}}M_{\mu,\nu}[a(1+x^{\frac{1}{2}})]$
8.70	$a^{-z}\Gamma(2z)W_{\mu-z,\nu+z}(a)$ $\mathrm{Re}\, z > 0$	$\frac{1}{2}(1+x^{\frac{1}{2}})^{\nu-\frac{1}{2}}e^{-\frac{1}{2}ax^{\frac{1}{2}}}W_{\mu,\nu}[a(1+x^{\frac{1}{2}})]$
8.71	$a^{-z}\Gamma(\frac{1}{2}-\mu-\nu-2z)$ $\cdot\Gamma(2z)W_{\mu+z,\nu+z}(a)$ $0 < \mathrm{Re}\, z < \frac{1}{4}-\frac{1}{2}\mathrm{Re}(\mu+\nu)$ $y = \log(1/x)$	$\frac{1}{2}\Gamma(\frac{1}{2}-\mu-\nu)(1+x^{\frac{1}{2}})^{\nu-\frac{1}{2}}$ $\cdot e^{\frac{1}{2}ax^{\frac{1}{2}}}W_{\mu,\nu}[a(1+x^{\frac{1}{2}})]$

	$\Phi(z) = \int_0^\infty x^{z-1}\phi(x)\,dx$	$\phi(x)$
8.72	$\Gamma(2z)[\Gamma(1+2\nu-2z)]^{-1}$ $\cdot a^{-z}M_{\mu-z,\nu-z}(a)$ $0 < \mathrm{Re}\ z < \frac{1}{4}+\mathrm{Re}(\nu+\mu)$	$\frac{1}{2}\Gamma(\mu+\nu+\frac{1}{2})[\Gamma(1+2\nu)]^{-1}(1+x^{\frac{1}{2}})^{-\nu-\frac{1}{2}}$ $\cdot e^{-\frac{1}{2}ax^{\frac{1}{2}}}M_{\mu,\nu}[a(1+x^{\frac{1}{2}})]$
8.73	$a^{-z}\Gamma(2z)W_{\mu-z,\nu+z}(a)$ $\mathrm{Re}\ z > 0$	$\frac{1}{2}(1+x^{\frac{1}{2}})^{\nu-\frac{1}{2}}e^{-\frac{1}{2}ax^{\frac{1}{2}}}W_{\mu,\nu}[a(1+x^{\frac{1}{2}})]$
8.74	$a^{-z}\Gamma(2z)\Gamma(\frac{1}{2}-\mu-\nu-z)$ $\cdot W_{\mu+z,\nu+z}(a)$ $0 < \mathrm{Re}\ z < \frac{1}{4}-\frac{1}{2}\mathrm{Re}(\mu+\nu)$	$\frac{1}{2}\Gamma(\frac{1}{2}-\mu-\nu)(1+x^{\frac{1}{2}})^{\nu-\frac{1}{2}}e^{\frac{1}{2}ax^{\frac{1}{2}}}$ $\cdot W_{\mu,\nu}[a(1+x^{\frac{1}{2}})]$
8.75	$a^{-z}\Gamma(\frac{1}{2}+\nu-\mu-2z)$ $[\Gamma(1-2z)]^{-1}W_{z-\nu}(a)W_{z+\mu}(a)$ $\mathrm{Re}\ z < \frac{1}{4}+\frac{1}{2}\mathrm{Re}(\nu-\mu)$	$0 \qquad x < 1$ $\frac{1}{2}[\Gamma(1+2\mu)]^{-1}(x^{\frac{1}{2}}-1)^{\mu-\frac{1}{2}}e^{-\frac{1}{2}ax^{\frac{1}{2}}}$ $\cdot M_{\nu,\mu}[a(x^{\frac{1}{2}}-1)] \qquad x > 1$
8.76	$\Gamma(\nu+z)\Gamma(1+\nu-z)$ $\cdot M_{z-\frac{1}{2},\nu}(a)M_{z-\frac{1}{2},\nu}(b)$ $-\mathrm{Re}\ \nu < \mathrm{Re}\ z < 1+\mathrm{Re}\ \nu$ $y = \log(1/x)$	$(ab)^{\frac{1}{2}}\Gamma^2(1+2\nu)(1+x)^{-1}J_{2\nu}[\frac{2(abx)^{\frac{1}{2}}}{1+x}]$ $\cdot \exp[\frac{1}{2}(a+b)(\frac{1-x}{1+x})]$

	$\Phi(z) = \int_0^\infty x^{z-1}\phi(x)\,dx$	$\phi(x)$
8.77	$\Gamma(\nu+z)\Gamma(1+\nu-z)$ $\cdot M_{\frac{1}{2}-z,\nu}(a)M_{z-\frac{1}{2},\nu}(b)$ $-\mathrm{Re}\ \nu < \mathrm{Re}\ z < 1+\mathrm{Re}\ \nu$	$(ab^{\frac{1}{2}}\Gamma^2(1+2\nu)(1+x)^{-1}I_{2\nu}[\frac{2(abx)^{\frac{1}{2}}}{1+x}]$ $\cdot\exp[\frac{1}{2}(a-b)(\frac{x-1}{x+1})]$ $a > b$
8.78	$\Gamma(\nu+z)$ $\cdot M_{z-\frac{1}{2},\nu}(b)W_{\frac{1}{2}-z,\nu}(a)$ $\mathrm{Re}\ z > -\mathrm{Re}\ \nu,\ a > b$	$(ab)^{\frac{1}{2}}\Gamma(1+2\nu)(1-x)^{-1}J_{2\nu}[\frac{2(abx)^{\frac{1}{2}}}{1-x}]$ $\cdot\exp[-\frac{1}{2}(a-b)(\frac{1+x}{1-x})],\quad x < 1$ $0 \quad x > 1$
8.79	$\Gamma(1+\nu-z)$ $\cdot M_{z-\frac{1}{2},\nu}(b)W_{z-\frac{1}{2},\nu}(a)$ $\mathrm{Re}\ z < 1+\mathrm{Re}\ \nu,\ a > b$	$0 \quad x < 1$ $(ab)^{\frac{1}{2}}\Gamma(1+2\nu)(x-1)^{-1}I_{2\nu}[\frac{2(abx)^{\frac{1}{2}}}{x-1}]$ $\cdot\exp[-\frac{1}{2}(a+b)(\frac{x+1}{x-1})] \quad x > 1$
8.80	$\Gamma(1+\nu-z)\Gamma(1-\nu-z)$ $\cdot W_{z-\frac{1}{2},\nu}(a)W_{z-\frac{1}{2},\nu}(b)$ $\mathrm{Re}\ z < 1\pm\mathrm{Re}\ \nu$ $y = \log(1/x)$	$0 \quad x < 1$ $2(ab)^{\frac{1}{2}}(x-1)^{-1}K_{2\nu}[\frac{2(abx)^{\frac{1}{2}}}{x-1}]$ $\cdot\exp[-\frac{1}{2}(a+b)(\frac{x+1}{x-1})] \quad x > 1$

Appendix. List of Notations and Definitions

Abbreviations: ε_n = Neumann's number

$\varepsilon_0 = 1, \quad \varepsilon_n = 2, \quad n = 1, 2, 3, \cdots$

$\gamma = 0.57721\cdots$ Euler's constant

$$(\alpha)_n = \alpha(\alpha+1)\cdots(\alpha+n-1) = \frac{\Gamma(\alpha+n)}{\Gamma(\alpha)}; \quad (\alpha)_0 = 1$$

$$\binom{\alpha}{n} = \alpha(\alpha-1)\cdots(\alpha-n+1)/n! = \frac{(-1)^n}{n!}\Gamma(n-\alpha)/\Gamma(-\alpha)$$

$$= \Gamma(1+\alpha)[n!\Gamma(1+\alpha-n)]^{-1}$$

1. Elementary functions

Trigonometric and inverse trigonometric functions:

$\sin x, \quad \cos x, \quad \tan x = \frac{\sin x}{\cos x}, \quad \cot x = \frac{\cos x}{\sin x}$

$\sec x = \frac{1}{\cos x}, \quad \csc x = \frac{1}{\sin x}, \quad \arcsin x, \quad \arccos x,$

$\arctan x, \quad \text{arccot}\, x.$

Hyperbolic and inverse hyperbolic functions:

$\sinh x = \frac{1}{2}(e^x - e^{-x}), \quad \cosh x = \frac{1}{2}(e^x + e^{-x}), \quad \sinh^{-1}x, \quad \cosh^{-1}x,$

$\tanh x = \frac{\sinh x}{\cosh x}, \quad \coth x = \frac{\cosh x}{\sinh x}, \quad \tanh^{-1}x, \quad \coth^{-1}x,$

$\text{sech}\, x = \frac{1}{\cosh x}, \quad \text{csch}\, x = \frac{1}{\sinh x}.$

2. Orthogonal polynomials

Legendre polynomials $P_n(x)$.

$$P_n(x) = 2^{-n}(n!)^{-1}\frac{d^n}{dx^n}(x^2-1)^n = {}_2F_1(-n, n+1; 1; \tfrac{1}{2}-\tfrac{1}{2}x)$$

Gegenbauer's polynomials $C_n^\nu(x)$

$$C_n^\nu(x) = [n!\Gamma(2\nu)]^{-1}\Gamma(2\nu+n)\,{}_2F_1(-n,2\nu+n;\tfrac{1}{2}+\nu;\tfrac{1}{2}-\tfrac{1}{2}x)$$

$$=(-2)^{-n}(1-x^2)^{\frac{1}{2}-\nu}(2\nu)_n[n!(\nu+\tfrac{1}{2})_n]^{-1}\frac{d^n}{dx^n}[(1-x^2)^{n+\nu-\frac{1}{2}}]$$

Chebycheff polynomials $T_n(x)$, $U_n(x)$

$$T_n(x) = \cos(n\arccos x) = {}_2F_1(-n,n;\tfrac{1}{2};\tfrac{1}{2}-\tfrac{1}{2}x) = \tfrac{1}{2}n\lim_{\nu=0}\Gamma(\nu)C_n^\nu(x)$$

$$= \tfrac{1}{2}\{[x+i(1-x^2)^{\frac{1}{2}}]^n + [x-i(1-x^2)^{\frac{1}{2}}]^n\}$$

$$U_n(x) = (1-x^2)^{-\frac{1}{2}}\sin[(n+1)\arccos x] = C_n^1(x)$$

$$= x(n+1)\,{}_2F_1(\tfrac{1}{2}-\tfrac{1}{2}n,\tfrac{3}{2}+\tfrac{1}{2}n;\tfrac{3}{2};1-x^2)$$

$$= -\tfrac{1}{2}i(1-x^2)^{-\frac{1}{2}}\{[x+i(1-x^2)^{\frac{1}{2}}]^n-[x-i(1-x^2)^{\frac{1}{2}}]^n\}$$

Jacobi polynomials $P_n^{\alpha,\beta}(x)$

$$P_n^{(\alpha,\beta)}(x) = [n!\Gamma(1+\alpha)]^{-1}\Gamma(1+\alpha+n)\,{}_2F_1(-n,n+\alpha+\beta+1;\alpha+1;\tfrac{1}{2}-\tfrac{1}{2}x)$$

$$= (-1)^n 2^{-n}(n!)^{-1}(1-x)^{-\alpha}(1+x)^{-\beta}\frac{d^n}{dx^n}[(1-x)^{\alpha+n}(1+x)^{\beta+n}]$$

Laguerre polynomials

$$L_n^\alpha(x) = (n!)^{-1}x^{-\alpha}e^x\frac{d^n}{dx^n}(e^{-x}x^{n+\alpha})$$

$$= [n!\Gamma(1+\alpha)]^{-1}\Gamma(\alpha+1+n)\,{}_1F_1(-n;\alpha+1;x)$$

$$L_n^0(x) = L_n(x)$$

Hermite polynomials

$$He_n(x) = (-1)^n \exp(\tfrac{1}{2}x^2) \frac{d^n}{dx^n} \exp(-\tfrac{1}{2}x^2)$$

$$H_n(x) = (-1)^n e^{x^2} \frac{d^n}{dx^n} e^{-x^2} = 2^{\frac{1}{2}n} He_n(2^{\frac{1}{2}}x)$$

$$He_{2n}(x) = (-1)^n 2^{-n} (n!)^{-1} (2n)! \, {}_1F_1(-n;\tfrac{1}{2};\tfrac{1}{2}x^2)$$

$$He_{2n+1}(x) = x(-1)^n 2^{-n} (n!)^{-1} (2n+1)! \, {}_1F_1(-n;\tfrac{3}{2};\tfrac{1}{2}x^2)$$

3. The Gamma function and related functions

$$\Gamma(z) = \int_0^\infty e^{-t} t^{z-1} dt \quad , \qquad \text{Re } z > 0$$

ψ-function

$$\psi(z) = \frac{d}{dz} \log z = \frac{\Gamma'(z)}{\Gamma(z)}$$

Beta function $B(x,y)$

$$B(x,y) = \frac{\Gamma(x)\Gamma(y)}{\Gamma(x+y)}$$

4. Legendre functions

(Definition according to Hobson)

$$p_\nu^\mu(z) = [\Gamma(1-\mu)]^{-1} \left(\frac{z+1}{z-1}\right)^{\frac{1}{2}\mu} {}_2F_1(-\nu,\nu+1;1-\mu;\tfrac{1}{2}-\tfrac{1}{2}z)$$

$$e^{-i\pi\mu} q_\nu^\mu(z) = 2^{-\nu-1} [\Gamma(\tfrac{3}{2}+\nu)]^{-1} \pi^{\frac{1}{2}} \Gamma(1+\nu+\mu) z^{-\nu-\mu-1} (z^2-1)^{\frac{1}{2}\mu} .$$

$$\cdot \, {}_2F_1(\tfrac{1}{2}\nu+\tfrac{1}{2}\mu+\tfrac{1}{2}, \tfrac{1}{2}\nu+\tfrac{1}{2}\mu+1; \nu+\tfrac{3}{2}; z^{-2})$$

z is a point in the complex z-plane cut along the real z-axis from $-\infty$ to $+1$

$$(z^2-1)^{\frac{1}{2}\mu} = (z-1)^{\frac{1}{2}\mu}(z+1)^{\frac{1}{2}\mu}; -\pi < \arg z < \pi, -\pi < \arg(z\pm 1) < \pi$$

$$P_\nu^\mu(x) = [\Gamma(1-\mu)]^{-1}\left(\frac{1+x}{1-x}\right)^{\frac{1}{2}\mu} {}_2F_1(-\nu,\nu+1;1-\mu;\tfrac{1}{2}-\tfrac{1}{2}z), \ -1 < x < 1$$

$$Q_\nu^\mu(x) = \tfrac{1}{2}e^{-i\pi\mu}[e^{-\frac{1}{2}i\pi\mu}q_\nu^\mu(x+i0) + e^{\frac{1}{2}i\pi\mu}q_\nu^\mu(x-i0)], \ -1 < x < 1$$

$$p_\nu^0(z) = p_\nu(z); \qquad q_\nu^0(z) = q_\nu(z)$$

$$P_\nu^0(z) = P_\nu(z); \qquad Q_\nu^0(z) = Q_\nu(z)$$

5. Bessel functions

$$J_\nu(z) = (\tfrac{1}{2}z)^\nu \sum_{n=0}^{\infty} \frac{(-1)^n (\frac{z}{2})^{2n}}{n!\Gamma(\nu+n+1)}$$

$$J_{-\nu}(z) = J_\nu(z)\cos(\pi\nu) - Y_\nu(z)\sin(\pi\nu);$$

$$Y_{-\nu}(z) = J_\nu(z)\sin(\pi\nu) + Y_\nu(z)\cos(\pi\nu);$$

$$Y_\nu(z) = \cot(\pi\nu)J_\nu(z) - \csc(\pi\nu)J_{-\nu}(z)$$

$$H_\nu^{(1)}(z) = J_\nu(z)+iY_\nu(z); \ H_\nu^{(2)}(z) = J_\nu(z) - iY_\nu(z)$$

6. Modified Bessel functions

$$I_\nu(z) = e^{-\frac{1}{2}i\pi\nu}J_\nu(ze^{i\frac{1}{2}\pi}) = (\tfrac{1}{2}z)^\nu \sum_{n=0}^{\infty} \frac{(\frac{1}{2}z)^{2n}}{n!\Gamma(\nu+n+1)}$$

$$K_\nu(z) = \tfrac{1}{2}\pi \csc(\pi\nu)[I_{-\nu}(z)-I_\nu(z)]$$

$$= \tfrac{1}{2}i\pi e^{\frac{1}{2}i\pi\nu}H_\nu^{(1)}(ze^{i\frac{1}{2}\pi}) = -\tfrac{1}{2}i\pi e^{-i\frac{1}{2}\pi\nu}H_\nu^{(2)}(ze^{-i\frac{1}{2}\pi\nu})$$

7. Anger-Weber functions

$$\mathbf{J}_\nu(z) = \pi^{-1}\int_0^\pi \cos(z\sin t - \nu t)\,dt$$

$$\mathbf{J}_n(z) = J_n(z), \quad n = 0,1,2,\cdots; \qquad \mathbf{E}_0(z) = -\mathbf{H}_0(z)$$

$$\mathbf{J}_{\frac{1}{2}}(z) = (\tfrac{1}{2}\pi z)^{-\frac{1}{2}}\{[C(z)-S(z)]\cos z+[C(z)+S(z)]\sin z\} = \mathbf{E}_{-\frac{1}{2}}(z)$$

$$\mathbf{J}_{-\frac{1}{2}}(z) = (\tfrac{1}{2}\pi z)^{-\frac{1}{2}}\{[C(z)+S(z)]\cos z-[C(z)-S(z)]\sin z\} = \mathbf{E}_{\frac{1}{2}}(z)$$

8. Struve functions

$$\mathbf{H}_\nu(z) = \sum_{n=0}^{\infty} \frac{(-1)^n(\tfrac{1}{2}z)^{\nu+2n+1}}{\Gamma(n+3/2)\Gamma(\nu+n+3/2)}$$

$$\mathbf{L}_\nu(z) = -ie^{-\frac{1}{2}i\pi\nu}\mathbf{H}_\nu(ze^{i\frac{1}{2}\nu\pi}) = \sum_{n=0}^{\infty} \frac{(\tfrac{1}{2}z)^{\nu+2n+1}}{\Gamma(n+3/2)\Gamma(\nu+n+3/2)}$$

9. Lommel functions

$$s_{\mu,\nu}(z) = \frac{z^{\mu+1}}{(\mu-\nu+1)(\mu+\nu+1)}\,{}_1F_2(1;\tfrac{1}{2}\mu-\tfrac{1}{2}\nu+3/2,\tfrac{1}{2}\mu+\tfrac{1}{2}\nu+3/2;-\tfrac{1}{4}z^2)$$

$$\mu\pm\nu \neq -1,\ -2,\ -3,\ \cdots$$

$$S_{\mu,\nu}(z) = s_{\mu,\nu}(z)-2^{\mu-1}\Gamma(\tfrac{1}{2}+\tfrac{1}{2}\mu-\tfrac{1}{2}\nu)\Gamma(\tfrac{1}{2}+\tfrac{1}{2}\mu+\tfrac{1}{2}\nu).$$

$$\cdot\{\sin[\tfrac{1}{2}\pi(\nu-\mu)]J_\nu(z)+\cos[\tfrac{1}{2}\pi(\nu-\mu)]Y_\nu(z)\}$$

$$s_{\nu,\mu}(z) = s_{\nu,-\mu}(z); \qquad S_{\mu,\nu}(z) = S_{\mu,-\nu}(z)$$

Special cases:

$$s_{\nu,\nu}(z) = \Pi^{\frac{1}{2}}2^{\nu-1}\Gamma(\tfrac{1}{2}+\nu)\mathbf{H}_\nu(z)$$

$$S_{\nu,\nu}(z) = \pi^{\frac{1}{2}}2^{\nu-1}\Gamma(\tfrac{1}{2}+\nu)[\mathbf{H}_\nu(z) - Y_\nu(z)]$$

$s_{\nu+1,\nu}(z) = z^{\nu}-2^{\nu}\Gamma(1+\nu)J_{\nu}(z); \qquad S_{\nu+1,\nu} = z^{\nu}$

$\lim_{\mu\to\nu} \cdot \frac{s_{\mu-1,\nu}(z)}{\Gamma(\mu-\nu)} = 2^{\nu-1}\Gamma(\nu)J_{\nu}(z)$

$s_{o,\nu}(z) = \tfrac{1}{2}\pi\csc(\pi\nu)[\mathbf{J}_{\nu}(z) - \mathbf{J}_{-\nu}(z)]$

$S_{o,\nu}(z) = \tfrac{1}{2}\pi\csc(\pi\nu)[\mathbf{J}_{\nu}(z) - \mathbf{J}_{-\nu}(z) - J_{\nu}(z) + J_{-\nu}(z)]$

$s_{-1,\nu}(z) = -\tfrac{1}{2}\pi\nu^{-1}\csc(\pi\nu)[\mathbf{J}_{\nu}(z)+\mathbf{J}_{-\nu}(z)]$

$S_{-1,\nu}(z = \tfrac{1}{2}\pi\nu^{-1}\csc(\pi\nu)[J_{\nu}(z)+J_{-\nu}(z)-\mathbf{J}_{\nu}(z)-\mathbf{J}_{-\nu}(z)]$

$s_{1,\nu}(z) = 1-\tfrac{1}{2}\pi\nu\csc(\pi\nu)[\mathbf{J}_{\nu}(z)+\mathbf{J}_{-\nu}(z)]$

$S_{1,\nu}(z) = 1+\tfrac{1}{2}\pi\nu\csc(\pi\nu)[J_{\nu}(z)+J_{-\nu}(z)-\mathbf{J}_{\nu}(z)-\mathbf{J}_{-\nu}(z)]$

$S_{\frac{1}{2},\frac{1}{2}}(z) = z^{-\frac{1}{2}}, \qquad S_{3/2,\frac{1}{2}}(z) = z^{\frac{1}{2}}$

$S_{-\frac{1}{2},\frac{1}{2}}(z) = z^{-\frac{1}{2}}[\sin z\ \mathrm{Ci}(z)-\cos z\ \mathrm{si}(z)]$

$S_{-3/2,\frac{1}{2}}(z) = -z^{-\frac{1}{2}}[\sin z\ \mathrm{si}(z)+\cos z\ \mathrm{Ci}(z)]$

Kelvin's functions

$J_{\nu}(ze^{i\,3/4\,\pi}) = \mathrm{ber}_{\nu}(z)+i\,\mathrm{bei}_{\nu}(z)$

$J_{\nu}(ze^{-i\,3/4\,\pi}) = \mathrm{ber}_{\nu}(z)-i\,\mathrm{bei}_{\nu}(z)$

$$K_\nu(ze^{i\frac{\pi}{4}}) = \mathrm{ker}_\nu(z) + i\,\mathrm{kei}_\nu(z)$$

$$K_\nu(ze^{-i\frac{\pi}{4}}) = \mathrm{ker}_\nu(z) - i\,\mathrm{kei}_\nu(z)$$

$$\mathrm{ber}_o(z) = \mathrm{ber}(z), \quad \mathrm{bei}_o(z) = \mathrm{bei}(z),$$

$$\mathrm{ker}_o(z) = \mathrm{ker}(z), \quad \mathrm{kei}_o(z) = \mathrm{kei}(z)$$

Neumann polynomials

$$O_n(x) = \tfrac{1}{4} \sum_{m=0}^{\leq\frac{1}{2}n} n(n-m-1)!(\tfrac{1}{2}x)^{2m-n-1}/m! \; ; \qquad O_o(x) = x^{-1}$$

Schläfli polynomials

$$S_n(x) = \sum_{m=0}^{\leq\frac{1}{2}n} (n-m-1)!(\tfrac{1}{2}z)^{2m-n}/m! \; ; \qquad S_o(x) = 0$$

10. Gauss' hypergeometric series

$${}_2F_1(a,b;c;z) = \frac{\Gamma(c)}{\Gamma(a)\Gamma(b)} \sum_{n=0}^{\infty} \frac{\Gamma(a+n)\Gamma(b+n)}{\Gamma(c+n)} \frac{z^n}{n!}, \quad |z| < 1$$

11. Confluent hypergeometric functions

Kummer's functions

$${}_1F_1(a;c;z) = \frac{\Gamma(c)}{\Gamma(a)} \sum_{n=0}^{\infty} \frac{\Gamma(a+n)}{\Gamma(c+n)} \frac{z^n}{n!} = z^{-\frac{1}{2}c} e^{\frac{1}{2}z} M_{\frac{1}{2}c-a,\frac{1}{2}c-\frac{1}{2}}(z)$$

$${}_1F_1(a;c;z) = e^z\,{}_1F_1(c-a;c;-z)$$

Whittaker functions

$$M_{k,\mu}(z) = z^{\mu+\frac{1}{2}}e^{-\frac{1}{2}z}\ {}_1F_1(\tfrac{1}{2}+\mu-k;2\mu+1;z)$$

$$= z^{\mu+\frac{1}{2}}e^{\frac{1}{2}z}\ {}_1F_1(\tfrac{1}{2}+\mu+k;2\mu+1;-z)$$

$$W_{k,\mu}(z) = \frac{\Gamma(-2\mu)}{\Gamma(\frac{1}{2}-\mu-k)} M_{k,\mu}(z) + \frac{\Gamma(2\mu)}{\Gamma(\frac{1}{2}+\mu-k)} M_{k,\mu}(z)$$

$$W_{k,-\mu}(z) = W_{k,\mu}(z)$$

Parabolic cylinder function

$$D_\alpha(z) = 2^{\frac{1}{4}+\frac{1}{2}\alpha}z^{-\frac{1}{2}}W_{\frac{1}{4}+\frac{1}{2}\alpha,\frac{1}{4}}(\tfrac{1}{2}z^2)$$

$$D_n(z) = e^{-\frac{1}{4}z^2}He_n(z)\ , \qquad n = 0,\ 1,\ 2,\ \cdots$$

$$D_{-1}(z) = (\tfrac{1}{2}\pi)^{\frac{1}{2}}e^{\frac{1}{4}z^2}\ \mathrm{Erfc}(2^{-\frac{1}{2}}z)$$

$$D_{-\frac{1}{2}}(z) = (2\pi z^{-1})^{-\frac{1}{2}}\ K_{\frac{1}{4}}(\tfrac{1}{4}z^2)$$

Error integrals

$$\mathrm{Erf}(x) = 2\pi^{-\frac{1}{2}}\int_0^x e^{-t^2}dt = 2\pi^{-\frac{1}{2}}x\ {}_1F_1(\tfrac{1}{2};\tfrac{3}{2};-x^2)$$

$$= 2(\pi x)^{-\frac{1}{2}}e^{-\frac{1}{2}x^2}\ M_{-\frac{1}{4},\frac{1}{4}}(x^2)$$

$$\mathrm{Erfc}(x) = 1-\mathrm{Erf}(x) = 2\pi^{-\frac{1}{2}}\int_x^\infty e^{-t^2}dt = (\pi x)^{-\frac{1}{2}}e^{-\frac{1}{2}x^2}W_{-\frac{1}{4},\frac{1}{4}}(x)$$

$$\mathrm{Erf}(x^{\frac{1}{2}}e^{i\frac{\pi}{4}}) = C(x) + S(x) + i[C(x) - S(x)]$$

$$\mathrm{Erfc}(x^{\frac{1}{2}}e^{i\frac{\pi}{4}}) = 1-C(x) - S(x) + i[S(x) - C(x)]$$

Fresnel's integrals

$$C(x) = (2\pi)^{-\frac{1}{2}} \int_0^x t^{-\frac{1}{2}} \cos t\, dt; S(x) = (2\pi)^{-\frac{1}{2}} \int_0^x t^{-\frac{1}{2}} \sin t\, dt$$

Exponential integrals

$$-\mathrm{Ei}(-z) = -\gamma - \log z - \sum_{n=1}^{\infty} (n\cdot n!)^{-1}(-z)^n = z^{-\frac{1}{2}} e^{-\frac{1}{2}z} W_{-\frac{1}{2},0}(z)$$

$$-\mathrm{Ei}(-x) = \int_x^{\infty} t^{-1} e^{-t} dt, \qquad x > 0$$

$$\overline{\mathrm{Ei}}(z) = \gamma + \log z + \sum_{n=1}^{\infty} (n\cdot n!)^{-1} z^n$$

$$\overline{\mathrm{Ei}}(x) = \tfrac{1}{2}[\mathrm{Ei}(-xe^{i\pi}) + \mathrm{Ei}(-xe^{-i\pi})] = -\mathrm{P\cdot V\cdot} \int_{-x}^{\infty} t^{-1} e^{-t} dt, \quad x > 0$$

$$\mathrm{Ei}(-ze^{\pm i\pi}) = \pm i\pi + \overline{\mathrm{Ei}}(z); \; \overline{\mathrm{Ei}}(ze^{\pm i\pi}) = \pm i\pi + \mathrm{Ei}(-z)$$

$$\mathrm{Ei}(-ze^{\pm\frac{1}{2}i\pi}) = \mathrm{Ci}(z) \mp i[\tfrac{1}{2}\pi - \mathrm{Si}(z)]; \overline{\mathrm{Ei}}(ze^{\pm\frac{1}{2}i\pi}) = \mathrm{Ci}(z) \pm i[\tfrac{1}{2}\pi + \mathrm{Si}(z)]$$

$$\mathrm{Ei}(-xe^{\pm\frac{1}{2}i\pi}) = \mathrm{Ci}(x) \pm i\,\mathrm{si}(x); \overline{\mathrm{Ei}}(xe^{\pm\frac{1}{2}i\pi}) = \mathrm{Ci}(x) \pm i[\pi + \mathrm{si}(x)]$$

$$x > 0$$

Sine and cosine integral

$$\mathrm{Si}(z) = \sum_{n=0}^{\infty} (-1)^n [(2n+1)(2n+1)!]^{-1} z^{2n+1} = \int_0^z t^{-1} \sin t\, dt$$

$$\mathrm{Ci}(z) = \gamma + \log z + \sum_{n=1}^{\infty} (-1)^n [2n(2n)!]^{-1} z^{2n}$$

$$\mathrm{Ci}(x) = -\int_x^{\infty} t^{-1} \cos t\, dt, \; \mathrm{si}(x) = -\int_x^{\infty} t^{-1} \sin t\, dt = \mathrm{Si}(x) - \frac{\pi}{2}$$

$$x > 0$$

$$\mathrm{Ci}(x) = \gamma+\log x - \int_0^x t^{-1}(1-\cos t)dt, \qquad x > 0$$

$$\mathrm{Ci}(z)=\tfrac{1}{2}[\mathrm{Ei}(-ze^{\frac{1}{2}i\pi})+\mathrm{Ei}(-ze^{-\frac{1}{2}i\pi})]=\tfrac{1}{2}[\overline{\mathrm{E}}\mathrm{i}(ze^{\frac{1}{2}i\pi})+\overline{\mathrm{E}}\mathrm{i}(ze^{-\frac{1}{2}i\pi})]$$

$$\mathrm{Si}(z) = \tfrac{1}{2}\pi+\tfrac{1}{2}i[\mathrm{Ei}(-ze^{\frac{1}{2}i\pi}) - \mathrm{Ei}(-ze^{-\frac{1}{2}i\pi})]$$

Incomplete gamma function

$$\gamma(\nu,x) = \int_0^x t^{\nu-1}e^{-t}dt=\nu^{-1}x^{\nu}{}_1F_1(\nu,\nu+1;-x), \qquad \mathrm{Re}\ \nu > 0$$

$$= \nu^{-1}x^{\frac{1}{2}\nu-\frac{1}{2}}e^{-\frac{1}{2}x}M_{\frac{1}{2}\nu-\frac{1}{2},\frac{1}{2}\nu}(x)$$

$$\Gamma(\nu,x) = \Gamma(\nu)-\gamma(\nu,x) = \int_x^{\infty} t^{\nu-1}e^{-t}dt=x^{\frac{1}{2}\nu-\frac{1}{2}}e^{-\frac{1}{2}x}W_{\frac{1}{2}\nu-\frac{1}{2},\frac{1}{2}\nu}(x)$$

$$\Gamma(\tfrac{1}{2},z^2) = \pi^{\frac{1}{2}}\mathrm{Erfc}(z); \quad \Gamma(0,z) = -\mathrm{Ei}(-z)$$

$$\Gamma(\tfrac{1}{2},z^2) = \pi^{\frac{1}{2}}\mathrm{Erf}(z); \quad \gamma(1,z) = 1-e^{-z}, \quad \Gamma(1,z) = e^{-z}$$

12. Particular cases of Whittaker's functions

$$M_{-\frac{1}{4},\frac{1}{4}}(z) = \tfrac{1}{2}\pi^{\frac{1}{2}}z^{\frac{1}{4}}e^{\frac{1}{2}z}\mathrm{Erf}(z^{\frac{1}{2}})$$

$$M_{\frac{1}{4},\frac{1}{4}}(z) = -i\tfrac{1}{2}\pi^{\frac{1}{2}}z^{\frac{1}{4}}e^{-\frac{1}{2}z}\mathrm{Erf}(iz^{\frac{1}{2}})$$

$$M_{k,0}(z) = z^{\frac{1}{2}}e^{-\frac{1}{2}z}L_{k-\frac{1}{2}}(z)$$

$$M_{0,\mu}(z) = 2^{2\mu}\Gamma(1+\mu)z^{\frac{1}{2}}I_\mu(\tfrac{1}{2}z)$$

$$M_{\mu+\frac{1}{2},\mu}(z) = z^{\mu+\frac{1}{2}}e^{-\frac{1}{2}z}$$

$$M_{k,k+\frac{1}{2}}(z) = (2k+1)z^{-k}e^{\frac{1}{2}z}\ \gamma(2k+1,z)$$

$$W_{-\frac{1}{4},\frac{1}{4}}(z) = \pi^{\frac{1}{2}}z^{\frac{1}{4}}e^{\frac{1}{2}z}\ \mathrm{Erfc}(z^{\frac{1}{2}})$$

$$W_{k,\frac{1}{4}}(z) = 2^{\frac{1}{4}-k}z^{\frac{1}{4}}\ D_{2k-\frac{1}{2}}[(2z)^{\frac{1}{2}}]$$

$$W_{0,\mu}(z) = \pi^{-\frac{1}{2}}z^{\frac{1}{2}}\ K_\mu(\tfrac{1}{2}z)$$

$$W_{-\frac{1}{2},0}(z) = -z^{\frac{1}{2}}e^{\frac{1}{2}z}\ \mathrm{Ei}(-z)$$

$$W_{k,\frac{1}{2}+k}(z) = z^{-k}e^{\frac{1}{2}z}\ \Gamma(2k+1,z)$$

$$W_{k,k-\frac{1}{2}}(z) = z^{\mu+\frac{1}{2}}e^{-\frac{1}{2}z}$$

13. Elliptic integrals and elliptic theta functions

Complete elliptic integrals

$$K(k) = \int_0^{\frac{1}{2}\pi} (1-k^2\sin^2 x)^{-\frac{1}{2}}dx = \tfrac{1}{2}\pi\ {}_2F_1(\tfrac{1}{2},\tfrac{1}{2};1;k^2)$$

$$E(k) = \int_0^{\frac{1}{2}\pi} (1-k^2\sin^2 x)^{\frac{1}{2}}dx = \tfrac{1}{2}\pi\ {}_2F_1(-\tfrac{1}{2},\tfrac{1}{2};1;k^2)$$

Theta functions

$$\theta_1(z|t) = (\pi t)^{-\frac{1}{2}} \sum_{n=-\infty}^{\infty} (-1)^n \exp[-(z+n-\tfrac{1}{2})^2/t]$$

$$= 2\sum_{n=0}^{\infty} (-1)^n \exp[-\pi^2 t(n+\tfrac{1}{2})^2]\sin[(2n+1)\pi z]$$

$$\theta_2(z|t) = (\pi t)^{-\frac{1}{2}} \sum_{n=-\infty}^{\infty} (-1)^n \exp[-(z+n)^2/t]$$

$$= 2 \sum_{n=0}^{\infty} \exp[-\pi^2 t(n+\tfrac{1}{2})^2] \cos[(2n+1)\pi z]$$

$$\theta_3(z|t) = (\pi t)^{-\frac{1}{2}} \sum_{n=-\infty}^{\infty} \exp[-(z+n)^2/t]$$

$$= \sum_{n=0}^{\infty} \varepsilon_n \exp(-\pi^2 tn^2)\cos(2\pi nz)$$

$$\theta_4(z|t) = (\pi t)^{-\frac{1}{2}} \sum_{n=-\infty}^{\infty} \exp[-(z+n+\tfrac{1}{2})^2/t]$$

$$= \sum_{n=0}^{\infty} (-1)^n \varepsilon_n \exp(-\pi^2 tn^2)\cos(2\pi nz)$$

Modified theta functions

$$\hat{\theta}_1(z|t) = (\pi t)^{-\frac{1}{2}}\{ \sum_{n=0}^{\infty} (-1)^n \exp[-(z+n+\tfrac{1}{2})^2/t]$$

$$- \sum_{n=-1}^{-\infty} (-1)^n \exp[-(z+n+\tfrac{1}{2})^2/t]\}$$

$$\hat{\theta}_2(z|t) = (\pi t)^{-\frac{1}{2}}\{ \sum_{n=0}^{\infty} (-1)^n \exp[-(z+n)^2/t]$$

$$- \sum_{n=-1}^{-\infty} (-1)^n \exp[-(z+n)^2/t]\}$$

$$\hat{\theta}_3(z|t) = (\pi t)^{-\frac{1}{2}}\{ \sum_{n=0}^{\infty} \exp[-(z+n)^2/t]$$

$$- \sum_{n=-1}^{-\infty} \exp[-(z+n)^2/t]\}$$

$$\hat{\theta}_4(z|t) = (\pi t)^{-\frac{1}{2}} \left\{ \sum_{n=0}^{\infty} \exp[-(z+n+\tfrac{1}{2})^2/t] - \sum_{n=-1}^{-\infty} \exp[-(z+n+\tfrac{1}{2})^2/t] \right\}$$

14. Generalized hypergeometric functions

$${}_pF_q(a_1,a_2,\cdots a_p;b_1,b_2,\cdots b_q;z) = \sum_{n=0}^{\infty} \frac{(a_1)_n\cdots(a_p)_n}{(b_1)_n\cdots(b_q)_n}\frac{z^n}{n!}$$

$$p,q = 0,\ 1,\ 2,\ \cdots$$

$|z|<1$ if $p = q+1$, $|z|<\infty$ if $p\leqq q$; divergent otherwise.

15. Meijer's G-function

$$G_{p,q}^{m,n}\left(x \,\middle|\, \begin{matrix} a_1,\cdots,a_p \\ b_1,\cdots,b_q \end{matrix} \right) = (2\pi i)^{-1} \int_L \frac{A(z)}{B(z)}\, x^z\, dz$$

where

$$A(z) = \prod_{k=1}^{m} \Gamma(b_k-z) \prod_{k=1}^{n} \Gamma(1-a_k+z)$$

$$B(z) = \prod_{k=m+1}^{q} \Gamma(1-b_k+z) \prod_{k=n+1}^{p} \Gamma(a_k-z)$$

L is a path separating the poles of $\Gamma(b_1-z)\cdots$, $\Gamma(b_m-z)$ from those of $\Gamma(1-a_1+z)\cdots\Gamma(1-a_n+z)$

(see Erdélyi et. al. Higher transcendental functions, Vol. 1, Sec. 5.3; 1953, McGraw Hill).

16. Miscellaneous functions

Riemann's zeta function

$$\zeta(z) = \sum_{n=1}^{\infty} n^{-z} \qquad \text{Re } z > 1$$

Hurwitz's zeta function

$$\zeta(z,a) = \sum_{n=0}^{\infty} (n+a)^{-z} \qquad \text{Re } z > 1$$

Lerch's zeta function

$$Y(z,s,a) = \sum_{n=0}^{\infty} (a+n)^{-s} z^n \qquad |z| < 1$$

Unit step function

$$H(t) = 1, \quad t > 0; \quad H(t) = 0, \qquad t < 0$$

List of Functions

Symbol	Name of the Function	Listed under
$C(x)$	Fresnel's integral	11
$Ci(x)$	Cosine integral	11
$C_n^{\nu}(x)$	Gegenbauer's polynomial	2
$D_{\nu}(z)$	Parabolic cylinder function	11
$E(k)$	Complete elliptic integral	13
$Ei(-x)$, $\overline{E}i(x)$	Exponential integrals	11
$Erf(z)$, $Erfc(z)$	Error integrals	11

Symbol	Name of the Function	Listed under
$\mathbf{E}_\nu(z)$	Anger-Weber function	7
${}_mF_n(z)$	Hypergeometric functions	10,11,12,14
$G^{m,n}_{p,q}\left(x\,\middle\vert\,\begin{matrix}a_1,\cdots,a_p\\ b_1,\cdots,b_q\end{matrix}\right)$	Meijer's G-function	15
$H(x)$	Unit step function	16
$He_n(x)$	Hermite's polynomial	2
$H^{(1),(2)}_\nu(z)$	Hankel's functions	5
$\mathbf{H}_\nu(z)$	Struve's function	8
$I_\nu(z)$	Modified Bessel function	6
$J_\nu(z)$	Bessel function	5
$\mathbf{J}_\nu(z)$	Anger-Weber function	7
$K(k)$	Complete elliptic integral	13
$K_\nu(z)$	Modified Hankel function	6
$L_\nu(z)$	Laguerre's function	11
$L^\alpha_n(x)$	Laguerre's polynomial	2
$\mathbf{L}_\nu(z)$	Struve's function	8
$M_{k,\mu}(z)$, $W_{k,\mu}(z)$	Whittaker's functions	11
$O_n(z)$	Neumann polynomials	9
$P_n(x)$	Legendre's polynomials	2

Symbol	Name of the Function	Listed under
$P_n^{(\alpha,\beta)}(x)$	Jacobi's polynomials	2
$p_\nu^\mu(z)$, $P_\nu^\mu(x)$, $q_\nu^\mu(z)$, $Q_\nu^\mu(x)$	Legendre functions	4
$S(x)$	Fresnel's integral	11
$S_n(z)$	Schläfli polynomials	9
$si(x)$, $Si(x)$	Sine integrals	11
$s_{\mu,\nu}(z)$, $S_{\mu,\nu}(z)$	Lommel's functions	9
$T_n(x)$, $U_n(x)$	Chebycheff's polynomials	2
$W_{\mu,\nu}(z)$	Whittaker's function	11
$Y(z,s,a)$	Lerch's zeta function	16
$Y_\nu(z)$	Neumann's function	5
$B(x,y)$	Beta function	3
$\Gamma(z)$	Gamma function	3
$\Gamma(\nu,z)$, $\gamma(\nu,z)$	Incomplete gamma functions	11

Symbol	Name of the Function	Listed under
$\psi(z)$	Psi function	3
$\zeta(z)$	Riemann's zeta function	16
$\zeta(z,a)$	Hurwitz's zeta function	16
$\theta_1(z\|t)$, $\theta_2(z\|t)$, $\theta_3(z\|t)$, $\theta_4(z\|t)$	Elliptic theta functions	13
$\hat{\theta}_1(z\|t)$, $\hat{\theta}_2(z\|t)$, $\hat{\theta}_3(z\|t)$, $\hat{\theta}_4(z\|t)$	Modified elliptic theta functions	13

G. Doetsch: Introduction to the Theory and Application of the Laplace Transformation

Translated by W. Nader
51 figures and a table of Laplace transforms
VII, 326 pages. 1974
Cloth DM 68,–; US $27.80
ISBN 3-540-06407-9
Prices are subject to change without notice

In anglo-american literature there exist numerous books, devoted to the application of the Laplace transformation in technical domains such as electrotechnics, mechanics etc. Chiefly, they treat problems which, in mathematical language, are governed by ordinary and partial differential equations, in various physically dressed forms. The theoretical foundations of the Laplace transformation are presented usually only in a simplified manner, presuming special properties with respect to the transformed functions, which allow easy proofs.
By contrast, the present book intends principally to develop those parts of the theory of the Laplace transformation, which are needed by mathematicians, physicists and engineers in their daily routine work, but in complete generality and with detailed, exact proofs. The applications to other mathematical domains and to technical problems are inserted, when the theory is adequately developed to present the tools necessary for their treatment.

Springer-Verlag
Berlin · Heidelberg · New York
München Johannesburg London Madrid New Delhi
Paris Rio de Janeiro Sydney Tokyo Utrecht Wien